Linear Operators and Their Essential Pseudospectra

Linear Operators and Their Essential Pseudospectra

Aref Jeribi, PhD

Department of Mathematics,

University of Sfax, Tunisia

E-mail: aref.jeribi@fss.rnu.tn

Apple Academic Press Inc.
3333 Mistwell Crescent
Oakville, ON L6L 0A2 Canada

Apple Academic Press Inc.
9 Spinnaker Way
Waretown, NJ 08758 USA

First issued in paperback 2021

Exclusive worldwide distribution by CRC Press, a member of Taylor & Francis Group
No claim to original U.S. Government works
Printed in the United States of America on acid-free paper

ISBN 13: 978-1-77463-400-4 (pbk)
ISBN 13: 978-1-77188-699-4 (hbk)

Library and Archives Canada Cataloguing in Publication

Jeribi, Aref, author
Linear operators and their essential pseudospectra / Aref Jeribi, PhD (Department of Mathematics, University of Sfax, Tunisia).

Includes bibliographical references and index.
Issued in print and electronic formats.
ISBN 978-1-77188-699-4 (hardcover).--ISBN 978-1-351-04627-5 (PDF)

1. Linear operators. 2. Spectral theory (Mathematics). I. Title.

QA329.2.J47 2018 515'.7246 C2018-900406-1 C2018-900407-X

CIP data on file with US Library of Congress

Apple Academic Press also publishes its books in a variety of electronic formats. Some content that appears in print may not be available in electronic format. For information about Apple Academic Press products, visit our website at **www.appleacademicpress.com** and the CRC Press website at **www.crcpress.com**

To

my mother Sania, my father Ali,
my wife Fadoua, my children Adam and Rahma,
my brothers Sofien and Mohamed Amin,
my sister Elhem, and
all members of my extended family

Contents

About the Author

Aref Jeribi, PhD
Professor, Department of Mathematics, University of Sfax, Tunisia

Aref Jeribi, PhD, is Professor in the Department of Mathematics at the University of Sfax, Tunisia. He is the author of the book *Spectral Theory and Applications of Linear Operators and Block Operator Matrices* (2015) and co-author of the book *Nonlinear Functional Analysis in Banach Spaces and Banach Algebras: Fixed Point Theory under Weak Topology for Nonlinear Operators and Block Operator Matrices with Applications* (CRC Press, 2015). He has published many journal articles in international journals. His areas of interest include spectral theory, matrice operators, transport theory, Gribov operator, Bargman space, fixed point theory, Riesz basis, and linear relations.

Preface

This book is intended to provide a fairly comprehensive study of spectral theory of linear operators defined on Banach spaces. The central items of interest include various essential spectra, but we also consider some of their generalizations that have been studied in recent years.

As the spectral theory of operators is an important part of functional analysis and has numerous applications in many parts of mathematics and, we require, in this book, some modest prerequisites from functional analysis and operator theory are necessary that the reader can find in the classical texts of functional analysis. We therefore have the hope that this book is accessible to newcomers and graduate students of mathematics with a standard background in analysis.

A considerable part of the content of this book corresponds to research activities developed in collaboration with some colleagues as well as with some of my graduate students over the course of several years. This book is considered as an attempt to organize the available material, most of which exists only in the form of research papers scattered throughout the literature. For this reason, it has been a great pleasure for me to organize the material in this book.

In recent years, spectral theory has witnessed an explosive development. In this book, we present a survey of results concerning various types of essential spectra and pseudospectra in a unified, axiomatic way, and we also gathered several topics that are new but that relate only to those concepts and methods emanating from other parts of the book.

The book covers an important list of topics in spectral theory and will be an excellent choice. It is well written, giving a survey of results concerning various types of spectra in a unified, axiomatic way. The main topics include essential spectra, essential pseudospectra, structured essential pseudospectra, and their relative sets.

We do hope that this book will be very useful for researchers, since it represents not only a collection of a previously heterogeneous material but is also an innovation through several extensions.

Of course, it is impossible for a single book to cover such a huge field of research. In making personal choices for inclusion of material, we tried to give useful complementary references in this research area, hence probably neglecting some relevant works. We would be very grateful to receive any comments from readers and researchers, providing us with some information concerning some missing references.

We would like to thank Salma Charfi for the improvement she has made in the introduction of this book. We are indebted to her. We would also like to thank Aymen Ammar for the improvement he has made throughout this book, and we are very grateful to him. Moreover, we should mention that the thesis work, performed under my direction, by my former students and present colleagues Asrar Elleuch, Mohamemd Zerai Dhahri, Bilel Boukettaya, Kamel Mahfoudhi, and Faten Bouzayeni, obtained results that have helped us in writing this book. Finally, we apologize in case we have forgotten to quote any author who has contributed, directly or indirectly, to this work.

—Aref Jeribi, PhD

Chapter 1

Introduction

Recently, the interest in description of essential spectra remains high because of the abundance of practical applications that help scientists to deal with information overload. This book is intended to provide an important list of topics in spectral theory of linear operators defined on Banach spaces. Central items of interest include Fredholm operators and various characterizations of essential spectra.

In this book, a survey of the state of the art of research related to essential spectra of closed, densely defined, and linear operators subjected to various perturbations is outlined. As important supersets of the essential spectra, the description of essential pseudospectra in this book are interesting objects by themselves, since they carry more information and have a better convergence and approximation properties than essential spectra. We are also interested in giving a much better insight into the essential pseudospectra, by studying some sets called structured essential pseudospectra. Further, a significant amount of research has been done in this book to treat important characterizations of S-essential spectra and S-structured essential pseudospectra.

In this book, we also turn our attention to the important concept of condition pseudospectrum which carries more information than spectrum and pseudospectrum, especially, about transient instead of just asymptotic behavior of dynamical systems.

Now, let us describe its contents in the following sections.

1.1 ESSENTIAL SPECTRA AND RELATIVE ESSENTIAL SPECTRA

The theory of the essential spectra of linear operators in Banach spaces is a modern section of the spectral analysis. It has numerous applications in many parts of mathematics and physics including matrix theory, function theory, complex analysis, differential and integral equations, and control theory.

The original definition of the essential spectrum goes back to H. Weyl [158] around 1909, when he defined the essential spectrum for a self-adjoint operator A on a Hilbert space as the set of all points of the spectrum of A that are not isolated eigenvalues of finite algebraic multiplicity and he proved that the addition of a compact operator to A does not affect the essential spectrum. Irrespective of whether A is bounded or not on a Banach space X, there are many ways to define the essential spectrum, most of them are enlargement of the continuous spectrum. Hence, we can find several definitions of the essential spectrum in the literature, which coincide for self-adjoint operators on Hilbert spaces (see, for example, [79, 137]).

On the other hand, the concept of essential spectra was introduced and studied by many mathematicians and we can cite the contributions of H. Weyl and his collaborators (see, for instance [1–3, 48, 82, 83, 108, 120, 141, 154, 158]). Further important characterizations concerning essential spectra and their applications to transport operators are established by A. Jeribi and his collaborators (see Refs. [1, 4–9, 11–14, 19, 26, 31, 34, 36, 37, 40, 41, 43–47, 51, 53, 59, 60, 84–88, 90, 94, 95, 97–100, 114–116, 123]).

Among these essential spectra, the following sets are defined for a closed densely defined linear operator A:

$$
\begin{aligned}
\sigma_{e1}(A) &:= \{\lambda \in \mathbb{C} \text{ such that } \lambda - A \notin \Phi_+(X)\},\\
\sigma_{e2}(A) &:= \{\lambda \in \mathbb{C} \text{ such that } \lambda - A \notin \Phi_-(X)\},\\
\sigma_{e3}(A) &:= \sigma_{e1}(A) \bigcap \sigma_{e2}(A),\\
\sigma_{e4}(A) &:= \{\lambda \in \mathbb{C} \text{ such that } \lambda - A \notin \Phi(X)\},\\
\sigma_{e5}(A) &:= \bigcap_{K \in \mathscr{K}(X)} \sigma(A+K),\\
\sigma_{e6}(A) &:= \mathbb{C} \backslash \Big\{\lambda \in \mathbb{C} \text{ such that } \lambda - A \text{ is Fredholm, } i(\lambda - A) = 0,\\
&\qquad\qquad \text{and all scalars near } \lambda \text{ are in } \rho(A)\Big\},\\
\sigma_{e7}(A) &:= \bigcap_{K \in \mathscr{K}(X)} \sigma_{ap}(A+K),\\
\sigma_{e8}(A) &:= \bigcap_{K \in \mathscr{K}(X)} \sigma_{\delta}(A+K),\\
\sigma_{ewl}(A) &:= \bigcap_{K \in \mathscr{K}(X)} \sigma_{l}(A+K),\\
\sigma_{ewr}(A) &:= \bigcap_{K \in \mathscr{K}(X)} \sigma_{r}(A+K),
\end{aligned}
$$

where $\Phi_+(X)$, $\Phi_-(X)$, and $\Phi(X)$ denote the sets of upper semi-Fredholm, lower semi-Fredholm, and Fredholm operators, respectively; $\rho(A)$, $\sigma(A)$, and $i(A)$ denote the resolvent set, the spectrum, and the index of A, respectively; $\mathscr{K}(X)$ denotes the set of all compact linear operators on X, and

$$
\begin{aligned}
\sigma_{ap}(A) &:= \Big\{\lambda \in \mathbb{C} \text{ such that } \inf_{\|x\|=1,\ x \in \mathscr{D}(A)} \|(\lambda - A)x\| = 0\Big\},\\
\sigma_{\delta}(A) &:= \Big\{\lambda \in \mathbb{C} \text{ such that } \lambda - A \text{ is not surjective}\Big\},\\
\sigma_{l}(A) &:= \{\lambda \in \mathbb{C} \text{ such that } \lambda - A \text{ is not left invertible}\},\\
\sigma_{r}(A) &:= \{\lambda \in \mathbb{C} \text{ such that } \lambda - A \text{ is not right invertible}\}.
\end{aligned}
$$

The sets $\sigma_{e1}(\cdot)$ and $\sigma_{e2}(\cdot)$ are the Gustafson and Weidmann essential spectra, respectively [79]; $\sigma_{e3}(\cdot)$ is the Kato essential spectrum [108]; the subset $\sigma_{e4}(\cdot)$ is the Wolf essential spectrum [160]; $\sigma_{e5}(\cdot)$ is the Schechter essential spectrum [86–88]; and $\sigma_{e6}(\cdot)$ denotes the Browder essential spectrum [48, 96]. The subset $\sigma_{e7}(\cdot)$ was introduced by V. Racočević [133] and designates the essential approximate point spectrum, and $\sigma_{e8}(\cdot)$ is the essential defect spectrum and was introduced by C. Shmoeger [143]. $\sigma_{ewl}(\cdot)$ and $\sigma_{ewr}(\cdot)$ are the left and the right of Weyl essential spectra, respectively.

Considerable attention has also been devoted by A. Jeribi [1,93,113] to give a new characterization of the Schechter essential spectrum named the Jeribi essential spectrum, $\sigma_j(A)$, of $A \in \mathscr{C}(X)$ (the set of closed densely defined linear operators on X) defined by

$$\sigma_j(A) = \bigcap_{K \in \mathscr{W}^*(X)} \sigma(A+K)$$

where $\mathscr{W}^*(X)$ stands for each one of the sets $\mathscr{W}(X)$ (the set of weakly compact operators) or $\mathscr{S}(X)$ (the set of strictly singular operators).

Recently, an important progress has been made by A. Ammar, B. Boukettaya, and A. Jeribi [20], who were interested in studying the stability problem of the left and right Weyl operator sets; and they defined the left and right Jeribi essential spectra for a closed densely defined operator A as:

$$\sigma_j^l(A) = \bigcap_{K \in \mathscr{W}^*(X)} \sigma_l(A+K)$$

and

$$\sigma_j^r(A) = \bigcap_{K \in \mathscr{W}^*(X)} \sigma_r(A+K).$$

At the first sight, $\sigma_{ewl}(A)$ (resp. $\sigma_{ewr}(A)$) and $\sigma_j^l(A)$ (resp. $\sigma_j^r(A)$) seem to be not equal. However, the authors has proved that $\sigma_{ewl}(A) = \sigma_j^l(A)$ and $\sigma_{ewr}(A) = \sigma_j^r(A)$ in L_1-spaces or X satisfies the Dunford-Pettis property; and hence $\sigma_{ewl}(A)$ and $\sigma_{ewr}(A)$ may be viewed as an extension of $\sigma_j^l(A)$ and $\sigma_j^r(A)$, respectively.

On the other hand, the question of stability of the left and the right essential Weyl spectra was well-treated in Ref. [20] by using the concept of polynomially Riesz operators in order to give a refinement on the definition of these essential spectra via this concept and to show that compactness condition can be relaxed in a very general Banach space setting.

Motivated by the notion of measure of noncompactness, A. Ammar, M. Z. Dhahri and A. Jeribi [23] are interested in the description of the essential approximate point spectrum and the essential defect spectrum of a closed densely defined linear operator by means of upper semi-Fredholm and lower semi-Fredholm operators, respectively.

In literature, the spectral theory of linear operators have been enriched by S. Goldberg [75], who presented a development of some powerful methods for the study of the convergence of a sequence of linear operators in a Banach space and the investigation of the convergence compactly to zero. Recently, A. Ammar and A. Jeribi are concerned in work [32], not only with the case of bounded linear operators $(A_n)_n$ converging compactly to a bounded operator A, but also with the case of a sequence of closed linear operators $(A_n)_n$ converging in the generalized sense to a closed linear operator A.

The notion of generalized convergence was approached as a generalization of convergence in norm for possibly unbounded linear operators, as well as a reliable method for comparing operators. Further, this notion make the authors generalize the results of T. Kato [108], which essentially represents the convergence between their graphs in a certain distance. The obtained results were exploited in Ref. [27] to examine the relationship between the various essential spectrum of A_n and A, where $(A_n)_n$ converges in the generalized sense to A.

In the last years, there have been many studies of the operators pencils, $\lambda S - A$, $\lambda \in \mathbb{C}$ (operator-valued functions of a complex argument) (see, for example, [119, 147]). It is known that many problems of mathematical physics (e.g., quantum theory, transport theory, $\cdots$) are reduced to the study of the essential spectra of $\lambda S - A$. For this, it seems interesting to study the following S-essential spectra:

$$
\begin{aligned}
\sigma_{e1,S}(A) &:= \{\lambda \in \mathbb{C} \text{ such that } \lambda S - A \notin \Phi_+(X)\},\\
\sigma_{e2,S}(A) &:= \{\lambda \in \mathbb{C} \text{ such that } \lambda S - A \notin \Phi_-(X)\},\\
\sigma_{e3,S}(A) &:= \{\lambda \in \mathbb{C} \text{ such that } \lambda S - A \notin \Phi_+(X) \bigcup \Phi_-(X)\},\\
\sigma_{e4,S}(A) &:= \{\lambda \in \mathbb{C} \text{ such that } \lambda S - A \notin \Phi(X)\},\\
\sigma_{e5,S}(A) &:= \{\lambda \in \mathbb{C} \text{ such that } \lambda S - A \notin \mathscr{W}(X)\},\\
\sigma_{b,S}(A) &:= \{\lambda \in \mathbb{C} \text{ such that } \lambda S - A \notin \mathscr{B}(X)\},\\
\sigma_{el,S}(A) &:= \{\lambda \in \mathbb{C} \text{ such that } \lambda S - A \notin \Phi_l(X)\},\\
\sigma_{er,S}(A) &:= \{\lambda \in \mathbb{C} \text{ such that } \lambda S - A \notin \Phi_r(X)\},\\
\sigma_{ewl,S}(A) &:= \{\lambda \in \mathbb{C} \text{ such that } \lambda S - A \notin \mathscr{W}_l(X)\},\\
\sigma_{ewr,S}(A) &:= \{\lambda \in \mathbb{C} \text{ such that } \lambda S - A \notin \mathscr{W}_r(X)\},\\
\sigma_{bl,S}(A) &:= \{\lambda \in \mathbb{C} \text{ such that } \lambda S - A \notin \mathscr{B}^l(X)\},\\
\sigma_{br,S}(A) &:= \{\lambda \in \mathbb{C} \text{ such that } \lambda S - A \notin \mathscr{B}^r(X)\},\\
\sigma_{eap,S}(A) &:= \Big\{\lambda \in \mathbb{C} \text{ such that } \lambda S - A \notin \Phi_+(X) \text{ and } i(\lambda S - A) \leq 0\Big\},\\
\sigma_{e\delta,S}(A) &:= \Big\{\lambda \in \mathbb{C} \text{ such that } \lambda S - A \notin \Phi_-(X) \text{ and } i(\lambda S - A) \geq 0\Big\},
\end{aligned}
$$

where $\mathscr{B}(X)$ is the set of Riesz-Schauder operators; $\Phi_l(X)$ is the set of left Fredholm operators; $\Phi_r(X)$ is the set of right Fredholm operators; $\mathscr{W}_l(X)$ is the set of left Weyl operators; $\mathscr{W}_r(X)$ is the set of right Weyl operators; $\mathscr{B}^l(X)$ is the set of left Browder operators; and $\mathscr{B}^r(X)$ is the set of right Browder operators. They can be ordered as

$$\sigma_{e3,S}(A) = \sigma_{e1,S}(A) \bigcap \sigma_{e2,S}(A) \subseteq \sigma_{e4,S}(A) \subseteq \sigma_{e5,S}(A) \subseteq \sigma_{b,S}(A),$$

$$\sigma_{el,S}(A) \subseteq \sigma_{ewl,S}(A) \subseteq \sigma_{bl,S}(A),$$

$$\sigma_{er,S}(A) \subseteq \sigma_{ewr,S}(A) \subseteq \sigma_{br,S}(A).$$

In Ref. [29], the authors proved that

$$\sigma_{e4,S}(A) = \sigma_{el,S}(A) \bigcup \sigma_{er,S}(A), \ \sigma_{e1,S}(A) \subset \sigma_{el,S}(A) \text{ and } \sigma_{e2,S}(A) \subset \sigma_{er,S}(A).$$

Note that if $S = I$, we recover the usual definition of the essential spectra of a bounded linear operator A defined in the first section of this introduction. These relative essential spectra drew the attention of Jeribi and his collaborators in Refs. [1, 10, 20, 29, 101].

More precisely, the authors A. Ammar, B. Boukettaya, and A. Jeribi pursued the analysis started in Refs. [1, 163, 164] for S-left-right essential spectra and they studied the invariance of $\sigma_{ei,S}(\cdot)$ $(i = l,\ r,\ wl,\ wr)$ by some class of perturbations and extended a part of the results obtained in Ref. [98] to a large class of perturbation operators $\mathscr{R}(X)$, which contains $\mathscr{F}(X)$ for the S-left and S-right spectrum, where $\mathscr{R}(X)$ (resp. $\mathscr{F}(X)$) denotes the set of Riesz operators (resp. Fredholm perturbations). They applied their obtained results to describe the S-essential spectra of an integro-differential operator with abstract boundary conditions acting in the Banach space.

At the same time, an important progress has been made in order to describe various relative essential spectra and we can cite in this context the contribution of the work of A. Ammar, M. Z. Dhahri and A. Jeribi in Ref. [25], where the authors are inspired by the work [10] and studied some types of S-essential spectra of linear bounded operators on a Banach space X. More precisely, in Ref. [25], we find a detailed treatment of some subsets of S-essential spectra of closed linear operators by means of the measure of non-strict-singularity.

When dealing with block operator matrices, we recall that the papers [39, 52, 54–56, 103, 111, 120, 146] are concerned with the study of the $\mathscr{I}$-essential spectra of operators defined by a 2×2 block operator matrix that

$$\mathscr{L}_0 = \begin{pmatrix} A & B \\ C & D \end{pmatrix}$$

acts on the product $X \times Y$ of Banach spaces, where $\mathscr{I}$ is the identity operator defined on the product space $X \times Y$ by

$$\mathscr{I} = \begin{pmatrix} I & 0 \\ 0 & I \end{pmatrix}.$$

Inspired by the works [39, 111, 120, 146], the aim of the authors in Ref. [101] was to generalize the previous work and they considered a bounded operator S formally defined on the product Banach space $X \times Y$ as

$$S = \begin{pmatrix} M_1 & M_2 \\ M_3 & M_4 \end{pmatrix}.$$

A considerable attention has been also devoted in Ref. [24] in order to give the characterization of the S-essential spectra of the 2×2 matrix operator $\mathscr{L}_0$ acting on a Banach space are given by using the notion of measure of non-strict-singularity. These results are considered as generalizations of the paper of N. Moalla in Ref. [122], where S-essential spectra of some 2×2 operator matrices on $X \times X$ are discussed with $S = I$.

1.2 ESSENTIAL PSEUDOSPECTRA

In 1967, J. M. Varah [155] introduced the first idea of pseudospectra. In 1986, J. H. Wilkinson [159] came up with the modern interpretation of pseudospectrum where he defined it for an arbitrary matrix norm induced by a vector norm. Throughout the 1990s, L. N. Trefethen [135, 150–152] not only initiated the study of pseudospectrum for matrices and operators, but also he talked of approximate eigenvalues and pseudospectrum and used this notion to study interesting problems in mathematical physics. By the same token, several authors worked on this field. For example, we may refer to E. B. Davies [58], A. Harrabi [80], A. Jeribi [1], E. Shargorodsky [145], and M. P. H. Wolff [161] who had introduced the term approximation pseudospectrum for linear operators.

Pseudospectra are interesting objects by themselves since they carry more information than spectra, especially about transient instead of just asymptotic behavior of dynamical systems. Also, they have better convergence and approximation properties than spectra. The definition of pseudospectra of a closed densely defined linear operator A, for every $\varepsilon > 0$, is given by:

$$\sigma_\varepsilon(A) := \sigma(A) \bigcup \left\{ \lambda \in \mathbb{C} \text{ such that } \|(\lambda - A)^{-1}\| > \frac{1}{\varepsilon} \right\},$$

or by

$$\Sigma_{\varepsilon}(A) := \sigma(A) \bigcup \left\{ \lambda \in \mathbb{C} \text{ such that } \|(\lambda - A)^{-1}\| \geq \frac{1}{\varepsilon} \right\}.$$

By convention, we write $\|(\lambda - A)^{-1}\| = \infty$ if $\lambda \in \sigma(A)$, (spectrum of A). For $\varepsilon > 0$, it can be shown that $\sigma_{\varepsilon}(A)$ is a larger set and is never empty. The pseudospectra of A are a family of strictly nested closed sets, which grow to fill the whole complex plane as $\varepsilon \to \infty$ (see [80, 151, 152]). From these definitions, it follows that the pseudospectra associated with various ε are nested sets. Then, for all $0 < \varepsilon_1 < \varepsilon_2$, we have

$$\sigma(A) \subset \sigma_{\varepsilon_1}(A) \subset \sigma_{\varepsilon_2}(A) \text{ and } \sigma(A) \subset \Sigma_{\varepsilon_1}(A) \subset \Sigma_{\varepsilon_2}(A),$$

and that the intersections of all the pseudospectra are the spectrum,

$$\bigcap_{\varepsilon > 0} \sigma_{\varepsilon}(A) = \sigma(A) = \bigcap_{\varepsilon > 0} \Sigma_{\varepsilon}(A).$$

In [58], E. B. Davies has defined another equivalent of pseudospectrum. One is in terms of perturbations of the spectrum, that is, for any closed operator A, we have

$$\sigma_{\varepsilon}(A) := \bigcup_{\|D\| < \varepsilon} \sigma(A + D).$$

Inspired by the notion of pseudospectra, A. Ammar and A. Jeribi in their works [29], thought to extend these results for essential spectra of densely closed linear operators on a Banach space. They declared the new concept of the essential pseudospectra of densely closed linear operators on a Banach space. Because of the existence of several essential spectra, they were interested in focusing their study on the pseudo-Browder essential spectrum. This set was shown to be characterized in the way one would expect by analogy with essential numerical range. As a consequence, the authors in Ref. [29] located the pseudo-Browder essential spectrum between the essential spectra and the essential numerical range.

More precisely, for $A \in \mathscr{C}(X)$ and for every $\varepsilon > 0$, they defined the pseudo-Browder essential spectrum as follows:

$$\sigma_{e6,\varepsilon}(A) = \sigma_{e6}(A) \bigcup \left\{ \lambda \in \mathbb{C} \text{ such that } \|R_b(A,\lambda)\| > \frac{1}{\varepsilon} \right\},$$

where $R_b(A,\lambda)$ is the Browder resolvent of A. The aim of this concept is to study the existence of eigenvalues far from the Browder essential spectrum and also to search the instability of the Browder essential spectrum under every perturbation. Their study of pseudo-Browder essential spectrum enabled them to determine and localize the Browder essential spectrum of a closed, densely defined linear operator on a Banach space.

In Refs. [1, 21, 29, 30], A. Jeribi and his collaborators pursued their studies about essential pseudospectra and defined the following sets:

$$\begin{array}{rcl}
\sigma_{e1,\varepsilon}(A) & := & \{\lambda \in \mathbb{C} \text{ such that } \lambda - A \notin \Phi^{\varepsilon}_{+}(X)\}, \\
\sigma_{e2,\varepsilon}(A) & := & \{\lambda \in \mathbb{C} \text{ such that } \lambda - A \notin \Phi^{\varepsilon}_{-}(X)\}, \\
\sigma_{e3,\varepsilon}(A) & := & \{\lambda \in \mathbb{C} \text{ such that } \lambda - A \notin \Phi^{\varepsilon}_{\pm}(X)\}, \\
\sigma_{e4,\varepsilon}(A) & := & \{\lambda \in \mathbb{C} \text{ such that } \lambda - A \notin \Phi^{\varepsilon}(X)\}, \\
\sigma_{e5,\varepsilon}(A) & := & \displaystyle\bigcap_{K \in \mathscr{K}(X)} \sigma_{\varepsilon}(A+K), \\
\sigma_{eap,\varepsilon}(A) & := & \sigma_{e1,\varepsilon}(A) \bigcup \{\lambda \in \mathbb{C} : i(\lambda - A - D) > 0, \ \forall \, \|D\| < \varepsilon\}, \\
\sigma_{e\delta,\varepsilon}(A) & := & \sigma_{e2,\varepsilon}(A) \bigcup \{\lambda \in \mathbb{C} : i(\lambda - A - D) < 0, \ \forall \, \|D\| < \varepsilon\},
\end{array}$$

where $\Phi^{\varepsilon}_{+}(X)$ (resp. $\Phi^{\varepsilon}_{-}(X)$) denotes the set of upper (resp. lower) pseudo semi-Fredholm operator, $\Phi^{\varepsilon}_{\pm}(X)$ denotes the set of pseudo semi-Fredholm operator, and $\Phi^{\varepsilon}(X)$ denotes the set of pseudo Fredholm operators. Note that if ε tends to 0, we recover the usual definition of the essential spectra of a closed linear operator A defined in the first section of this introduction.

In [161], M. P. H. Wolff has given a motivation to study the essential approximation pseudospectrum. In [33], A. Ammar, A. Jeribi and K. Mahfoudhi showed that the notion of essential approximation pseudospectrum can be extended by devoting our studies to the essential approximation spectrum. For $\varepsilon > 0$ and $A \in \mathscr{C}(X)$, they define

$$\begin{aligned}\sigma_{eap,\varepsilon}(A) &= \bigcap_{K \in \mathscr{K}(X)} \sigma_{ap,\varepsilon}(A+K),\\ \Sigma_{eap,\varepsilon}(A) &= \bigcap_{K \in \mathscr{K}(X)} \Sigma_{ap,\varepsilon}(A+K),\end{aligned}$$

where

$$\sigma_{ap,\varepsilon}(A) := \left\{\lambda \in \mathbb{C} \text{ such that } \inf_{x \in \mathscr{D}(A),\ \|x\|=1} \|(\lambda - A)x\| < \varepsilon\right\},$$

and

$$\Sigma_{ap,\varepsilon}(A) := \left\{\lambda \in \mathbb{C} \text{ such that } \inf_{x \in \mathscr{D}(A),\ \|x\|=1} \|(\lambda - A)x\| \leq \varepsilon\right\}.$$

In their work, the authors measure the sensitivity of the set $\sigma_{ap}(A)$ with respect to additive perturbations of A by an operator $D \in \mathscr{L}(X)$ of a norm less than ε. So, they define the approximation pseudospectrum of A by

$$\sigma_{ap,\varepsilon}(A) = \bigcup_{\|D\|<\varepsilon} \sigma_{ap}(A+D),$$

and they give a characterization of the essential approximation pseudospectrum $\sigma_{eap,\varepsilon}(\cdot)$ which nicely blends these properties of the essential and the approximation pseudospectra, and accordingly the authors are interested in the following essential approximation spectrum

$$\sigma_{eap}(A) := \bigcap_{K \in \mathscr{K}(X)} \sigma_{ap}(A+K).$$

As a consequence, the authors obtain the following result

$$\sigma_{eap,\varepsilon}(A) = \bigcup_{\|D\|<\varepsilon} \sigma_{eap}(A+D).$$

1.3 STRUCTURED ESSENTIAL PSEUDOSPECTRA AND RELATIVE STRUCTURED ESSENTIAL PSEUDOSPECTRA

Structured pseudospectra are very useful tools in control theory and other fields [81]. In particular, D. Hinrichsen and B. Kelb [81] presented in 1993 a graphical method to determine and visualize the spectral value sets of a matrix A.

We also refer the reader to E. B. Davies [58] who defined the structured pseudospectra, or spectral value sets of a closed densely defined linear operator A on a Banach space X by

$$\sigma(A,B,C,\varepsilon) = \bigcup_{\|D\|<\varepsilon} \sigma(A+CDB),$$

where $B \in \mathscr{L}(X,Y)$ (the set of bounded linear operators from X into Y) and $C \in \mathscr{L}(Z,X)$.

The notion of structured pseudospectra drew the attention of A. Elleuch and A. Jeribi who have recently been concerned in [67,68] with the study of essential spectra of an operator A which is subjected to structured perturbation of the form $A \rightsquigarrow A+CDB$, where B, C are given bounded operators and D is unknown disturbance operator satisfies $\|D\| < \varepsilon$ for a given $\varepsilon > 0$. More specifically, inspired by [1, 86–89, 91, 112], where the authors proved the invariance of the Schechter essential spectrum of a closed densely defined linear operators, A. Elleuch and A. Jeribi, in [67, 68], extended the obtained results for the case of structured Ammar-Jeribi essential pseudospectrum. For this analysis, they needed to characterize the structured pseudospectra of bounded operator and they defined the following structured essential pseudospectra of a closed, densely defined linear operator A with respect to the perturbation structure (B,C), where $B \in \mathscr{L}(X,Y)$, $C \in \mathscr{L}(Z,X)$ and uncertainty level $\varepsilon > 0$,

$$\sigma_{e4}(A,B,C,\varepsilon) := \bigcup_{\|D\|<\varepsilon} \sigma_{e4}(A+CDB),$$

$$\sigma_{e5}(A,B,C,\varepsilon) := \bigcup_{\|D\|<\varepsilon} \sigma_{e5}(A+CDB),$$

$$\sigma_{e6}(A,B,C,\varepsilon) := \bigcup_{\|D\|<\varepsilon} \sigma_{e6}(A+CDB).$$

The subset $\sigma_{e4}(A,B,C,\varepsilon)$ is the structured Wolf essential pseudospectrum, $\sigma_{e5}(A,B,C,\varepsilon)$ is the structured Ammar-Jeribi essential pseudospectrum, and $\sigma_{e6}(A,B,C,\varepsilon)$ denotes the structured Browder essential pseudospectrum.

The main purpose of the authors in Ref. [67] consists in studying the structured essential pseudospectra of the sum of two operators in order to improve and extend the results of Refs. [1, 12, 144] which were done for various essential spectra. In addition, the authors described the behavior of structured Ammar-Jeribi essential pseudospectrum and they presented a fine description of this set by means of compact and Fredholm perturbations.

Further, they used in Ref. [68] some criteria in order to prove the stability of the structured Wolf and the structured Ammar-Jeribi essential pseudospectra. They also used the notion of the measure of noncompactness to give a new characterization of the structured Ammar-Jeribi essential pseudospectrum and they presented some properties of the structured Ammar-Jeribi and Browder essential pseudospectra by means of the measure of non-strict-singularity.

On the other hand, based on the definition of the S-essential spectra and the structured S-pseudospectra, A. Elleuch and A. Jeribi, in [67, 68], introduced and characterized the structured S-pseudospectrum of a closed, densely defined linear operator on a Banach space X. They also established a relationship between the structured S-essential pseudospectra and the S-essential spectra of a bounded linear operator under some conditions.

1.4 CONDITION PSEUDOSPECTRUM

There are several generalizations of the concept of the spectrum in literature such as Ransford spectrum [134], pseudospectrum [1, 80, 152], and condition spectrum [109]. It is natural to ask whether there are any results similar for operator in infinite dimensional Banach spaces for these sets. The concept of condition pseudospectrum is very important since it carries more information than spectra and pseudospectra, especially, about transient instead of just asymptotic behavior of dynamical systems. Also, condition pseudospectrums have better convergence and approximation properties than spectra and pseudospectra. For a detailed study of the properties of the condition spectrum in finite dimensional space or in Banach algebras, we may refer to S. H. Kulkarni and D. Sukumar in Ref. [109], M. Karow in [106] and G. K. Kumar and S. H. Lui in Ref. [110].

The aim of the authors A. Ammar, A. Jeribi and K. Mahfoudhi in [35] is to show some properties of condition pseudospectra of a linear operator A in Banach spaces and reveal the relation between their condition pseudospectrum and the pseudospectrum (resp. numerical range). One of the central questions consists in characterizing the condition pseudospectra of all bounded linear operators acting in a Banach space. More precisely, the authors consider, in their work, one more such extension in terms of the condition number. In a complex Banach space, the condition pseudospectrum of a bounded linear operator A in infinite dimensional Banach spaces is given for every $0 < \varepsilon < 1$ by:

$$\Sigma^{\varepsilon}(A) := \sigma(A) \bigcup \left\{ \lambda \in \mathbb{C} \text{ such that } \|\lambda - A\| \|(\lambda - A)^{-1}\| > \frac{1}{\varepsilon} \right\},$$

with the convention that $\|\lambda - A\|\|(\lambda - A)^{-1}\| = \infty$, if $\lambda - A$ is not invertible.

The authors summarize some properties and useful results of the condition pseudospectrum and they characterize it in order to improve lots recent results of Refs. [106, 109, 110] in infinite dimensional Banach spaces.

1.5 OUTLINE OF CONTENTS

Going on to a brief indication of the contents of the individual chapters, we mention, first of all that our book consists of 12 chapters.

We recall in Chapter 2 some definitions, notations, and basic information on both bounded and unbounded Fredholm operators. We also introduce the concept of semigroup theory, compactly convergence, measure of noncompactness, gap topology, and quasi-inverse operator. Basic concepts and classical results are summarized in the first two sections. In the subsequent sections we study the approximate point spectrum, which is one of the most important examples of a spectrum in Banach algebras. The approximate point spectrum is closely related with the notions of removable and non-removable ideals.

We continue Chapter 3 by giving some fundamentals about characterization of various spectra and essential spectra such as left and right Jeribi essential spectra, S-resolvent set, S-spectra, S-essential spectra, pseudospectra and structured pseudospectra.

The aim of Chapter 4 is to investigate the perturbation of unbounded linear operators via the concept of γ-relative boundedness introduced in [92]. In fact, this chapter is a survey of the paper [22] where the authors try to generalize the some results of [61] by using the new concept of the gap and the γ-relative boundedness in order to establish some sufficient conditions imposed to three unbounded linear operators and to obtain the closedness of their algebraic sum. The obtained results are useful to study some specific properties of both 2×2 and 3×3 block operator matrices.

Chapter 5 addresses the characterization of essential spectra such as left and right Jeribi and Weyl essential spectra. We also study the stability of essential approximate point spectrum and essential defect spectrum of linear operators and we treat the elegant interaction between essential spectra of closed linear operators and the notion of convergence in the generalized sense.

In Chapters 6 and 7, we focus on the study of the stability of S-left and S-right essential spectra. Moreover, we give the characterization of the Schechter S-essential spectrum by means of compact and Fredholm perturbations and also via the concept of measure of non-strict-singularity. Finally, illustrate the obtained results to matrix operator.

We concentrate ourselves exclusively in Chapter 8 on the characterization of S-pseudospectra, structured S-pseudospectra, and structured S-essential pseudospectra. Inspired by the definition of essential spectra, we define and characterize in Chapter 9 the structured Ammar-Jeribi, Browder and Wolf, essential pseudospectra and in particular, we treat the relationship between structured Jeribi and structured Ammar-Jeribi essential pseudospectra.

Chapter 10 focuses on a new description of the structured essential pseudospectra. Not only do we give a characterization of the structured Ammar-Jeribi essential pseudospectrum by Kuratowski measure of noncompactness but also we investigate the structured Browder essential pseudospectrum by means of measure of non-strict-singularity.

The aim of Chapter 11 is to investigate the essential approximation pseudospectrum of closed, densely defined linear operators on a Banach space. We refine the notion of the ε-pseudospectrum and we focus on the characterization of the approximation pseudospectrum and the essential approximation pseudospectrum. We also establish some properties of essential pseudospectra. Further, we study the pseudospectrum of block operator matrix.

Finally, in Chapter 12, we summarize some properties and useful results of the condition pseudospectrum and we characterize it in infinite dimensional Banach spaces.

Chapter 2

Fundamentals

The aim of this chapter is to introduce the basic concepts, notations, and elementary results which are used throughout the book.

2.1 OPERATORS

Let X and Y be two Banach spaces. A mapping A which assigns to each element x of a set $\mathscr{D}(A) \subset X$ a unique element $y \in Y$ is called an operator (or transformation), we use the notation $A(X \longrightarrow Y)$. The set $\mathscr{D}(A)$ on which A acts is called the domain of A.

2.1.1 Linear Operators

The operator A is called linear, if $\mathscr{D}(A)$ is a subspace of X, and if

$$A(\lambda x + \beta y) = \lambda Ax + \beta Ay$$

for all scalars λ, β and all elements x, y in $\mathscr{D}(A)$. By an operator A from X into Y, we mean a linear operator with a domain $\mathscr{D}(A) \subset X$. We denote by $N(A)$ its null space, and $R(A)$ its range. For injective A, the inverse $A^{-1}(Y \longrightarrow X)$ is an operator with $\mathscr{D}(A^{-1}) = R(A)$ and $R(A^{-1}) = \mathscr{D}(A)$. The nullity, $\alpha(A)$, of A is defined as the dimension of $N(A)$ and the deficiency, $\beta(A)$, of A is defined as the codimension of $R(A)$ in Y.

2.1.2 Bounded Operators

The operator A is called bounded if there is a constant M such that

$$\|Ax\| \leq M\|x\|, \quad x \in X.$$

The norm of such an operator is defined by

$$\|A\| = \sup_{x \neq 0} \frac{\|Ax\|}{\|x\|}.$$

For an introduction to the theory of bounded or unbounded linear operators, we refer to the books of E. B. Davies [58], N. Dunford and J. T. Schwartz [65], I. C. Gohberg, S. Goldberg and M. A. Kaashoek [73], A. Jeribi [1–3] and T. Kato [108].

2.1.3 Closed and Closable Operators

The graph $G(A)$ of a linear operator A on $\mathscr{D}(A) \subset X$ into Y is the set

$$\{(x, Ax) \text{ such that } x \in \mathscr{D}(A)\}$$

in the product space $X \times Y$. Then, A is called a closed linear operator when its graph $G(A)$ constitutes a closed linear subspace of $X \times Y$. Hence, the notion of a closed linear operator is an extension of the notion of a bounded linear operator. The operator B will be called A-closed if $x_n \to x$, $Ax_n \to y$, $Bx_n \to z$ for $\{x_n\} \subset \mathscr{D}(A)$ implies that $x \in \mathscr{D}(B)$ and $Bx = z$. A sequence $(x_n)_n \subset \mathscr{D}(A)$ will be called A-convergent to $x \in X$ if both $(x_n)_n$ and $(Ax_n)_n$ are Cauchy sequences and $x_n \to x$. A linear operator A on $\mathscr{D}(A) \subset X$ into Y is said to be closable if A has a closed extension. It is equivalent to the condition that the graph $G(A)$ is a submanifold (or subspace) of a closed linear manifold (or space) which is at the same time a graph. It follows that A is closable if, and only if, the closure $\overline{G(A)}$ of $G(A)$ is a graph. We are thus led to the criterion: A is closable if, and only if, no element of the form $(0, x)$, $x \neq 0$ is the limit of elements of the form (x, Ax). In other words, A is closable if, and only if, $(x_n)_n \in \mathscr{D}(A)$, $x_n \to 0$ and $Ax_n \to x$ imply $x = 0$. When A is closable, there is a closed operator $\overline{A}$ with $G(\overline{A}) = \overline{G(A)}$. $\overline{A}$

is called the closure of A. It follows immediately that $\overline{A}$ is the smallest closed extension of A, in the sense that any closed extension of A is also an extension of $\overline{A}$. Since $x \in \mathscr{D}(\overline{A})$ is equivalent to $(x, \overline{A}x) \in \overline{G(A)}$, $x \in X$ belongs to $\mathscr{D}(\overline{A})$ if, and only if, there exists a sequence $(x_n)_n$ that is A-convergent to x. In this case, we have

$$\overline{A}x = \lim_{n \to \infty} Ax_n.$$

The operator B will be called A-closable if $x_n \to 0$, $Ax_n \to 0$, $Bx_n \to z$ implies $z = 0$.

By $\mathscr{C}(X,Y)$, we denote the set of all closed, densely defined linear operators from X into Y, and by $\mathscr{L}(X,Y)$ the Banach space of all bounded linear operators from X into Y. If $X = Y$, the sets $\mathscr{L}(X,Y)$ and $\mathscr{C}(X,Y)$ are replaced, respectively, by $\mathscr{L}(X)$ and $\mathscr{C}(X)$.

It is easy to see that a bounded operator defined on the whole Banach space X is closed. The inverse is also true and follows from the closed graph theorem which is the following theorem.

Theorem 2.1.1 *If X, Y are Banach spaces, and A is a closed linear operator from X into Y, with $\mathscr{D}(A) = X$, then $A \in \mathscr{L}(X,Y)$.* ◊

Theorem 2.1.2 *If A is a one-to-one closed linear operator from X into Y, then $R(A)$ is closed in Y if, and only if, A^{-1} is a bounded linear operator from Y into X.* ◊

Theorem 2.1.3 *Let X, Y be Banach spaces, and let A be a closed linear operator from X into Y. Then, a necessary and sufficient condition that $R(A)$ be closed in Y is that*

$$dist(x, N(A)) \leq C\|Ax\|, \;\; x \in \mathscr{D}(A)$$

holds, where C is a constant and $dist(x, N(A))$ denotes the distance between x and $N(A)$. ◊

We will recall the Hahn-Banach theorem for normed spaces:

Theorem 2.1.4 *Let X be a normed space and let $x_0 \neq 0$ be any element of X. Then, there exists a bounded linear functional f on X such that $\|f\| = 1$ and $f(x_0) = \|x_0\|$.* ◇

We will recall the Liouville theorem.

Theorem 2.1.5 *Let $f(z)$ be an analytic function on the complex plane. If $f(z)$ is bounded, then $f(z)$ is constant.* ◇

2.1.4 Adjoint Operator

Let $A \in \mathscr{L}(X,Y)$. For each $y' \in Y^*$ (the adjoint space of Y), the expression $y'(Ax)$ assigns a scalar to each $x \in X$. Thus, it is a functional $F(x)$. Clearly, F is linear. It is also bounded since

$$\begin{aligned} |F(x)| &= |y'(Ax)| \\ &\leq \|y'\|\|Ax\| \\ &\leq \|y'\|\|A\|\|x\|. \end{aligned}$$

Thus, there is an $x' \in X^*$ (the adjoint space of X) such that

$$y'(Ax) = x'(x), \quad x \in X. \tag{2.1}$$

This functional is unique, for any other functional satisfying (2.1) would have to coincide with x' on each $x \in X$. Thus, (2.1) can be written in the form

$$y'(Ax) = A^*y'(x). \tag{2.2}$$

The operator A^* is called the adjoint of A, depending on the mood one is in. If A, B are bounded operators defined everywhere, it is easily checked that $(BA)^* = A^*B^*$. We just follow the definition of adjoint for bounded operator (see (2.2)) for define the adjoint of unbounded operator.

By an operator A from X into Y, we mean a linear operator with a domain $\mathscr{D}(A) \subset X$. We want

$$A^*y'(x) = y'(Ax), \quad x \in \mathscr{D}(A).$$

Thus, we say that $y' \in \mathscr{D}(A^*)$ if there is an $x' \in X^*$ such that

$$x'(x) = y'(Ax), \; x \in \mathscr{D}(A). \tag{2.3}$$

Then, we define A^*y' to be x'. In order that this definition make sense, we need x' to be unique, i.e., that

$$x'(x) = 0$$

for all $x \in \mathscr{D}(A)$ should imply that $x' = 0$. Thus, it is true if $\mathscr{D}(A)$ is dense in X. To summarize, we can define A^* for any linear operator from X into Y provided $\mathscr{D}(A)$ is dense in X. We take $\mathscr{D}(A^*)$ to be the set of those $y' \in Y^*$ for which there is an $x' \in X^*$ satisfying (2.3). This x' is unique, and we set

$$A^*y' = x'.$$

If A, B are only densely defined, $\mathscr{D}(AB)$ need not be dense, and consequently, $(BA)^*$ need not exist. If $\mathscr{D}(BA)$ is dense, it follows that

$$\mathscr{D}(A^*B^*) \subset \mathscr{D}((BA)^*)$$

and

$$(BA)^*z' = A^*B^*z',$$

for all $z' \in \mathscr{D}(A^*B^*)$. Let N be a subset of a normed vector space X. A functional $x' \in X^*$, is called an annihilator of N if $x'(x) = 0$, $x \in N$. The set of all annihilators of N is denoted by $N^\perp$. The subspace $N^\perp$ is closed. For any subset R of X^*, we call $x \in X$ an annihilator of R if

$$x'(x) = 0,$$

$x' \in R$. We denote the set of such annihilators of R by ${}^\perp R$. The subspace ${}^\perp R$ is closed. If M is a closed subspace of X, then

$${}^\perp(M^\perp) = M.$$

Lemma 2.1.1 *(M. Schechter [140, Lemma 4.14, p. 93]) If $x'_1, \cdots, x'_m$ are linearly independent vectors in X^*, then there are vectors $x_1, \cdots, x_m$ in X, such that*

$$x'_j(x_k) = \delta_{jk} := \begin{cases} 1 & \text{if } j = k \\ 0 & \text{if } j \neq k \end{cases} \quad 1 \leq j, k \leq m. \tag{2.4}$$

Moreover, if $x_1, \cdots, x_m$ are linearly independent vectors in X, then there are vectors $x'_1, \cdots, x'_m$ in X^*, such that (2.4) holds. ♢

We will recall a simple lemma due to F. Riesz.

Lemma 2.1.2 *Let M be a closed subspace of a normed vector space X. If M is not the whole of X, then for each number θ satisfying $0 < \theta < 1$, there is an element $x_\theta \in X$, such that $\|x_\theta\| = 1$ and*

$$dist(x_\theta, M) := \inf_{x \in M} \|x_\theta - x\| \geq \theta,$$

where $dist(x_\theta, M)$ denotes the distance between x_θ and M. ♢

2.1.5 Direct Sum

It is obvious that the sum $M+N$ of two linear subspaces M and N of a vector space X is again a linear subspace. If $M \bigcap N = \{0\}$, then this sum is called the direct sum of M and N, and will be denoted by $M \bigoplus N$. In this case, for every $z = x + y$ in $M+N$, the components x, y are uniquely determined. A subspace $M \subset X$ is said to be complemented, if there exists a closed subspace $N \subset X$, such that $X = M \bigoplus N$. If $X = M \bigoplus N$, then N is called an algebraic complement of M. A particularly important class of endomorphisms are the so-called projections. If $X = M \bigoplus N$ and $x = y + z$, with $x \in M$ and $y \in N$, define $P : X \longrightarrow M$ by $Px := y$. The linear map P projects X onto M along N. Clearly, $I - P$ projects X onto N along M and, we have

$$P(X) = N(I-P) = M,$$

$$N(P) = (I-P)(X) = N,$$

with $P^2 = P$, i.e., P is an idempotent operator. Suppose now that X is a Banach space. If $X = M \bigoplus N$ and the projection P is continuous, then M

is said to be complemented and N is said to be a topological complement of M. Note that each complemented subspace is closed, but the converse is not true, for instance c_0, the Banach space of all sequences which converge to 0, is a not complemented closed subspace of l_∞, where l_∞ denotes the Banach space of all bounded sequences, see [126].

Lemma 2.1.3 *(V. Müller [124, Lemma 2, p. 156]) Let $T \in \mathscr{L}(X,Y)$ and let $F \subset Y$ be a finite-dimensional subspace. Suppose that $R(T)+F$ is closed. Then, $R(T)$ is closed. In particular, if $codimR(T) < \infty$, then $R(T)$ is closed.*

Proof. Let $F_0 = R(T)\bigcap F$ and choose a subspace F_1 such that

$$F = F_0 \bigoplus F_1.$$

Let

$$S : \left(X_{|N(T)}\right) \bigoplus F_1 \longrightarrow Y$$

be the operator defined by

$$S(x+N(T))\bigoplus f = Tx+f.$$

Then,

$$R(S) = R(T)+F$$

and S is one-to-one. Hence, S is bounded below, and consequently, the space

$$R(T) = S\left(X_{|N(T)}\bigoplus\{0\}\right)$$

is closed. Q.E.D.

2.1.6 Resolvent Set and Spectrum

Definition 2.1.1 *Let A be a closable linear operator in a Banach space X. The resolvent set and the spectrum of A are, respectively, defined as*

$$\begin{aligned} \rho(A) &= \{\lambda \in \mathbb{C} \text{ such that } \lambda - A \text{ is injective and } (\lambda - A)^{-1} \in \mathscr{L}(X)\}, \\ \sigma(A) &= \mathbb{C}\backslash\rho(A) \end{aligned}$$

and the point spectrum, continuous, and the residual spectrum are defined as

$$\begin{aligned}\sigma_p(A) &= \{\lambda \in \mathbb{C} \text{ such that } \lambda - A \text{ is not injective}\},\\ \sigma_c(A) &= \{\lambda \in \mathbb{C} \text{ such that } \lambda - A \text{ is injective, } \overline{R(\lambda - A)} = X, \text{ and } \\ &\qquad R(\lambda - A) \neq X\},\\ \sigma_R(A) &= \{\lambda \in \mathbb{C} \text{ such that } \lambda - A \text{ is injective, } \overline{R(\lambda - A)} \neq X\}. \quad \diamondsuit\end{aligned}$$

Remark 2.1.1 *Note that, if $\rho(A) \neq \emptyset$, then A is closed. In fact, if $\lambda \in \rho(A)$, then $(\lambda - A)^{-1}$ is closed, which is also valid for $\lambda - A$. Then, according to the closed graph theorem (see Theorem 2.1.1), we deduce that*

$$\rho(A) = \{\lambda \in \mathbb{C} \text{ such that } \lambda - A \text{ is bijective}\}$$

and hence,

$$\sigma(A) = \sigma_p(A) \bigcup \sigma_c(A) \bigcup \sigma_R(A). \qquad \diamondsuit$$

Theorem 2.1.6 *Let A, B be elements of $\mathcal{L}(X)$. Then,*

$$\sigma(AB)\backslash\{0\} = \sigma(BA)\backslash\{0\}. \qquad \diamondsuit$$

2.1.7 Compact Operators

Definition 2.1.2 *Let X and Y be two Banach spaces. A bounded linear operator K from X into Y is said to be compact (or completely continuous) if it transforms every bounded set of X in a relatively compact set of Y. In a similar way, K is said to be compact if for any bounded sequence $(x_n)_n$ of X, the sequence $(Kx_n)_n$ has a convergent sequence in Y.* $\diamondsuit$

We denote by $\mathcal{K}(X,Y)$ the set of all compact linear operators from X into Y. If $X = Y$, then the set $\mathcal{K}(X,X)$ is replaced by $\mathcal{K}(X)$.

Theorem 2.1.7 *Let $K \in \mathcal{L}(X,Y)$. Then,*

(i) If $(K_n)_n$ is compact and converge in norm to K, then K is compact.

(ii) K^* is compact if, and only if, K is compact.

(iii) $\mathcal{K}(X,Y)$ is a two-sided ideal of $\mathcal{L}(X,Y)$. $\diamondsuit$

Lemma 2.1.4 *For any n elements $a_1, \cdots, a_n$ of a complex Banach space X, the set E, of their linear combinations, forms a subspace of X of dimension $\leq n$.* ◇

Definition 2.1.3 *A bounded operator A on X is said to be finite rank if* $\dim R(A) < \infty$. ◇

Lemma 2.1.5 *Operators of finite rank are compact.* ◇

2.1.8 A-Defined, A-Bounded, and A-Compact Operators

Let $A \in \mathscr{C}(X,Y)$. The graph norm of A is defined by

$$\|x\|_A = \|x\| + \|Ax\|, \; x \in \mathscr{D}(A). \tag{2.5}$$

From the closedness of A, it follows that $\mathscr{D}(A)$ endowed with the norm $\|\cdot\|_A$ is a Banach space. Let us denote by X_A, the space $\mathscr{D}(A)$ equipped with the norm $\|\cdot\|_A$. Clearly, the operator A satisfies

$$\|Ax\| \leq \|x\|_A$$

and consequently,

$$A \in \mathscr{L}(X_A, X).$$

Let $J : X \longrightarrow Y$ be a linear operator on X. If $\mathscr{D}(A) \subset \mathscr{D}(J)$, then J will be called A-defined. If J is A-defined, we will denote by $\widehat{J}$ its restriction to $\mathscr{D}(A)$. Moreover, if $\widehat{J} \in \mathscr{L}(X_A, Y)$, we say that J is A-bounded. We can easily check that, if J is closed (or closable), then J is A-bounded. Thus, we have the obvious relation

$$\left\{ \begin{array}{l} \alpha(\widehat{A}) = \alpha(A), \;\; \beta(\widehat{B}) = \beta(B), \;\;\; R(\widehat{A}) = R(A), \\ \alpha(\widehat{A}+\widehat{B}) = \alpha(A+B), \\ \beta(\widehat{A}+\widehat{J}) = \beta(A+J) \;\text{ and }\; R(\widehat{A}+\widehat{J}) = R(A+J), \end{array} \right. \tag{2.6}$$

where $\alpha(A) := \dim N(A)$ and $\beta(A) := \operatorname{codim} R(A)$.

Definition 2.1.4 *Let X and Y be two Banach spaces, $A \in \mathscr{C}(X,Y)$ and let J be an A-defined linear operator on X. We say that J is A-compact if $\widehat{J} \in \mathscr{K}(X_A, Y)$.* ◇

We can define the A-compact by the sequence.

Definition 2.1.5 *Let X, Y and Z be Banach spaces, and let A and S be two linear operators from X into Y and from X into Z, respectively. S is called relatively compact with respect to A (or A-compact), if* $\mathscr{D}(A) \subset \mathscr{D}(S)$ *and for every bounded sequence* $(x_n)_n \in \mathscr{D}(A)$ *such that* $(Ax_n)_n \subset Y$ *is bounded, the sequence* $(Sx_n)_n \subset Z$ *contains a convergent subsequence.*

Let $A\mathscr{K}(X,Y)$ denotes the set of A-compact on X. If $X = Y$ we write $A\mathscr{K}(X)$ for $A\mathscr{K}(X,X)$.

We can define the iterates $A^2, A^3, \cdots$ of A. If $n > 1$, $\mathscr{D}(A^n)$ is the set

$$\Big\{x \in X \text{ such that } x, Ax, \cdots, A^{n-1}x \in \mathscr{D}(A)\Big\},$$

and

$$A^n x = A(A^{n-1}x).$$

Definition 2.1.6 *Let X be a Banach space. We consider the following linear operators* $A : \mathscr{D}(A) \subset X \longrightarrow X$ *and* $K : \mathscr{D}(K) \subset X \longrightarrow X$. *We say that K commutes with A if*

(i) $\mathscr{D}(A) \subset \mathscr{D}(K)$,
(ii) $Kx \in \mathscr{D}(A)$ *whenever* $x \in \mathscr{D}(A)$, *and*
(iii) $KAx = AKx$ *for* $x \in \mathscr{D}(A^2)$. ♢

Definition 2.1.7 *Let X, Y and Z be Banach spaces, and let A and S be two linear operators from X into Y and from X into Z, respectively. S is called relatively bounded with respect to A (or A-bounded), if* $\mathscr{D}(A) \subset \mathscr{D}(S)$ *and there exist two constants* $a_S \geq 0$, *and* $b_S \geq 0$, *such that*

$$\|Sx\| \leq a_S\|x\| + b_S\|Ax\|, \quad x \in \mathscr{D}(A). \tag{2.7}$$

The infimum δ *of all* b_S *that (2.7) holds for some* $a_S \geq 0$ *is called relative bound of S with respect to A (or A-bounded of S).* ♢

Clearly, a compact operator is always A-compact, and an A-compact operator is always A-bounded.

Definition 2.1.8 *An operator B is said to be A-pseudo-compact if*

$$\|x_n\| + \|Ax_n\| + \|Bx_n\| \leq c,$$

for all $\{x_n\} \in \mathscr{D}(A)$ implies that $\{Bx_n\}$ has a convergent subsequence, where c is a constant. $\diamondsuit$

2.1.9 Weakly Compact and A-Weakly Compact Operators

Definition 2.1.9 *An operator $A \in \mathscr{L}(X,Y)$ is said to be weakly compact, if $A(B)$ is relatively weakly compact in Y, for every bounded subset $B \subset X$.* $\diamondsuit$

The family of weakly compact operators from X into Y is denoted by $\mathscr{W}(X,Y)$. If $X = Y$, then the family of weakly compact operators on X, $\mathscr{W}(X) := \mathscr{W}(X,X)$, is a closed two-sided ideal of $\mathscr{L}(X)$ containing $\mathscr{K}(X)$ (cf. [65, 75]).

Definition 2.1.10 *Let X and Y be two Banach spaces and let S and A be two linear operators from X into Y. The operator S is said to be relatively weakly compact with respect to A* (or A-weakly compact) *if $\mathscr{D}(A) \subset \mathscr{D}(S)$ and, for every bounded sequence $(x_n)_n \in \mathscr{D}(A)$ such that $(Ax_n)_n$ is bounded, the sequence $(Sx_n)_n$ contains a weakly convergent subsequence.* $\diamondsuit$

2.1.10 Dunford-Pettis Property

Definition 2.1.11 *Let X be a Banach space. We say that X possesses the Dunford-Pettis property (for short, property DP) if, for each Banach space Y, every weakly compact operator $A : X \longrightarrow Y$ takes weakly compact sets in X into norm compact sets of Y.* $\diamondsuit$

It is well known that if X has the property DP, then

$$\mathcal{W}(X)\mathcal{W}(X) \subset \mathcal{K}(X). \tag{2.8}$$

Indeed, Let A_1 and $A_2 \in \mathcal{W}(X)$. If U is a bounded subset of X, then $A_1(U)$ is relatively weakly compact. Accordingly, since X has the DP property, then $A_2(A_1(U))$ is a relatively compact subset of X. That is, $A_2A_1 \in \mathcal{K}(X)$. It is well known that any L_1-space has the property DP [64]. Also, if Ω is a compact Hausdorff space, then $C(\Omega)$ has the property DP [78]. For further examples we refer to [62]. Actually, if X is a separable Hilbert space, then $\mathcal{K}(X)$ is the unique proper closed two-sided ideal of $\mathcal{L}(X)$. This result hold true for the space l_p, $1 \leq p \leq \infty$ and c_0.

2.1.11 Strictly Singular Operators

Definition 2.1.12 *Let X and Y be two Banach spaces. An operator $A \in \mathcal{L}(X,Y)$ is called strictly singular if, for every infinite-dimensional subspace M, the restriction of A to M is not a homeomorphism.* ◇

Let $\mathcal{S}(X,Y)$ denotes the set of strictly singular operators from X into Y.

The concept of strictly singular operators was introduced by T. Kato in his pioneering paper [108] as a generalization of the notion of compact operators. For a detailed study of the properties of strictly singular operators, we may refer to [75, 108]. For our own use, let us recall the following four facts. The set $\mathcal{S}(X,Y)$ is a closed subspace of $\mathcal{L}(X,Y)$. If $X = Y$, then $\mathcal{S}(X) := \mathcal{S}(X,X)$ is a closed two-sided ideal of $\mathcal{L}(X)$ containing $\mathcal{K}(X)$. If X is a Hilbert space, then

$$\mathcal{K}(X) = \mathcal{S}(X)$$

and the class of weakly compact operators on L_1-spaces (resp. $C(K)$-spaces with K being a compact Haussdorff space) is, nothing else but, the family of strictly singular operators on L_1-spaces (resp. $C(K)$-spaces) (see [128, Theorem 1]). In fact, if $X = L_1(\Omega, \Sigma, \mu)$, where (Ω, Σ, μ) is a positive measure space or $X = C(K)$ with K is a compact Hausdorff space, then

$$\mathscr{W}(X) = \mathscr{S}(X). \tag{2.9}$$

For the detailed study of the basic properties of the classes $\mathscr{S}(X)$, we refer the reader to [127, Theorem 1].

Lemma 2.1.6 *(A. Pelczynski [127]) Let (Ω, Σ, μ) be a positive measure space and let $p \geq 1$. If X is isomorphic to one of the spaces $L_p(\Omega, \Sigma, d\mu)$, then*

$$\mathscr{S}(X)\mathscr{S}(X) \subset \mathscr{K}(X). \qquad \diamondsuit$$

Definition 2.1.13 *Let X and Y be two Banach spaces, $A \in \mathscr{C}(X,Y)$ and let J be an A-defined linear operator on X. We say that J is A-strictly singular if $\widehat{J} \in \mathscr{S}(X_A, Y)$.* $\diamondsuit$

Let $A\mathscr{S}(X,Y)$ denotes the set of A-strictly singular from X into Y. If $X = Y$ we write $A\mathscr{S}(X)$ for $A\mathscr{S}(X,X)$.

Remark 2.1.2 *If J is A-defined and strictly singular, then J is A-strictly singular.* $\diamondsuit$

2.1.12 Strictly Cosingular

Let X be a Banach space. If N is a closed subspace of X, we denote by π_N^X the quotient map $X \longrightarrow X/N$. The codimension of N, codim(N), is defined to be the dimension of the vector space X/N.

Definition 2.1.14 *Let X and Y be two Banach spaces. An operator $F \in \mathscr{L}(X,Y)$ is called strictly cosingular from X into Y if there exists no closed subspace N of Y with codim(N) $= \infty$ such that $\pi_N^Y F : X \longrightarrow Y/N$ is surjective.* $\diamondsuit$

Let $\mathscr{S}\mathscr{C}(X,Y)$ denotes the set of strictly cosingular operators from X into Y. This class of operators was introduced by A. Pelczynski [128]. If $X = Y$, the family of strictly cosingular operators on X, $\mathscr{S}\mathscr{C}(X) := \mathscr{S}\mathscr{C}(X,X)$, is a closed two-sided ideal of $\mathscr{L}(X)$ (see [156]).

Definition 2.1.15 *Let X and Y be two Banach spaces, $A \in \mathscr{C}(X,Y)$ and let J be an A-defined linear operator on X. We say that J is A-strictly cosingular if $\widehat{J} \in \mathscr{S}\mathscr{C}(X_A,Y)$.* ♢

Let $A\mathscr{S}\mathscr{C}(X,Y)$ denotes the set of A-strictly cosingular from X into Y. If $X = Y$, we write $A\mathscr{S}\mathscr{C}(X)$ for $A\mathscr{S}\mathscr{C}(X,X)$.

Remark 2.1.3 *If J is A-defined and strictly cosingular, then J is A-strictly cosingular.* ♢

2.1.13 Perturbation Function

The family of infinite dimensional subspaces of X is denoted by $\mathscr{I}(X)$. For a linear subspace M of X, we denote by i_M the canonical injection of M into X and by Ai_M the restriction of A to M with the usual convention $Ai_M = Ai_{(M \bigcap \mathscr{D}(A))}$.

Definition 2.1.16 *Let $A : \mathscr{D}(A) \subset X \longrightarrow Y$ be a linear operator. The minimum modulus of A is defined as follows*

$$\gamma_1(A) = \sup\Big\{\alpha \geq 0 \text{ such that } \alpha \; dist(x,N(A)) \leq \|Ax\| \; \forall \, x \in \mathscr{D}(A)\Big\},$$

where $dist(x,N(A))$ denotes the distance between x and $N(A)$. ♢

Definition 2.1.17 *Let X, Y be normed spaces and let $A : \mathscr{D}(A) \subset X \longrightarrow Y$ be a linear operator. The surjection modulus $q(A)$ is defined by*

$$q(A) = \sup\{\alpha \geq 0 \text{ such that } A(B_{\mathscr{D}(A)}) \supset B_Y\},$$

where $B_{\mathscr{D}(A)}$ (resp. B_Y) denotes the closed unit ball in $\mathscr{D}(A)$ (resp. Y), that is, the set of all $x \in \mathscr{D}(A)$ (resp. $x \in Y$) satisfying $\|x\| \leq 1$. ♢

Definition 2.1.18 *We define a perturbation function to be a function ψ assigning, to each pair of normed spaces X, Y and a linear operator $A : \mathscr{D}(A) \subset X \longrightarrow Y$, a number $\psi(A) \in [0,\infty]$, verifying the following properties:*

(i) $\psi(\lambda A) = |\lambda|\psi(A)$, *for all* $\lambda \in \mathbb{C}$.

(ii) $\psi(A+S) = \psi(A)$, *where* S *is a precompact operator.*

(iii) $\psi(A) \leq \|A\|$ *whenever* $A \in \mathscr{L}(X,Y)$. *Otherwise* $\|A\| = \infty$.

(iv) $q(A) \leq \psi(A)$ *whenever* $\dim \mathscr{D}(A^*) = \infty$.

(v) $\psi(Ai_M) \leq \psi(A)$, $M \in \mathscr{I}(\mathscr{D}(A))$, *where* $\mathscr{I}(\mathscr{D}(A))$ *denotes the family of infinite dimensional subspaces of* $\mathscr{D}(A)$. ◇

2.1.14 Measure of Non-Strict-Singularity

Definition 2.1.19 *Let* X, Y *be two normed spaces and* ψ *be a perturbation function. For a linear operator* $A : \mathscr{D}(A) \subset X \longrightarrow Y$, *we define the functions* $\Gamma_\psi(\cdot)$ *and* $\Delta_\psi(\cdot)$ *in the following way:*
If $\dim \mathscr{D}(A) < \infty$, *then* $\Gamma_\psi(A) = 0$. *However, if* $\dim \mathscr{D}(A) = \infty$, *then*

$$\Gamma_\psi(A) = \inf\Big\{\psi(Ai_M) : M \in \mathscr{I}(\mathscr{D}(A))\Big\} \tag{2.10}$$

and

$$\Delta_\psi(A) = \sup\Big\{\Gamma_\psi(Ai_M) : M \in \mathscr{I}(\mathscr{D}(A))\Big\}. \tag{2.11}$$

$\Delta_\psi(\cdot)$ *is called the measure of non-strict-singularity.* ◇

Theorem 2.1.8 *(T. Alvarez [18, Theorem 2.7]) Let* $T \in \mathscr{L}(X,Y)$. *Then,* T *is strictly singular operator if, and only if,* $\Delta_\psi(T) = 0$. ◇

2.1.15 Semigroup Theory

Definition 2.1.20 *Let* X *be a Banach space. A strongly continuous semigroup is an operator-valued function* $T(t)$ *from* $\mathbb{R}_+$ *into* $\mathscr{L}(X)$ *that satisfies the following properties*

$$\begin{aligned} T(t+s) &= T(t)T(s) \text{ for } t,\ s \geq 0 \\ T(0) &= I \\ \|T(t)z_0 - z_0\| &\to 0 \text{ as } t \to 0^+ \text{ for all } z_0 \in X. \end{aligned}$$

We shall subsequently use the standard abbreviation C_0*-semigroup for a strongly continuous semigroup.* ◇

Definition 2.1.21 *The infinitesimal generator A of a C_0-semigroup on a Banach space X is defined by*

$$Az = \lim_{t \to 0^+} \frac{1}{t}(T(t) - I)z$$

whenever the limit exists; the domain of A, $\mathscr{D}(A)$, being the set of elements in X for which the limit exists. $\diamondsuit$

2.2 FREDHOLM AND SEMI-FREDHOLM OPERATORS

2.2.1 Definitions

Let X and Y be two Banach spaces. By an operator A from X into Y we mean a linear operator with domain $\mathscr{D}(A) \subset X$ and range $R(A) \subset Y$. The operator A is called left invertible if A is injective and $R(A)$ is closed and a complemented subspaces of X. The operator A is called right invertible if A is onto and $N(A)$ is a complemented subspaces of X. An operator $A \in \mathscr{C}(X,Y)$ is said to be a Fredholm operator from X into Y if

(i) $\alpha(A) := \dim N(A)$ is finite,

(ii) $R(A)$ is closed in Y, and

(iii) $\beta(A) := \text{codim} R(A)$ is finite.

The set of upper semi-Fredholm operators is defined by

$$\Phi_+(X,Y) := \Big\{ A \in \mathscr{C}(X,Y) \text{ such that } \alpha(A) < \infty \text{ and } R(A) \text{ is closed in } Y \Big\}.$$

The set of lower semi-Fredholm operators from X into Y is defined by

$$\Phi_-(X,Y) := \Big\{ A \in \mathscr{C}(X,Y) \text{ such that } \beta(A) < \infty \text{ and } R(T) \text{ is closed in } Y \Big\}.$$

The set of semi-Fredholm operators from X into Y is defined by

$$\Phi_\pm(X,Y) := \Phi_+(X,Y) \bigcup \Phi_-(X,Y).$$

The set of Fredholm operators from X into Y is defined by

$$\Phi(X,Y) := \Phi_+(X,Y) \bigcap \Phi_-(X,Y).$$

The set of left invertible and right invertible operators are denoted by $\mathcal{G}_l(X)$ and $\mathcal{G}_r(X)$, respectively. Note that A is invertible if A is left and right invertible. The set of left Fredholm operators is defined by

$$\Phi_l(X,Y) = \Big\{A \in \mathcal{C}(X,Y) \text{ such that } R(A) \text{ is closed, complemented subspaces and } \alpha(A) < \infty\Big\}$$

and the set of right Fredholm operators is defined by

$$\Phi_r(X,Y) = \Big\{A \in \mathcal{C}(X,Y) \text{ such that } R(A) \text{ is closed, } N(A) \text{ is complemented subspaces and } \beta(A) < \infty\Big\}.$$

Thus, we have the following inclusions

$$\Phi(X,Y) \subseteq \Phi_l(X,Y) \subseteq \Phi_+(X,Y)$$

and

$$\Phi(X,Y) \subseteq \Phi_r(X,Y) \subseteq \Phi_-(X,Y).$$

Let $A \in \mathcal{C}(X,Y)$ and $S \in \mathcal{L}(X,Y)$. A complex number λ is in $\Phi_{+A,S}$, $\Phi_{-A,S}$, $\Phi^l_{A,S}$, $\Phi^r_{A,S}$, $\Phi_{\pm A,S}$ or $\Phi_{A,S}$, if $\lambda S - A$ is in $\Phi_+(X,Y)$, $\Phi_-(X,Y)$, $\Phi_l(X,Y)$, $\Phi_r(X,Y)$, $\Phi_\pm(X,Y)$ or $\Phi(X,Y)$, respectively. Let $\Phi^b(X,Y)$, $\Phi^b_+(X,Y)$, and $\Phi^b_-(X,Y)$ denote the sets $\Phi(X,Y) \bigcap \mathcal{L}(X,Y)$, $\Phi_+(X,Y) \bigcap \mathcal{L}(X,Y)$, and $\Phi_-(X,Y) \bigcap \mathcal{L}(X,Y)$, respectively. If $X = Y$, then $\Phi_+(X,Y)$, $\Phi_-(X,Y)$, $\Phi_\pm(X,Y)$, $\Phi(X,Y)$, $\Phi_l(X,Y)$, $\Phi_r(X,Y)$, $\Phi^b(X,Y)$, $\Phi^b_+(X,Y)$ and $\Phi^b_-(X,Y)$ are replaced by $\Phi_+(X)$, $\Phi_-(X)$, $\Phi_\pm(X)$, $\Phi(X)$, $\Phi_l(X)$, $\Phi_r(X)$, $\Phi^b(X)$, $\Phi^b_+(X)$ and $\Phi^b_-(X)$, respectively. A number complex λ is in Φ_{+A}, Φ_{-A}, Φ_{lA}, Φ_{rA} or Φ_A if $\lambda - A$ is in $\Phi_+(X)$, $\Phi_-(X)$, $\Phi_l(X)$, $\Phi_r(X)$ or $\Phi(X)$, respectively.

Theorem 2.2.1 *Φ_A and $\rho(A)$ are open sets of $\mathbb{C}$. Hence, $\sigma(A)$ is a closed set.* ◇

If A is a semi-Fredholm operator, then the index of A is defined by

$$i(A) := \alpha(A) - \beta(A).$$

Clearly, if $A \in \Phi(X,Y)$, then

$$i(A) < \infty.$$

If $A \in \Phi_l(X,Y) \backslash \Phi(X,Y)$, then

$$i(A) = -\infty$$

and, if $A \in \Phi_r(X,Y) \backslash \Phi(X,Y)$, then

$$i(A) = +\infty.$$

The set of left Weyl operators is defined by

$$\mathscr{W}_l(X) = \Big\{ A \in \Phi_l(X) \text{ such that } i(A) \leq 0 \Big\}$$

and, the set of right Weyl operators is defined by

$$\mathscr{W}_r(X) = \Big\{ A \in \Phi_r(X) \text{ such that } i(A) \geq 0 \Big\}.$$

The set of Weyl operators is defined by

$$\mathscr{W}_l(X) \bigcap \mathscr{W}_r(X) = \Big\{ A \in \Phi(X) \text{ such that } i(A) = 0 \Big\}.$$

Let $A \in \mathscr{C}(X)$ and $S \in \mathscr{L}(X)$. A number complex λ is in $\mathscr{W}^l_{A,S}$ or $\mathscr{W}^r_{A,S}$ if $\lambda S - A$ is in $\mathscr{W}_l(X)$ or $\mathscr{W}_r(X)$, respectively.

Theorem 2.2.2 *Assume that T^* exists. Then, $R(T)$ is closed if, and only if, $R(T^*)$ is closed. In this case we have $R(T)^{\perp} = N(T^*)$, $N(T)^{\perp} = R(T^*)$, $\alpha(T^*) = \beta(T)$, and $\beta(T^*) = \alpha(T)$.* ◊

Definition 2.2.1 *Let $\varepsilon > 0$ and $A \in \mathscr{C}(X)$.*

(i) A is called an upper (resp. lower) pseudo semi-Fredholm operators if $A + D$ is an upper (resp. lower) semi-Fredholm operator for all $D \in \mathscr{L}(X)$ such that $||D|| < \varepsilon$.

(ii) A is called a pseudo semi-Fredholm operator if $A + D$ is a semi-Fredholm operator for all $D \in \mathscr{L}(X)$ such that $||D|| < \varepsilon$.

(iii) A is called a pseudo Fredholm operator if $A + D$ is a Fredholm operator for all $D \in \mathscr{L}(X)$ such that $||D|| < \varepsilon$. ◊

We denote by $\Phi^{\varepsilon}(X)$ the set of pseudo Fredholm operators, by $\Phi^{\varepsilon}_{\pm}(X)$ the set of pseudo semi-Fredholm operator and by $\Phi^{\varepsilon}_{+}(X)$ (resp. $\Phi^{\varepsilon}_{-}(X)$) the set of upper (resp. lower) pseudo semi-Fredholm operator. A complex number λ is in $\Phi^{\varepsilon}_{\pm A}$, Φ^{ε}_{+A}, Φ^{ε}_{-A} or Φ^{ε}_{A} if $\lambda - A$ is in $\Phi^{\varepsilon}_{\pm}(X)$, $\Phi^{\varepsilon}_{+}(X)$, $\Phi^{\varepsilon}_{-}(X)$ or $\Phi^{\varepsilon}(X)$, respectively.

2.2.2 Basics on Bounded Fredholm Operators

Theorem 2.2.3 *(M. Schechter [137, Theorem 5.4, p. 103]) Let X and Y be two Banach spaces, and let $A \in \Phi^{b}(X,Y)$. Then, there is a closed subspace X_0 of X, such that*

$$X = X_0 \bigoplus N(A)$$

and, a subspace Y_0 of Y of dimension $\beta(A)$ such that

$$Y = R(A) \bigoplus Y_0.$$

Moreover, there is an operator $A_0 \in \mathscr{L}(Y,X)$, such that $N(A_0) = Y_0$, $R(A_0) = X_0$,

$$A_0 A = I \text{ on } X_0,$$

and

$$AA_0 = I \text{ on } R(A).$$

In addition,

$$A_0 A = I - F_1 \text{ on } X,$$

and

$$AA_0 = I - F_2 \text{ on } Y,$$

where $F_1 \in \mathscr{L}(X)$ with $R(F_1) = N(A)$ and $F_2 \in \mathscr{L}(Y)$ with $R(F_2) = Y_0$. $\diamondsuit$

Theorem 2.2.4 *(M. Schechter [137, Theorem 5.5, p. 105]) Let* $A \in \mathcal{L}(X,Y)$*. Suppose that there exist* A_1*,* $A_2 \in \mathcal{L}(Y,X)$*,* $F_1 \in \mathcal{K}(X)$ *and* $F_2 \in \mathcal{K}(Y)$ *such that*

$$A_1 A = I - F_1 \text{ on } X$$

and

$$AA_2 = I - F_2 \text{ on } Y.$$

Then, $A \in \Phi^b(X,Y)$*.* ◇

Theorem 2.2.5 *(M. Schechter [139]) Let* $A \in \mathcal{L}(X,Y)$ *and* $B \in \mathcal{L}(Y,Z)$*, where* X*,* Y*, and* Z *are Banach spaces. If* A *and* B *are Fredholm operators, upper semi-Fredholm operators, lower semi-Fredholm operators, then* BA *is a Fredholm operator, an upper semi-Fredholm operator, a lower semi-Fredholm operator, respectively, and*

$$i(BA) = i(B) + i(A).$$ ◇

Theorem 2.2.6 *(M. Schechter [137, Theorem 5.13, p. 110]) Let* X*,* Y*, and* Z *be three Banach spaces. Assume that* $A \in \mathcal{L}(X,Y)$ *and* $B \in \mathcal{L}(Y,Z)$ *are such that* $BA \in \Phi^b(X,Z)$*. Then,* $A \in \Phi^b(X,Y)$ *if, and only if,* $B \in \Phi^b(Y,Z)$*.* ◇

Theorem 2.2.7 *(M. Schechter [137, Theorem 5.14, p. 111]) Let* X*,* Y*, and* Z *be three Banach spaces. Assume that* $A \in \mathcal{L}(X,Y)$ *and* $B \in \mathcal{L}(Y,Z)$ *are such that* $BA \in \Phi^b(X,Z)$*. If* $\alpha(B) < \infty$*, then* $A \in \Phi^b(X,Y)$ *and* $B \in \Phi^b(Y,Z)$*.* ◇

Theorem 2.2.8 *(M. Schechter [137, Theorem 7.3, p. 157]) Let* X *and* Y *be Banach spaces. If* $A \in \Phi^b(X,Y)$ *and* $B \in \Phi^b(Y,Z)$*, then* $BA \in \Phi^b(X,Z)$ *and*

$$i(BA) = i(B) + i(A).$$ ◇

Lemma 2.2.1 *(A. Lebow and M. Schechter [117, Lemma 4.3]) Let* X *and* Y *be Banach spaces. An operator* $A \in \mathcal{L}(X,Y)$ *is in* $\Phi_+^b(X,Y)$ *if, and only if,* $\alpha(A-K) < \infty$ *for all* $K \in \mathcal{K}(X,Y)$*.* ◇

Lemma 2.2.2 *(A. Lebow and M. Schechter [117, Lemma 4.5]) Let X and Y be Banach spaces. If* $A \in \Phi_+^b(X,Y)$ *(resp.* $\Phi_-^b(X,Y)$, $\Phi^b(X,Y)$*), then there exist an* $\eta > 0$ *such that* $B \in \Phi_+^b(X,Y)$ *(resp.* $\Phi_-^b(X,Y)$, $\Phi^b(X,Y)$*) with*

$$i(B) = i(A),$$

for all $B \in \mathscr{L}(X,Y)$ *satisfying* $\|B - A\| < \eta$. ◇

Theorem 2.2.9 *(V. Müller [125, Theorem 5, p. 156]) Let X, Y, and Z be three Banach spaces,* $A \in \mathscr{L}(X,Y)$ *and* $B \in \mathscr{L}(Y,Z)$*. Then,*

(i) if $A \in \Phi_-^b(X,Y)$ *and* $B \in \Phi_-^b(Y,Z)$*, then* $BA \in \Phi_-^b(X,Z)$*,*
(ii) if $A \in \Phi_+^b(X,Y)$ *and* $B \in \Phi_+^b(Y,Z)$*, then* $BA \in \Phi_+^b(X,Z)$*, and*
(iii) if $A \in \Phi^b(X,Y)$ *and* $B \in \Phi^b(Y,Z)$*, then* $BA \in \Phi^b(X,Z)$*.* ◇

Theorem 2.2.10 *(V. Müller [125, Theorem 6, p. 157]) Let X, Y, and Z be three Banach spaces,* $A \in \mathscr{L}(X,Y)$ *and* $B \in \mathscr{L}(Y,Z)$*. Then,*

(i) if $BA \in \Phi_+^b(X,Z)$*, then* $A \in \Phi_+^b(X,Y)$*,*
(ii) if $BA \in \Phi_-^b(X,Z)$*, then* $B \in \Phi_-^b(Y,Z)$*, and*
(iii) if $BA \in \Phi^b(X,Z)$*, then* $B \in \Phi_-^b(Y,Z)$ *and* $A \in \Phi_+^b(X,Y)$*.* ◇

Lemma 2.2.3 *Assume that* $A \in \mathscr{L}(X)$ *and there exist operators* B_0, $B_1 \in \mathscr{L}(X)$ *such that* B_0A *and* AB_1 *are in* $\Phi^b(X)$*. Then,* $A \in \Phi^b(X)$*.* ◇

Proof. By referring to Theorem 2.2.3, there are operators A_0, $A_1 \in \mathscr{L}(X)$ such that $A_0B_0A - I$ and $AB_1A_1 - I$ are in $\mathscr{K}(X)$. The result follows from Theorem 2.2.4. Q.E.D.

Theorem 2.2.11 *(V. Müller [125, Theorem 7]) For a bounded operator A on a Banach space X, the following assertions are equivalent:*

(i) A is a Fredholm operator having index 0,

(ii) there exist $K \in \mathscr{K}(X)$ *and an invertible operator* $S \in \mathscr{L}(X)$ *such that* $A = S + K$ *is invertible.* ◇

Theorem 2.2.12 *(P. Aiena [16, Corollary 1.52]) Suppose that* $T \in \mathscr{L}(X)$*. Then,* *(i)* $T \in \Phi_r(X)$ *if, and only if, the class rest* $\widehat{T} = T + \mathscr{K}(X)$ *is right invertible in the Calkin algebra* $\mathscr{L}(X)/\mathscr{K}(X)$.

(ii) $T \in \Phi_l(X)$ *if, and only if, the class rest* $\widehat{T} = T + \mathscr{K}(X)$ *is left invertible in* $\mathscr{L}(X)/\mathscr{K}(X)$.

(iii) $T \in \Phi^b(X)$ *if, and only if, the class rest* $\widehat{T} = T + \mathscr{K}(X)$ *is invertible in* $\mathscr{L}(X)/\mathscr{K}(X)$. ◊

Let $P(X)$ denote the set

$$P(X) := \{F \in \mathscr{L}(X) \text{ such that there exists } r \in \mathbb{N}^* \text{ satisfying } F^r \in \mathscr{K}(X)\}.$$

Lemma 2.2.4 *(A. Jeribi [84], see also [113]) Let* $A \in P(X)$ *and set* $A = I - F$. *Then,*

(i) $\dim[N(A)] < \infty$,

(ii) $R(A)$ *is closed, and*

(iii) $codim[R(A)] < \infty$. ◊

Proof. (i) Since $F \in P(X)$, there exists $r \in \mathbb{N}^*$ such that $F^r \in \mathscr{K}(X)$. Let $x \in N(A)$, then

$$F^r x = x$$

i.e., $x \in N(I - F^r)$ and, therefore

$$N(A) \subset N(I - F^r).$$

On the other hand, the identity I restricted to the kernel of $I - F^r$ is equal to F^r and, consequently compact. Hence, $N(I - F^r)$ is finite dimensional and, therefore

$$\dim[N(A)] < \infty.$$

(ii) Since A commutes with I, Newton's binomial formula gives

$$F^r = (I - A)^r = I + \sum_{k=1}^{r} (-1)^k C_r^k A^k. \tag{2.12}$$

Let E be a closed complement for $N(A)$, so that

$$X = N(A) \bigoplus E.$$

Thus, we obtain two linear continuous maps

$$A_{|E} : E \longrightarrow X$$

and

$$F_{|E} : E \longrightarrow X,$$

the restrictions of A and F to E. It is clear that the kernel of $A_{|E}$ is $\{0\}$. In order to conclude, it suffices to show that

$$A_{|E}(E) = A(E) = A(X)$$

is closed. For this, it suffices to show that the map

$$(A_{|E})^{-1} : A(E) \longrightarrow E$$

is continuous. By linearity, it even suffices to prove that $(A_{|E})^{-1}$ is continuous at 0. Suppose that this is not the case. Then, we can find a sequence $(x_n)_n$ in E such that

$$Ax_n \to 0,$$

but $(x_n)_n$ does not converge to 0. Selecting a suitable subsequence, we can assume without loss of generality that

$$\|x_n\| \geq \eta > 0$$

for all n. Then,

$$\frac{1}{\|x_n\|} < \frac{1}{\eta}$$

for all n and, consequently $A(\frac{x_n}{\|x_n\|})$ also converges to 0. Furthermore, $\frac{x_n}{\|x_n\|}$ has norm 1, and hence some subsequence of $F^r(\frac{x_n}{\|x_n\|})$ converges. It follows from (2.12) that $\frac{x_n}{\|x_n\|}$ has a converging subsequence to an element z in E verifying $\|z\| = 1$ and

$$F^r(z) = z.$$

On the other hand,

$$F^r = F - FA - F^2A - \cdots - F^{r-1}A$$

and so, we get

$$F^r(z) = F(z).$$

Hence, we infer that

$$F(z) - z = 0,$$

which implies that $z \in N(A)$. This contradicts the fact that

$$E \bigcap N(A) = \{0\}$$

(because $\|z\| = 1$) and completes the proof of (ii).

(iii) If $A(X)$ does not have finite codimension, we can find a sequence of closed subspaces

$$A(X) = M_0 \subset M_1 \subset M_2 \subset \cdots \subset M_n \subset \cdots$$

such that each M_n is closed and of codimension 1 in M_{n+1} just by adding one-dimensional spaces to $A(X)$ inductively. By Riesz's lemma (see Lemma 2.1.2), we can find, in each M_n, an element x_n such that $\|x_n\| = 1$ and,

$$\|x_n - y\| \geq 1 - \varepsilon$$

for all y in M_{n-1} with $0 < \varepsilon < 1$. Then, by using (2.12), together with the fact that

$$X \supset R(A) \supset R(A^2) \cdots \supset R(A^n) \supset \cdots,$$

for all $k < n$, we get

$$\begin{aligned} \|F^r x_n - F^r x_k\| &= \left\| x_n - \sum_{i=1}^{r} (-1)^i C_r^i A^i x_n - x_k + \sum_{i=1}^{r} (-1)^i C_r^i A^i x_k \right\| \\ &= \left\| x_n - x_k - \sum_{i=1}^{r} (-1)^i C_r^i A^i (x_n - x_k) \right\| \\ &\geq 1 - \varepsilon \end{aligned}$$

because

$$x_k + \sum_{i=1}^{r} (-1)^i C_r^i A^i (x_n - x_k) \in M_{n-1}.$$

This proves that the sequence $(F^r x_n)_n$ cannot have a convergent subsequence, and contradicts the compactness of F^r. Q.E.D.

Theorem 2.2.13 *Assume that the hypothesis of Lemma 2.2.4 holds. Then, $A = I - F$ is a Fredholm operator and*

$$i(A) = 0. \qquad \diamondsuit$$

Let X be a Banach space. We denote by $\mathscr{P}(X)$ the set defined by

$$\mathscr{P}(X) = \Big\{ A \in \mathscr{L}(X) \text{ such that there exists a polynomial } P(z) = \sum_{k=0}^{n} a_k z^k \text{ satisfying } P(1) \neq 0,\ P(1) - a_0 \neq 0 \text{ and } P(A) \in \mathscr{K}(X) \Big\}.$$

Lemma 2.2.5 *(A. Jeribi and K. Latrach [115, Lemma 2.2]) If $A \in \mathscr{P}(X)$, then $I + A \in \Phi(X)$ and*

$$i(I + A) = 0. \qquad \diamondsuit$$

2.2.3 Basics on Unbounded Fredholm Operators

Theorem 2.2.14 *(M. Schechter [137, Theorem 5.10]) Let $A \in \mathscr{C}(X,Y)$. If $A \in \Phi(X,Y)$ and $K \in \mathscr{K}(X,Y)$, then $A + K \in \Phi(X,Y)$ and*

$$i(A + K) = i(A). \qquad \diamondsuit$$

Theorem 2.2.15 *(M. Schechter [137, Theorem 7.9, p. 161]) Let X and Y be Banach spaces. For $A \in \Phi(X,Y)$, there is an $\eta > 0$ such that, for every $T \in \mathscr{L}(X,Y)$ satisfying $\|T\| < \eta$, one has $A + T \in \Phi(X,Y)$,*

$$i(A + T) = i(A)$$

and

$$\alpha(A + T) \leq \alpha(A). \qquad \diamondsuit$$

Let A be a closed linear from X into Y. Then, $N(A)$ is a closed subspace of X and, hence the quotient space $\widetilde{X} := X/N(A)$ is a Banach space with respect to the norm

$$\|\widetilde{x}\| = \text{dist}(x, N(A)) := \inf\{\|x-y\| \text{ such that } y \in N(A)\}.$$

Since $N(A) \subset \mathscr{D}(A)$, the quotient space $\mathscr{D}(\widetilde{A}) := \mathscr{D}(A)/N(A)$ is contained in $\widetilde{X}$. Defining $\widetilde{A}x = Ax$ for every $\widetilde{x} \in \mathscr{D}(\widetilde{A})$, it follows that $\widetilde{A}$ is a well defined closed linear operator with $\mathscr{D}(\widetilde{A}) \subset \widetilde{X}$ and $R(\widetilde{A}) = R(A)$. Since $\widetilde{A}$ is one-to-one, the inverse $\widetilde{A}^{-1}$ exists on $R(A)$. The reduced minimum modulus of A is defined by

$$\widetilde{\gamma}(A) := \begin{cases} \inf\limits_{x \notin N(A)} \dfrac{\|Ax\|}{\text{dist}(x, N(A))} & \text{if } \; A \neq 0, \\ \infty & \text{if } \; A = 0. \end{cases} \tag{2.13}$$

It is to remember that the reduced minimum modulus measures the closedness of the range of operators in the sense that $R(A)$ is closed if, and only if, $\widetilde{\gamma}(A) = \|\widetilde{A}^{-1}\|^{-1} > 0$.

Remark 2.2.1 *Let $A \in \mathscr{C}(X)$. Then, $\widetilde{\gamma}(A) = \gamma_1(A)$.* ◊

Theorem 2.2.16 *(T. Kato [107, Theorem 5.22, p. 236]) Let $T \in \mathscr{C}(X,Y)$ be semi-Fredholm (so that $\gamma = \widetilde{\gamma}(T) > 0$). Let A be a T-bounded operator from X into Y so that we have the inequality*

$$\|Ax\| \leq a\|x\| + b\|Tx\| \quad x \in \mathscr{D}(T) \subset \mathscr{D}(A),$$

where $a < (1-b)\gamma$. Then, $S = T + A$ belongs to $\mathscr{C}(X,Y)$, S is semi-Fredholm, $\alpha(S) \leq \alpha(T)$, $\beta(S) \leq \beta(T)$, and $i(S) = i(T)$. ◊

Lemma 2.2.6 *(F. Fakhfakh and M. Mnif [70, Lemma 3.1]) Let X be a Banach space, $A \in \mathscr{C}(X)$, and $K \in \mathscr{L}(X)$. If $\Delta_\psi(K) < \Gamma_\psi(A)$, then $A + K \in \Phi_+(X)$, $A \in \Phi_+(X)$, and*

$$i(A+K) = i(A),$$

where $\Delta_\psi(\cdot)$ is the measure of non-strict-singularity given in (2.11) and $\Gamma_\psi(\cdot)$ is given in (2.10). ◊

Lemma 2.2.7 *Let $A \in \Phi_+(X)$. Then, the following statements are equivalent*

(i) $i(A) \leq 0$,

(ii) A can be expressed in the form

$$A = U + K,$$

where $K \in \mathscr{K}(X)$ *and* $U \in \mathscr{C}(X)$ *an operator with closed range and* $\alpha(U) = 0$. ◇

This lemma is well known for bounded upper semi-Fredholm operators. The proof is a straightforward adaption of the proof of Theorem 3.9 in [162].

Theorem 2.2.17 *(M. Schechter [137, Theorem 7.1, p. 157]) Let X and Y be Banach spaces, and* $A \in \Phi(X,Y)$. *Then, there is an operator* $A_0 \in \mathscr{L}(Y,X)$, *such that* $N(A_0) = Y_0$, $R(A_0) = X_0 \bigcap \mathscr{D}(A)$,

$$A_0 A = I \text{ on } X_0 \bigcap \mathscr{D}(A),$$

and

$$AA_0 = I \text{ on } R(A)$$

i.e., there are operators $F_1 \in \mathscr{L}(X)$, $F_2 \in \mathscr{L}(Y)$, *such that*

$$A_0 A = I - F_1 \text{ on } \mathscr{D}(A),$$

$$AA_0 = I - F_2 \text{ on } Y,$$

$R(F_1) = N(A)$, $N(F_1) = X_0$, *and* $R(F_2) = Y_0$, $N(F_2) = R(A)$. ◇

Theorem 2.2.18 *(T. Kato [108, Theorem 5.31, p. 241]) Let X and Y be Banach spaces. Let* $A \in \mathscr{C}(X,Y)$ *be semi-Fredholm and let S be an A-bounded operator from X into Y. Then,* $\lambda S + A$ *is semi-Fredholm and* $\alpha(\lambda S + A)$, $\beta(\lambda S + A)$ *are constant for a sufficiently small* $|\lambda| > 0$. ◇

Theorem 2.2.19 *(M. Schechter [141, Theorem 7.3, p. 157]) Let X and Y be Banach spaces. If $A \in \Phi(X,Y)$ and $B \in \Phi(Y,Z)$, then $BA \in \Phi(X,Z)$ and $i(BA) = i(B) + i(A)$.* ♢

Theorem 2.2.20 *(M. Schechter [140, Theorem 3.8]) Let X, Y, Z be Banach spaces and suppose $B \in \Phi^b(Y,Z)$. Assume that A is a closed, densely defined linear operator from X into Y such that $BA \in \Phi(X,Z)$. Then, $A \in \Phi(X,Y)$.* ♢

Theorem 2.2.21 *(M. Schechter [137, Theorem 7.12, p. 162]) Let X, Y, and Z be Banach spaces. If $A \in \Phi(X,Y)$ and B is a densely defined closed linear operator from Y into Z such that $BA \in \Phi(X,Z)$, then $B \in \Phi(Y,Z)$.* ♢

Theorem 2.2.22 *(M. Schechter [137, Theorem 7.14, p. 164]) Let X, Y, and Z be Banach spaces and let A be a densely defined closed linear operator from X into Y. Suppose that $B \in \mathscr{L}(Y,Z)$ with $\alpha(B) < \infty$ and $BA \in \Phi(X,Z)$. Then,*

$$A \in \Phi(X,Y).$$ ♢

Proposition 2.2.1 *Let $A \in \mathscr{C}(X,Y)$ and let S be a non null bounded linear operator from X into Y. Then, we have the following results*

(i) $\Phi_{A,S}$ is open,

(ii) $i(\lambda S - A)$ is constant on any component of $\Phi_{A,S}$, and

(iii) $\alpha(\lambda S - A)$ and $\beta(\lambda S - A)$ are constant on any component of $\Phi_{A,S}$, except on a discrete set of points on which they have larger values. ♢

Proof. (i) Let $\lambda_0 \in \Phi_{A,S}$. Then, according to Theorem 2.2.15, there exists $\eta > 0$ such that, for all $\mu \in \mathbb{C}$ with $|\mu| < \frac{\eta}{\|S\|}$, the operator $\lambda_0 S - \mu S - A$ is a Fredholm operator,

$$i(\lambda_0 S - \mu S - A) = i(\lambda_0 S - A)$$

and

$$\alpha(\lambda_0 S - \mu S - A) \leq \alpha(\lambda_0 S - A).$$

Consider $\lambda \in \mathbb{C}$ such that

$$|\lambda - \lambda_0| < \frac{\eta}{\|S\|}.$$

Then, $\lambda S - A$ is a Fredholm operator,

$$i(\lambda S - A) = i(\lambda_0 S - A)$$

and

$$\alpha(\lambda S - A) \leq \alpha(\lambda_0 S - A).$$

In particular, this implies that $\Phi_{A,S}$ is open.

(ii) Let λ_1 and λ_2 be any two points in $\Phi_{A,S}$ which are connected by a smooth curve Γ whose points are all in $\Phi_{A,S}$. Since $\Phi_{A,S}$ is an open set, then for each $\lambda \in \Gamma$, there exists an $\varepsilon > 0$ such that, for all $\mu \in \mathbb{C}$, $|\lambda - \mu| < \varepsilon$, $\mu \in \Phi_{A,S}$ and

$$i(\mu S - A) = i(\lambda S - A).$$

By using the Heine-Borel theorem, there exist a finite number of such sets, which cover Γ. Since each of these sets overlaps with, at least, another set and since $i(\mu S - A)$ is constant on each one, we see that

$$i(\lambda_1 S - A) = i(\lambda_2 S - A).$$

(iii) Let λ_1 and λ_2 be any two points in $\Phi_{A,S}$ which are connected by a smooth curve Γ whose points are all in $\Phi_{A,S}$. Since $\Phi_{A,S}$ is an open set, then for each $\lambda \in \Gamma$, there is a sufficiently small $\varepsilon > 0$ such that, for all $\mu \in \mathbb{C}$, $|\lambda - \mu| < \varepsilon$, $\mu \in \Phi_{A,S}$ and by using Theorem 2.2.18, $\alpha(A + \mu S)$ and $\beta(A + \mu S)$ are constant for all $\mu \in \mathbb{C}$, $0 < |\lambda - \mu| < \varepsilon$. By referring to Heine-Borel's theorem, there is a finite number of such sets which cover Γ. Since each of these sets overlaps with, at least, another set, we see that $\alpha(\mu S + A)$ and $\beta(\mu S + A)$ are constant for all $\mu \in \Gamma$, except for a finite number of points of Γ. Q.E.D.

The following theorem is developed by M. Gonzalez and M. O. Onieva in [77].

Lemma 2.2.8 *(M. Gonzalez and M. O. Onieva [77, Theorem 2.3]) Let $A \in \mathscr{C}(X,Y)$, then*

(i) $A \in \Phi_l(X,Y)$ if, and only if, there exist $B \in \mathscr{L}(Y,X)$ and $K \in \mathscr{K}(X)$ such that

$$R(B)\bigcup R(K) \subset \mathscr{D}(A)$$

and

$$BA = I - K \text{ on } \mathscr{D}(A),$$

(ii) $A \in \Phi_r(X,Y)$ if, and only if, there exist $B \in \mathscr{L}(Y,X)$ and $K \in \mathscr{K}(Y)$ such that $R(B) \subset \mathscr{D}(A)$, BA and KA are continuous, and

$$AB = I - K. \qquad \diamondsuit$$

Lemma 2.2.9 *(M. Gonzalez and M. O. Onieva [77, Theorem 2.5]) Let $A \in \mathscr{C}(Y,Z)$ and $B \in \mathscr{C}(X,Y)$, then*

(i) if $A \in \Phi_l(Y,Z)$, $B \in \Phi_l(X,Y)$ and $\overline{\mathscr{D}(AB)} = X$, then $AB \in \Phi_l(X,Z)$ and

$$i(AB) = i(A) + i(B),$$

(ii) if $A \in \Phi_r(Y,Z)$, $B \in \Phi_r(X,Y)$ and AB is closed, then $AB \in \Phi_r(X,Z)$ and

$$i(AB) = i(A) + i(B). \qquad \diamondsuit$$

Lemma 2.2.10 *(M. Gonzalez and M. O. Onieva [77, Theorem 2.7]) If $A \in \Phi_l(X)$ (resp. $\Phi_r(X)$) and $K \in \mathscr{K}(X)$, then $A + K \in \Phi_l(X)$ (resp. $\Phi_r(X)$) and*

$$i(A+K) = i(A). \qquad \diamondsuit$$

2.3 PERTURBATION

2.3.1 Fredholm and Semi-Fredholm Perturbations

Definition 2.3.1 *Let X and Y be two Banach spaces, and let $F \in \mathscr{L}(X,Y)$. F is called a Fredholm perturbation, if $U + F \in \Phi(X,Y)$ whenever $U \in$*

$\Phi(X,Y)$. F is called an upper (resp. lower) Fredholm perturbation, if $U+F\in\Phi_+(X,Y)$ (resp. $U+F\in\Phi_-(X,Y)$) whenever $U\in\Phi_+(X,Y)$ (resp. $U\in\Phi_-(X,Y)$). ◇

The sets of Fredholm, upper semi-Fredholm and lower semi-Fredholm perturbations are, respectively, denoted by $\mathscr{F}(X,Y)$, $\mathscr{F}_+(X,Y)$, and $\mathscr{F}_-(X,Y)$. In general, we have

$$\mathscr{K}(X,Y)\subseteq\mathscr{F}_+(X,Y)\subseteq\mathscr{F}(X,Y)$$

$$\mathscr{K}(X,Y)\subseteq\mathscr{F}_-(X,Y)\subseteq\mathscr{F}(X,Y).$$

If $X=Y$, we may write $\mathscr{F}(X)$, $\mathscr{F}_+(X)$ and $\mathscr{F}_-(X)$ instead of $\mathscr{F}(X,X)$, $\mathscr{F}_+(X,X)$ and $\mathscr{F}_-(X,X)$, respectively. In Definition 2.3.1, if we replace $\Phi(X,Y)$, $\Phi_+(X,Y)$ and $\Phi_-(X,Y)$ by $\Phi^b(X,Y)$, $\Phi^b_+(X,Y)$ and $\Phi^b_-(X,Y)$, we obtain the sets $\mathscr{F}^b(X,Y)$, $\mathscr{F}^b_+(X,Y)$ and $\mathscr{F}^b_-(X,Y)$, respectively. These classes of operators were introduced and investigated in [73]. In particular, it was shown that $\mathscr{F}^b(X,Y)$ is a closed subset of $\mathscr{L}(X,Y)$ and $\mathscr{F}^b(X):=\mathscr{F}^b(X,X)$ is a closed two-sided ideal of $\mathscr{L}(X)$. In general, we have

$$\mathscr{K}(X,Y)\subseteq\mathscr{F}^b_+(X,Y)\subseteq\mathscr{F}^b(X,Y)$$

$$\mathscr{K}(X,Y)\subseteq\mathscr{F}^b_-(X,Y)\subseteq\mathscr{F}^b(X,Y).$$

In Ref. [72], it was shown that $\mathscr{F}^b(X)$ and $\mathscr{F}^b_+(X):=\mathscr{F}^b_+(X,X)$ are closed two-sided ideals of $\mathscr{L}(X)$. It is worth noticing that, in general, the structure ideal of $\mathscr{L}(X)$ is extremely complicated. Most of the results on ideal structure deal with the well-known closed ideals which have arisen from applied work with operators. We can quote, for example, compact operators, weakly compact operators, strictly singular operators, strictly cosingular operators, upper semi-Fredholm perturbations, lower semi-Fredholm perturbations, and Fredholm perturbations. In general, we have

$$\mathscr{K}(X) \subset \mathscr{S}(X) \subset \mathscr{F}_{+}^{b}(X) \subset \mathscr{F}^{b}(X) \subset \mathscr{J}(X), \text{ and}$$

$$\mathscr{K}(X) \subset \mathscr{SC}(X) \subset \mathscr{F}_{-}^{b}(X) \subset \mathscr{F}^{b}(X) \subset \mathscr{J}(X),$$

where $\mathscr{F}_{-}^{b}(X) := \mathscr{F}_{-}^{b}(X,X)$, and where $\mathscr{J}(X)$ denotes the set

$$\mathscr{J}(X) = \{F \in \mathscr{L}(X) \text{ such that } I - F \in \Phi^{b}(X) \text{ and } i(I-F) = 0\}.$$

Remark 2.3.1 *$\mathscr{J}(X)$ is not an ideal of $\mathscr{L}(X)$ (since $I \notin \mathscr{J}(X)$).* ♢

If X is isomorphic to an L_p-space with $1 \leq p \leq \infty$ or to $C(\Sigma)$, where Σ is a compact Hausdorff space, then

$$\mathscr{K}(X) \subset \mathscr{S}(X) = \mathscr{F}_{+}^{b}(X) = \mathscr{F}_{-}^{b}(X) = \mathscr{SC}(X) = \mathscr{F}^{b}(X).$$

Lemma 2.3.1 *(A. Jeribi and N. Moalla [98, Lemma 2.1]) Let $A \in \mathscr{C}(X,Y)$ and let $J : X \longrightarrow Y$ be a linear operator. Assume that $J \in \mathscr{F}(X,Y)$. Then,*

(i) if $A \in \Phi(X,Y)$, then $A + J \in \Phi(X,Y)$ and

$$i(A+J) = i(A).$$

Moreover,

(ii) if $A \in \Phi_{+}(X,Y)$ and $J \in \mathscr{F}_{+}(X,Y)$, then $A + J \in \Phi_{+}(X,Y)$ and

$$i(A+J) = i(A),$$

(iii) if $A \in \Phi_{-}(X,Y)$ and $J \in \mathscr{F}_{-}(X,Y)$, then $A + J \in \Phi_{-}(X,Y)$ and

$$i(A+J) = i(A),$$

(iv) if $A \in \Phi_{\pm}(X,Y)$ and $J \in \mathscr{F}_{+}(X,Y) \bigcap \mathscr{F}_{-}(X,Y)$, then $A + J \in \Phi_{\pm}(X,Y)$ and

$$i(A+J) = i(A).$$ ♢

Lemma 2.3.2 *Let X, Y and Z be three Banach spaces. Then,*

(i) if $F_1 \in \mathscr{F}^{b}(X,Y)$ and $A \in \mathscr{L}(Y,Z)$, then $AF_1 \in \mathscr{F}^{b}(X,Z)$, and

(ii) if $F_2 \in \mathscr{F}^{b}(Y,Z)$ and $B \in \mathscr{L}(X,Y)$, then $F_1 B \in \mathscr{F}^{b}(Y,Z)$. ♢

Definition 2.3.2 *An operator $A \in \mathscr{L}(X)$ is said to be polynomially Fredholm perturbation if there exists a nonzero complex polynomial $P(z) = \sum_{n=0}^{p} a_n z^n$ such that $P(A)$ is a Fredholm perturbation.* ♢

Lemma 2.3.3 *Let X and Y be two Banach spaces. Then,*

$$\mathscr{F}^b(X,Y) = \mathscr{F}(X,Y). \qquad \diamondsuit$$

An immediate consequence of Lemma 2.3.3 that $\mathscr{F}(X)$ is a closed two-sided ideal of $\mathscr{L}(X)$.

2.3.2 Semi-Fredholm Perturbations

In the beginning of this section, let us prove some results for semi-Fredholm perturbations.

Proposition 2.3.1 *Let X, Y and Z be three Banach spaces.*

(i) If the set $\Phi^b(Y,Z)$ is not empty, then

$$E_1 \in \mathscr{F}_+^b(X,Y) \text{ and } A \in \Phi^b(Y,Z) \text{ imply } AE_1 \in \mathscr{F}_+^b(X,Z)$$
$$E_1 \in \mathscr{F}_-^b(X,Y) \text{ and } A \in \Phi^b(Y,Z) \text{ imply } AE_1 \in \mathscr{F}_-^b(X,Z).$$

(ii) If the set $\Phi^b(X,Y)$ is not empty, then

$$E_2 \in \mathscr{F}_+^b(Y,Z) \text{ and } B \in \Phi^b(X,Y) \text{ imply } E_2B \in \mathscr{F}_+^b(X,Z)$$
$$E_2 \in \mathscr{F}_-^b(Y,Z) \text{ and } B \in \Phi^b(X,Y) \text{ imply } E_2B \in \mathscr{F}_-^b(X,Z). \qquad \diamondsuit$$

Proof. (i) Since $A \in \Phi^b(Y,Z)$, and using Theorem 2.2.3, it follows that there exist $A_0 \in \mathscr{L}(Z,Y)$ and $K \in \mathscr{K}(Z)$ such that

$$AA_0 = I - K.$$

From Lemma 2.2.5, we get $AA_0 \in \Phi^b(Z)$. Using Theorem 2.2.6, we have $A_0 \in \Phi^b(Z,Y)$, and so $A_0 \in \Phi_+^b(Z,Y)$ and $A_0 \in \Phi_-^b(Z,Y)$. Let $J \in \Phi_+^b(X,Z)$ (resp. $\Phi_-^b(X,Z)$), using Theorem 2.2.9, we deduce that $A_0J \in \Phi_+^b(X,Y)$ (resp. $\Phi_-^b(X,Y)$). This implies that $E_1 + A_0J \in \Phi_+^b(X,Y)$ (resp. $\Phi_-^b(X,Y)$). So, $A(E_1 + A_0J) \in \Phi_+^b(X,Z)$ (resp. $\Phi_-^b(X,Z)$). Now, using the relation

$$AE_1 + J - KJ = A(E_1 + A_0J)$$

together with the compactness of the operator KJ, we get $(AE_1 + J) \in \Phi_+^b(X,Z)$ (resp. $\Phi_-^b(X,Z)$). This implies that $AE_1 \in \mathscr{F}_+^b(X,Z)$ (resp.

$\mathscr{F}_-^b(X,Z)$).

(ii) The proof of (ii) is obtained in a similar way as the proof of the item (i). Q.E.D.

Theorem 2.3.1 *Let X, Y and Z be three Banach spaces. Then,*

(i) *If the set $\Phi^b(Y,Z)$ is not empty, then*

$$E_1 \in \mathscr{F}_+^b(X,Y) \, and \, A \in \mathscr{L}(Y,Z) \text{ imply } AE_1 \in \mathscr{F}_+^b(X,Z)$$
$$E_1 \in \mathscr{F}_-^b(X,Y) \, and \, A \in \mathscr{L}(Y,Z) \text{ imply } AE_1 \in \mathscr{F}_-^b(X,Z).$$

(ii) *If the set $\Phi^b(X,Y)$ is not empty, then*

$$E_2 \in \mathscr{F}_+^b(Y,Z) \, and \, B \in \mathscr{L}(X,Y) \text{ imply } E_2B \in \mathscr{F}_+^b(X,Z)$$
$$E_2 \in \mathscr{F}_-^b(Y,Z) \, and \, B \in \mathscr{L}(X,Y) \text{ imply } E_2B \in \mathscr{F}_-^b(X,Z). \qquad \diamondsuit$$

Proof. (i) Let $C \in \Phi^b(Y,Z)$ and $\lambda \in \mathbb{C}$. Let $A_1 = A - \lambda C$ and $A_2 = \lambda C$. For a sufficiently large λ, and using Lemma 2.2.2, we have $A_1 \in \Phi^b(Y,Z)$. From Proposition 2.3.1 (i), it follows that $A_1E_1 \in \mathscr{F}_+^b(X,Z)$ (resp. $\mathscr{F}_-^b(X,Z)$) and $A_2E_1 \in \mathscr{F}_+^b(X,Z)$ (resp. $\mathscr{F}_-^b(X,Z)$). This implies that

$$A_1E_1 + A_2E_1 = AE_1$$

is an element of $\mathscr{F}_+^b(X,Z)$ (resp. $\mathscr{F}_-^b(X,Z)$).

(ii) The proof may be achieved in a similar way as (i). It is sufficient to replace Proposition 2.3.1 (i) by Proposition 2.3.1 (ii). Q.E.D.

2.3.3 Riesz Operator

Let X be a Banach space. An operator $R \in \mathscr{L}(X)$ is said to be a Riesz operators if $\Phi_R = \mathbb{C} \backslash \{0\}$. This family of Riesz operators is denoted by $\mathscr{R}(X)$. We know that $\mathscr{R}(X)$ is not an ideal of $\mathscr{L}(X)$ (see [49]). In general, we have

$$\mathscr{K}(X) \subset \mathscr{S}(X) \subset \mathscr{F}(X) \subset \mathscr{R}(X),$$

where

$$\mathscr{R}(X) := \left\{ A \in \mathscr{L}(X) \text{ such that } \lambda - A \in \Phi^b(X) \text{ for each } \lambda \neq 0 \right\}$$

which is the class of all Riesz operators. The set $\mathscr{R}(X)$ is, not generally, a closed ideal of $\mathscr{L}(X)$.

Theorem 2.3.2 *(P. Aiena [15, Theorem 2.3]) Let X be a Banach space. Then,*

(i) if A, $B \in \mathscr{R}(X)$ and $AB - BA \in \mathscr{K}(X)$, then $A + B \in \mathscr{R}(X)$, and

(ii) if $A \in \mathscr{R}(X)$, $B \in \mathscr{L}(X)$ and $AB - BA \in \mathscr{K}(X)$, then AB and $BA \in \mathscr{R}(X)$. ♢

2.3.4 Some Perturbation Results

Definition 2.3.3 *Let X and Y be two Banach spaces.*

(i) An operator $A \in \mathscr{C}(X)$ is said to have a left Fredholm inverse if there are maps $R_l \in \mathscr{L}(X)$ and $F \in \mathscr{F}(X)$ such that $I_X + F$ extends $R_l A$. The operator R_l is called left Fredholm inverse of A.

(ii) An operator $A \in \mathscr{C}(X)$ is said to have a right Fredholm inverse if there are maps $R_r \in \mathscr{L}(X)$ such that $R_r(Y) \subset \mathscr{D}(A)$ and $AR_r - I_Y \in \mathscr{F}(X)$. The operator R_r is called right Fredholm inverse of A. ♢

The following theorem is well known for bounded upper or lower semi-Fredholm operators. The proof is a straightforward adaption of the proof in [162].

Theorem 2.3.3 *$T \in \Phi_+(X,Y)$ $(\Phi_-(X,Y))$ either without $\Phi_-(X,Y)$ $(\Phi_+(X,Y))$ or with it with index ≥ 0 (≤ 0) if, and only if, is of the form $U_1 + U_2$, where $R(U_1)$ is closed (U_1 has range Y) and U_2 is compact.* ♢

Theorem 2.3.4 *(S. Č. Živković-Zlatanović, D. S. Djordjević, and R. E. Harte [163, Theorem 8]) Let $A \in \mathscr{L}(X)$ and $E \in \mathscr{R}(X)$, then*

(i) if $A \in \Phi_l(X)$ and $AE - EA \in \mathscr{F}^b(X)$, then $A + E \in \Phi_l(X)$,

(ii) if $A \in \Phi_r(X)$ and $AE - EA \in \mathscr{F}^b(X)$, then $A + E \in \Phi_r(X)$,

(iii) if $A \in \mathscr{W}_l(X)$ and $AE - EA \in \mathscr{F}^b(X)$, then $A + E \in \mathscr{W}_l(X)$, and

(iv) if $A \in \mathscr{W}_r(X)$ and $AE - EA \in \mathscr{F}^b(X)$, then $A + E \in \mathscr{W}_r(X)$. ♢

A direct consequence of Theorem 2.3.4, the ideality of $\mathscr{F}(X)$ and the local constancy of the index, we infer the following result.

Corollary 2.3.1 *Let $A \in \mathscr{L}(X)$ and $E \in \mathscr{F}^b(X)$, then*

(i) if $A \in \Phi_l(X)$, then $A+E \in \Phi_l(X)$ and $i(A) = i(A+E)$, and
(ii) if $A \in \Phi_r(X)$, then $A+E \in \Phi_r(X)$ and $i(A) = i(A+E)$. ◊

2.3.5 A-Fredholm Perturbation

Definition 2.3.4 *Let X and Y be two Banach spaces, $A \in \mathscr{C}(X,Y)$, and let F be an arbitrary A-defined linear operator on X. We say that F is an A-Fredholm perturbation if $\widehat{F} \in \mathscr{F}^b(X_A,Y)$. The operator F is called an upper (resp. lower) A-Fredholm perturbation if $\widehat{F} \in \mathscr{F}^b_+(X_A,Y)$ (resp. $\widehat{F} \in \mathscr{F}^b_-(X_A,Y)$).* ◊

The sets of A-Fredholm, upper A-semi-Fredholm and lower A-semi-Fredholm perturbations are denoted by $A\mathscr{F}(X,Y)$, $A\mathscr{F}_+(X,Y)$, and $A\mathscr{F}_-(X,Y)$, respectively. If $X = Y$, we write $A\mathscr{F}(X)$, $A\mathscr{F}_+(X)$, and $A\mathscr{F}_-(X)$ for $A\mathscr{F}(X,X)$, $A\mathscr{F}_+(X,X)$, and $A\mathscr{F}_-(X,X)$, respectively.

Remark 2.3.2 *(i) If B is bounded (resp. compact, weakly compact, strictly singular, strictly cosingular) implies that B is A-bounded (resp. A-compact, A-weakly compact, A-strictly singular, A-strictly cosingular).*

(ii) Notice that the concept of A-compactness and A-Fredholmness are not connected with the operator A itself, but only with its domain.

(iii) Using Definition 2.3.4 *and* [74, page 69], *we have*

$$A\mathscr{K}(X) \subseteq A\mathscr{S}(X) \subseteq A\mathscr{F}_+(X) \subseteq A\mathscr{F}(X).$$

and

$$A\mathscr{K}(X) \subseteq A\mathscr{S}\mathscr{C}(X) \subseteq A\mathscr{F}_-(X) \subseteq A\mathscr{F}(X).$$

(iv) Let B be an arbitrary A-Fredholm perturbation operator, hence we can regard A and B as operators from X_A into X, they will be denoted by

$\widehat{A}$ and $\widehat{B}$, respectively, these operators belong to $\mathscr{L}(X_A, X)$. Furthermore, we have the obvious relations

$$\begin{cases} \alpha(\widehat{A}) = \alpha(A), \ \beta(\widehat{B}) = \beta(B), \ \ R(\widehat{A}) = R(A), \\ \alpha(\widehat{A}+\widehat{B}) = \alpha(A+B), \\ \beta(\widehat{A}+\widehat{B}) = \beta(A+B) \text{ and } R(\widehat{A}+\widehat{B}) = R(A+B). \end{cases} \tag{2.14}$$

◇

2.3.6 A-Compact Perturbations

Theorem 2.3.5 *(T. Kato [107, Theorem 5.26, p. 238]) Let X and Y be Banach spaces. Suppose that A is a semi-Fredholm operator and B is an A-compact operator from X into Y, then $A+B$ is also semi-Fredholm with*

$$i(A+B) = i(A).$$

◇

As a consequence of Theorem 2.3.5, we have the following:

Theorem 2.3.6 *(M. Schechter [140, Theorem 7.26, p. 172]) $\Phi_{A+K} = \Phi_A$ for all K which are A-compact, and*

$$i(A+K-\lambda) = i(A-\lambda)$$

for all $\lambda \in \Phi_A$. ◇

Theorem 2.3.7 *(M. Schechter [140, Theorem 2.12, p. 9]) Let X and Y be two Banach spaces. If A is a closed linear operator from X into Y, and if B is A-compact, then*

(i) $\|Bx\| \le c(\|Ax\| + \|x\|)$, $x \in \mathscr{D}(A)$,

(ii) $\|Ax\| \le c(\|(A+B)x\| + \|x\|)$, $x \in \mathscr{D}(A)$,

(iii) $A+B$ is a closed operator, and

(iv) B is $(A+B)$-compact. ◇

Theorem 2.3.8 *Let G be a connected open set in the complex plane, such that $\rho(A)\bigcap G \neq \emptyset$. Then, each point of $\sigma(A)\bigcap G$ is a pole of finite rank of $(\lambda - A)^{-1}$ if, and only if,*

$$\alpha(\lambda - A) = \beta(\lambda - A) < \infty$$

at each point λ of G. 

2.3.7 The Convergence Compactly

Definition 2.3.5 *Let $(T_n)_n$ be a sequence of bounded linear operators on X, $(T_n)_n$ is said to be converge to zero compactly, written $T_n \xrightarrow{c} 0$ if, for all $x \in X$, $T_n x \to 0$ and $(T_n x_n)_n$ is relatively compact for every bounded sequence $(x_n)_n \subset X$.* ♢

Clearly, $\|T_n\| \to 0$ implies that $T_n \xrightarrow{c} 0$.

Theorem 2.3.9 *(S. Goldberg [76, Theorem 4]) Let $(K_n)_n$ be a sequence of bounded linear operators converging to zero compactly i.e., $K_n \xrightarrow{c} 0$, and let T be a closed linear operator. If T is a semi-Fredholm operator, then there exists $n_0 \in \mathbb{N}$ such that for all $n \geq n_0$,*

(i) $T + K_n$ is semi-Fredholm,
(ii) $\alpha(T + K_n) \leq \alpha(T)$,
(iii) $\beta(T + K_n) \leq \beta(T)$, and
(iv) $i(T + K_n) = i(T)$. 

We recall the following results due to M. Schechter in [139].

Theorem 2.3.10 *Let $(T_n)_n$ be a sequence of bounded linear operators which converges (in norm) to a bounded operator T. Then,*

(i) if $(T_n)_n \in \mathcal{F}^b(X)$, then $T \in \mathcal{F}^b(X)$, and
(ii) if $(T_n)_n \in \mathcal{F}^b_-(X)$, then $T \in \mathcal{F}^b_-(X)$. ♢

Definition 2.3.6 *Let $(T_n)_n$ be a sequence of bounded linear operators on X and let $T \in \mathcal{L}(X)$. The sequence $(T_n)_n$ is said to be converge to T compactly, written $T_n \xrightarrow{c} T$, if $(T_n - T)_n$ converges to zero compactly.* ♢

If $(T_n)_n$ converges to T, then $(T_n)_n$ converges to T compactly. A generalization of the Theorem 2.3.10 is given by the following proposition.

Proposition 2.3.2 *(A. Ammar and A. Jeribi [32]) Let $(T_n)_n$ be a sequence of bounded linear operators which converges compactly to a bounded operator T. Then,*

(i) if $(T_n)_n \in \mathscr{F}^b(X)$, then $T \in \mathscr{F}^b(X)$,
(ii) if $(T_n)_n \in \mathscr{F}_+^b(X)$, then $T \in \mathscr{F}_+^b(X)$, and
(iii) if $(T_n)_n \in \mathscr{F}_-^b(X)$, then $T \in \mathscr{F}_-^b(X)$. ◊

2.4 ASCENT AND DESCENT OPERATORS

2.4.1 Bounded Operators

We will denote the set of nonnegative integers by $\mathbb{N}$ and if $A \in \mathscr{L}(X)$, we define the ascent and the descent of A, respectively, by

$$\text{asc}(A) := \min\Big\{n \in \mathbb{N} \text{ such that } N(A^n) = N(A^{n+1})\Big\},$$

and

$$\text{desc}(A) := \min\Big\{n \in \mathbb{N} \text{ such that } R(A^n) = R(A^{n+1})\Big\}.$$

If no such integer exists, we shall say that A has infinite ascent or infinite descent. First, we will recall the following result due to A. E. Taylor.

Proposition 2.4.1 *(A. E. Taylor [149, Theorem 3.6]) Let $A \in \mathscr{L}(X)$. If $asc(A)$ and $desc(A)$ are finite, then*

$$asc(A) = desc(A). \quad ◊$$

Proposition 2.4.2 *Let A be a bounded linear operator on a Banach space X. If $A \in \Phi^b(X)$ with $asc(A)$ and $desc(A)$ are finite. Then,*

$$i(A) = 0. \quad ◊$$

Proof. Since $asc(A)$ and $desc(A)$ are finite. Using Proposition 2.4.1, there exists an integer k such that

$$asc(A) = desc(A) = k.$$

Hence,

$$N(A^k) = N(A^{n+k})$$

and

$$R(A^k) = R(A^{n+k})$$

for all $n \in \mathbb{N}$. Therefore,

$$i(A^k) = i(A^{n+k}).$$

However, since $A \in \Phi^b(X)$ and by using Theorem 2.2.5, we deduce that

$$i(A^k) = ki(A) = i(A^{n+k}) = (n+k)i(A),$$

for all $n \geq 0$. Hence,

$$i(A) = 0.$$ Q.E.D.

Theorem 2.4.1 *(M. A. Kaashoek [104]) Suppose that* $\alpha(A) = \beta(A) < \infty$, *and that* $p = asc(A) < \infty$. *Then,*

(i) $desc(A) = asc(A)$,
(ii) $\alpha(A^i) = \beta(A^i) < \infty$ *for* $i = 0, 1, 2, \cdots$, *and*
(iii) $X = R(A^p) \bigoplus N(A^p)$. ◊

Theorem 2.4.2 *Suppose* λ_0 *is an isolated point of* $\sigma(A)$, *and let* m *be a positive integer. Then,* λ_0 *is a pole of order* m *of* $(\lambda - A)^{-1}$ *if, and only if,*

$$asc(\lambda_0 - A) = desc(\lambda_0 - A) = m$$

and $R[(\lambda_0 - A)^m]$ *is closed.* ◊

Lemma 2.4.1 *(S. R. Caradus [50]) For any non negative* k, *we have*

(i) $\alpha(A^k) \leq asc(A)\alpha(A)$, *and*
(ii) $\beta(A^k) \leq desc(A)\beta(A)$. ◊

Lemma 2.4.2 *(A. E. Taylor [149]) Suppose there exists a nonnegative integer N such that* $\alpha(A^k) \leq N$ *when* $k = 0, 1, 2 \cdots$. *Then,*

$$asc(A) \leq N. \qquad \diamondsuit$$

Theorem 2.4.3 *(V. Rakočević [132, Theorem 1]) Suppose that* $A, K \in \mathscr{L}(X)$ *and* $AK = KA$. *Then,*

(i) if $A \in \Phi_+^b(X)$, $asc(A) < \infty$ *and* $K \in \mathscr{F}_+^b(X)$, *then* $asc(A+K) < \infty$, *and*

(ii) if $A \in \Phi_-^b(X)$, $desc(A) < \infty$ *and* $K \in \mathscr{F}_-^b(X)$, *then* $desc(A+K) < \infty$. $\diamondsuit$

Theorem 2.4.4 *Let* $A \in \mathscr{L}(X)$ *and assume that there exists a nonzero complex polynomial*

$$P(z) = \sum_{k=0}^{n} a_k z^k$$

satisfying

$$P(A) \in \mathscr{R}(X).$$

If $P(\lambda) \neq 0$ *for some* $\lambda \in \mathbb{C}$, *then* $\lambda - A \in \Phi^b(X)$ *with* $asc(\lambda - A) < \infty$ *and* $desc(\lambda - A) < \infty$. $\diamondsuit$

Proof. Let $\lambda \in \mathbb{C}$ with $P(\lambda) \neq 0$. We have

$$P(\lambda) - P(A) = \sum_{k=1}^{n} a_k(\lambda^k - A^k).$$

Moreover, for any $k \in \{1, \cdots, n\}$

$$\lambda^k - A^k = (\lambda - A) \sum_{i=0}^{k-1} \lambda^i A^{k-1-i}.$$

Then,

$$P(\lambda) - P(A) = (\lambda - A)H(A) = H(A)(\lambda - A),$$

where

$$H(A) = \sum_{k=1}^{n} a_k \sum_{i=0}^{k-1} \lambda^i A^{k-1-i}.$$

Hence, for $n \in \mathbb{N}$,

$$(P(\lambda) - P(A))^n = (\lambda - A)^n H(A)^n = H(A)^n (\lambda - A)^n.$$

This implies that

$$N[(\lambda - A)^n] \subset N[(P(\lambda) - P(A))^n]$$

and

$$R[(P(\lambda) - P(A))^n] \subset R[(\lambda - A)^n]$$

for all $n \in \mathbb{N}$. The last inclusions gives

$$\bigcup_{n \in \mathbb{N}} N[(\lambda - A)^n] \subset \bigcup_{n \in \mathbb{N}} N[(P(\lambda) - P(A))^n] \tag{2.15}$$

and

$$\bigcap_{n \in \mathbb{N}} R[(P(\lambda) - P(A))^n] \subset \bigcap_{n \in \mathbb{N}} R[(\lambda - A)^n]. \tag{2.16}$$

On the other hand, since $P(\lambda) \neq 0$ and $P(A) \in \mathscr{R}(X)$, then $P(\lambda) - P(A) \in \Phi^b(X)$ with $\text{asc}(P(\lambda) - P(A)) < \infty$ and $\text{desc}(P(\lambda) - P(A)) < \infty$. Therefore, Proposition 2.4.1 gives

$$\text{asc}(P(\lambda) - P(A)) = \text{desc}(P(\lambda) - P(A)).$$

Let n_0 this quantity. Since $P(\lambda)$ commutes with $P(A)$, Newton's binomial formula gives

$$\begin{aligned}(P(\lambda) - P(A))^{n_0} &= \sum_{k=0}^{n_0} (-1)^k C_{n_0}^k P(\lambda)^{n_0 - k} P(A)^k \\ &= P(\lambda)^{n_0} - \mathscr{K},\end{aligned}$$

where

$$\mathscr{K} = P(A) \sum_{k=1}^{n_0} (-1)^{k-1} C_{n_0}^k P(\lambda)^{n_0 - k} P(A)^{k-1}$$

is a Riesz operator on X. Then,

$$\dim \bigcup_{n \in \mathbb{N}} N[(P(\lambda) - P(A))^n] = \dim N[(P(\lambda) - P(A))^{n_0}] < \infty$$

and

$$\operatorname{codim} \bigcap_{n \in \mathbb{N}} R[(P(\lambda) - P(A))^n] = \operatorname{codim} R[(P(\lambda) - P(A))^{n_0}] < \infty.$$

It follows from Eqs. (2.15) and (2.16) that

$$\dim \bigcup_{n \in \mathbb{N}} N[(\lambda - A)^n] < \infty$$

and

$$\operatorname{codim} \bigcap_{n \in \mathbb{N}} R[(\lambda - A)^n] < \infty.$$

Hence, $\operatorname{asc}(\lambda - A) < \infty$ and $\operatorname{desc}(\lambda - A) < \infty$. This implies that $\alpha(\lambda - A) < \infty$ and $\beta(\lambda - A) < \infty$ and consequently, by Lemma 2.1.3, $R(\lambda - A)$ is closed. This proves that $\lambda - A \in \Phi^b(X)$ with $\operatorname{asc}(\lambda - A) < \infty$ and $\operatorname{desc}(\lambda - A) < \infty$. Q.E.D.

Corollary 2.4.1 *Assume that the hypotheses of Theorem 2.4.4 hold. Then, $\lambda - A$ is a Fredholm operator on X of index zero.* ◇

The next corollary is a consequence of Theorem 2.4.4 and Corollary 2.4.1.

Corollary 2.4.2 *Let $A \in \mathscr{PR}(X)$ i.e., there exists a nonzero complex polynomial*

$$P(z) = \sum_{k=0}^{n} a_k z^k$$

satisfying

$$P(-1) \neq 0 \text{ and } P(A) \in \mathscr{R}(X).$$

Then, $I + A \in \Phi^b(X)$ and $i(I + A) = 0$. ◇

2.4.2 Unbounded Operators

To discuss the definitions of ascent and descent for an unbounded linear operator one must consider the case in which $\mathscr{D}(A)$ and $R(A)$ are in the same linear space X. Obviously,

$$N(A^n) \subset N(A^{n+1})$$

and

$$R(A^{n+1}) \subset R(A^n)$$

for all $n \geq 0$ with the convention $A^0 = I$ (the identity operator on X). Thus, if $N(A^k) = N(A^{k+1})$ (resp. $R(A^k) = R(A^{k+1})$), then $N(A^n) = N(A^k)$ (resp. $R(A^n) = R(A^k)$) for all $n \geq k$. Then, the smallest nonnegative integer n such that $N(A^n) = N(A^{n+1})$ (resp. $R(A^n) = R(A^{n+1})$) is called the ascent (resp. the descent) of A, and denoted by $asc(A)$ (resp. $desc(A)$). In case where n does not exist, we define $asc(A) = \infty$ (resp. $desc(A) = \infty$).

Proposition 2.4.3 *(M. A. Kaashoek and D. C. Lay [105]) Suppose that $\alpha(A) = \beta(A) < \infty$ and $asc(A) < \infty$. Then, there exists a bounded linear operator B defined on X with finite-dimensional range and such that*

(i) $BAx = ABx$ for $x \in \mathscr{D}(A)$, in particular, B commutes with A,
(ii) $0 \in \rho(A+B)$, i.e., $A+B$ has a bounded inverse defined on X, and
(iii) if C commutes with A, then $CBx = BCx$ for $x \in \mathscr{D}(A)$. ◇

Lemma 2.4.3 *(M. A. Kaashoek and D. C. Lay [105]) If C commutes with A, then $-C$ commutes with $A+C$.* ◇

Theorem 2.4.5 *(M. A. Kaashoek and D. C. Lay [105]) Suppose that $\alpha(A) = \beta(A) < \infty$ and $asc(A) < \infty$. Let C commute with A and suppose that C satisfies at least one of the following conditions:*

(i) C is a compact linear operator on X,
(ii) C is a Riesz operator, and
(iii) C is A^n-compact, $n \in \mathbb{N}^*$.

Then, $A+C$ is closed, $\alpha(A+C) = \beta(A+C) < \infty$, and $asc(A+C) = desc(A+C)) < \infty$. ◇

2.5 SEMI-BROWDER AND BROWDER OPERATORS

For $A \in \mathscr{C}(X)$, we define the generalized kernel of A by

$$N^{\infty}(A) = \bigcup_{k=1}^{\infty} N(A^k)$$

and the generalized range of A by

$$R^{\infty}(A) = \bigcap_{k=1}^{\infty} R(A^k).$$

Proposition 2.5.1 *(T. T. West [157]) Let $A \in \mathscr{L}(X)$.*

(i) If $\alpha(A) < \infty$, then A has finite ascent if, and only if,

$$\left(\bigcup_{n=1}^{\infty} N(A^n)\right) \cap \left(\bigcap_{n=1}^{\infty} A^n(X)\right) = \{0\}.$$

(ii) If $\beta(A) < \infty$, then A has finite descent if, and only if,

$$\bigcup_{n=1}^{\infty} N(A^n) \bigoplus \bigcap_{n=1}^{\infty} A^n(X) = X.$$

◇

2.5.1 Semi-Browder Operators

Two important classes of operators in Fredholm theory are given by the classes of semi-Fredholm operators which possess finite ascent or finite descent. We shall distinguish the class of all upper semi-Browder operators on a Banach space X that is defined by

$$\mathscr{B}_{+}(X) := \Big\{A \in \Phi_{+}(X) \text{ such that } \operatorname{asc}(A) < \infty\Big\},$$

and the class of all lower semi-Browder operators that is defined by

$$\mathscr{B}_{-}(X) := \Big\{A \in \Phi_{-}(X) \text{ such that } \operatorname{desc}(A) < \infty\Big\}.$$

The class of all Browder operators (known in the literature also as Riesz-Schauder operators) is defined by

$$\mathscr{B}(X) := \mathscr{B}_+(X) \bigcap \mathscr{B}_-(X).$$

The class of all left Browder operators is defined by

$$\mathscr{B}^l(X) = \left\{A \in \Phi_l(X) \text{ such that } \mathrm{asc}(A) < \infty\right\}$$

and the class of all right Browder operators is defined by

$$\mathscr{B}^r(X) = \left\{A \in \Phi_r(X) \text{ such that } \mathrm{desc}(A) < \infty\right\}.$$

Theorem 2.5.1 *(S. Č. Živković-Zlatanović, D. S. Djordjević, and R. E. Harte [163, Theorem 7]) Let A, $E \in \mathscr{L}(X)$. Then,*

(i) if $A \in \mathscr{B}^l(X)$ and $E \in \mathscr{R}(X)$ such that $AE = EA$, then $A+E \in \mathscr{B}^l(X)$, and

(ii) if $A \in \mathscr{B}^r(X)$ and $E \in \mathscr{R}(X)$ such that $AE = EA$, then $A+E \in \mathscr{B}^r(X)$.

$\diamondsuit$

Theorem 2.5.2 *(F. Fakhfakh and M. Mnif [70, Theorem 3.1]) Let X be a Banach space, $A \in \mathscr{C}(X)$ and $K \in \mathscr{L}(X)$ such that either $\rho(A)$ or $\rho(A+K) \neq \emptyset$. Assume that K commutes with A, and $\Delta_\psi(K) < \Gamma_\psi(A)$, where ψ is a perturbation function. Then,*

$$A \in \mathscr{B}_+(X) \text{ if, and only if, } A+K \in \mathscr{B}_+(X). \qquad \diamondsuit$$

Theorem 2.5.3 *(F. Fakhfakh and M. Mnif [70, Theorem 3.2]) Let X be a Banach space, ψ be a perturbation function, $A \in \mathscr{C}(X)$, $K \in \mathscr{L}(X)$, and neither $\rho(A)$ nor $\rho(A+K)$ is empty. If we suppose that K commutes with A and $\Delta_\psi(K) < \Gamma_\psi(A)$, then*

$$A \in \mathscr{B}(X) \text{ if, and only if, } A+K \in \mathscr{B}(X). \qquad \diamondsuit$$

Theorem 2.5.4 *(F. Leon-Saavedra and V. Müller [118, Theorem 10, p. 186]) An operator $A \in \mathscr{L}(X)$ is upper semi-Browder (lower semi-Browder, Browder, respectively) if, and only if, there exists a decomposition*

$$X = X_1 \bigoplus X_2$$

such that $\dim X_1 < \infty$, $AX_i \subset X_i$ $(i = 1, 2)$, $A_{|X_1}$ *is nilpotent and* $A_{|X_2}$ *is bounded below (onto, invertible, respectively).* ♢

2.5.2 Fredholm Operator with Finite Ascent and Descent

Theorem 2.5.5 *Let $A \in \mathscr{P}\mathscr{F}(X)$ i.e., there exists a nonzero complex polynomial*

$$P(z) = \sum_{i=0}^{n} a_i z^i$$

satisfying $P(A) \in \mathscr{F}^b(X)$. Let $\lambda \in \mathbb{C}$ with $P(\lambda) \neq 0$ and set $B = \lambda - A$. Then, B is a Fredholm operator on X with finite ascent and descent. ♢

We need the following lemma for the proof of the Theorem 2.5.5.

Lemma 2.5.1 *Let P be a complex polynomial and $\lambda \in \mathbb{C}$ such that $P(\lambda) \neq 0$. Then, for all $A \in \mathscr{L}(X)$ satisfying $P(A) \in \mathscr{F}^b(X)$, the operator $P(\lambda) - P(A)$ is a Fredholm operator on X with finite ascent and descent.* ♢

Proof. Put

$$\begin{aligned} B &= P(\lambda) - P(A) \\ &= P(\lambda)\left(I - \frac{P(A)}{P(\lambda)}\right) \\ &= P(\lambda)(I - F), \end{aligned}$$

where $F = \frac{P(A)}{P(\lambda)} \in \mathscr{F}^b(X)$. Let $C = I - F$, then

$$B = P(\lambda)C.$$

It is clear that $C + F \in \mathscr{B}(X)$ and so, we can write $C = C + F - F$ with $C + F \in \mathscr{B}(X)$ and $F \in \mathscr{F}^b(X)$. On the other hand, we have

$$(C + F)F = F(C + F).$$

By using Theorem 2.4.3, we deduce that $C \in \mathscr{B}(X)$ and therefore, $B \in \mathscr{B}(X)$.

Q.E.D.

Proof of Theorem 2.5.5 Let $\lambda \in \mathbb{C}$ with $P(\lambda) \neq 0$. We have

$$P(\lambda) - P(A) = \sum_{i=1}^{n} a_i(\lambda^i - A^i).$$

On the one hand, for any $i \in \{1, \cdots, n\}$, we have

$$\lambda^i - A^i = (\lambda - A) \sum_{j=0}^{k-1} \lambda^j A^{k-1-j}.$$

So,

$$P(\lambda) - P(A) = (\lambda - A)Q(A) = Q(A)(\lambda - A), \tag{2.17}$$

where

$$Q(A) = \sum_{i=1}^{n} a_i \sum_{j=0}^{k-1} \lambda^j A^{k-1-j}.$$

Let $p \in \mathbb{N}$, the Eq. (2.17) gives

$$\big(P(\lambda) - P(A)\big)^p = (\lambda - A)^p Q(A)^p = Q(A)^p (\lambda - A)^p.$$

Hence,

$$N[(\lambda - A)^p] \subset N\big[\big(P(\lambda) - P(A)\big)^p\big], \quad \forall p \in \mathbb{N}$$

and

$$R\big[\big(P(\lambda) - P(A)\big)^p\big] \subset R[(\lambda - A)^p], \quad \forall p \in \mathbb{N}.$$

This implies that

$$\bigcup_{p \in \mathbb{N}} N[(\lambda - A)^p] \subset \bigcup_{p \in \mathbb{N}} N\big[\big(P(\lambda) - P(A)\big)^p\big], \tag{2.18}$$

and

$$\bigcap_{p \in \mathbb{N}} R\big[\big(P(\lambda) - P(A)\big)^p\big] \subset \bigcap_{p \in \mathbb{N}} R[(\lambda - A)^p]. \tag{2.19}$$

On the other hand, since $P(\lambda) \neq 0$ and $P(A) \in \mathscr{F}^b(X)$, then by using Lemma 2.5.1, $P(\lambda) - P(A)$ is a Fredholm operator on X with finite ascent and descent. So, by Proposition 2.4.1, we have

$$asc\big(P(\lambda) - P(A)\big) = desc\big(P(\lambda) - P(A)\big).$$

Let p_0 this quantity, then

$$\dim \bigcup_{p\in\mathbb{N}} N\left[\left(P(\lambda)-P(A)\right)^p\right] = \dim N\left[\left(P(\lambda)-P(A)\right)^{p_0}\right] < \infty,$$

and

$$\operatorname{codim} \bigcap_{p\in\mathbb{N}} R\left[\left(P(\lambda)-P(A)\right)^p\right] = \operatorname{codim} R\left[\left(P(\lambda)-P(A)\right)^{p_0}\right] < \infty.$$

Using Eqs. (2.18) and (2.19), we have

$$\dim \bigcup_{p\in\mathbb{N}} N[(\lambda-A)^p] < \infty,$$

and

$$\operatorname{codim} \bigcap_{p\in\mathbb{N}} R[(\lambda-A)^p] < \infty.$$

Therefore, $\operatorname{asc}(\lambda-A) < \infty$ and $\operatorname{desc}(\lambda-A) < \infty$. We have also, $\alpha(\lambda-A) < \infty$ and $\beta(\lambda-A) < \infty$. Q.E.D.

Corollary 2.5.1 *Let $A \in \mathscr{P}\mathscr{F}(X)$, i.e., there exists a nonzero complex polynomial*

$$P(z) = \sum_{r=0}^{p} a_r z^r$$

satisfying $P(A) \in \mathscr{F}^b(X)$. Let $\lambda \in \mathbb{C}$ with $P(\lambda) \neq 0$. Then, the operator $\lambda - A$ is a Fredholm operator on X of index zero. ◇

Proof. The proof follows immediately from both Theorem 2.5.5 and Proposition 2.4.2. Q.E.D.

2.6 MEASURE OF NONCOMPACTNESS

The notion of measure of noncompactness turned out to be a useful tool in some topological problems, in functional analysis, and in operator theory (see [1, 17, 42, 63, 121, 129]).

2.6.1 Measure of Noncompactness of a Bounded Subset

In order to recall the measure of noncompactness, let $(X, \|\cdot\|)$ be an infinite-dimensional Banach space. We denote by $\mathcal{M}_X$ the family of all nonempty and bounded subsets of X, while N_X denotes its subfamily consisting of all relatively compact sets. Moreover, let us denote the convex hull of a set $A \subset X$ by $conv(A)$. Let us recall the following definition.

Definition 2.6.1 *A mapping $\mu : \mathcal{M}_X \longrightarrow [0, +\infty[$ is said to be a measure of noncompactness in the space X, if it satisfies the following conditions:*

(i) The family $Ker(\mu) := \{D \in \mathcal{M}_X$ such that $\mu(D) = 0\}$ is nonempty, and $Ker(\mu) \subset N_X$.

For $A, B \in \mathcal{M}_X$, we have the following:

(ii) If $A \subset B$, then $\mu(A) \leq \mu(B)$.

(iii) $\mu(\overline{A}) = \mu(A)$.

(iv) $\mu(\overline{conv(A)}) = \mu(A)$.

(v) $\mu(\lambda A + (1-\lambda)B) \leq \lambda\mu(A) + (1-\lambda)\mu(B)$, for all $\lambda \in [0, 1]$.

(vi) If $(A_n)_n$ is a sequence of sets from $\mathcal{M}_X$ such that $A_{n+1} \subset A_n$, $\overline{A}_n = A_n$ $(n = 1, 2, \cdots)$ and $\lim_{n\to+\infty} \mu(A_n) = 0$, then

$$A_\infty := \bigcap_{n=1}^{\infty} A_n$$

is nonempty and $A_\infty \in Ker(\mu)$. ◇

The family Ker(μ), described in Definition 2.6.1 (i), is called the kernel of the measure of noncompactness μ.

Definition 2.6.2 *A measure of noncompactness μ is said to be sublinear if, for all $A, B \in \mathcal{M}_X$, it satisfies the two following conditions:*

(i) $\mu(\lambda A) = |\lambda|\mu(A)$ for $\lambda \in \mathbb{R}$ (μ is said to be homogenous), and

(ii) $\mu(A+B) \leq \mu(A) + \mu(B)$ (μ is said to be subadditive). ◇

Definition 2.6.3 *A measure of noncompactness μ is referred to as a measure with maximum property if* $\max(\mu(A), \mu(B)) = \mu(A \bigcup B)$. ◇

Definition 2.6.4 *A measure of noncompactness μ is said to be regular if $Ker(\mu) = N_X$, sublinear and has a maximum property.* $\diamondsuit$

For $A \in \mathscr{M}_X$, the most important examples of measures of noncompactness ([121]) are:

- Kuratowski measure of noncompactness

$$\gamma(A) = \inf\{\varepsilon > 0 \,:\, A \text{ may be covered by a finite number of sets of diameter} \leq \varepsilon\}.$$

- Hausdorff measure of noncompactness

$$\overline{\gamma}(A) = \inf\{\varepsilon > 0 \,:\, A \text{ may be covered by a finite number of open balls of radius} \leq \varepsilon\}.$$

Note that these measures $\gamma(\cdot)$ and $\overline{\gamma}(\cdot)$ are regular. The relations between these measures are given by the following inequalities, which were obtained by J. Danes [57]

$$\overline{\gamma}(A) \leq \gamma(A) \leq 2\overline{\gamma}(A),$$

for any $A \in \mathscr{M}_X$. The following proposition gives some frequently used properties of the Kuratowski's measure of noncompactness.

Proposition 2.6.1 *Let A and A' be two bounded subsets of X. Then, we have the following properties:*

(i) $\gamma(A) = 0$ if, and only if, A is relatively compact.
(ii) If $A \subseteq A'$, then $\gamma(A) \leq \gamma(A')$.
(iii) $\gamma(A + A') \leq \gamma(A) + \gamma(A')$.
(iv) For every $\alpha \in \mathbb{C}$, $\gamma(\alpha A) = |\alpha|\,\gamma(A)$. $\diamondsuit$

2.6.2 Measure of Noncompactness of an Operator

Definition 2.6.5 *(i) Let $T : \mathscr{D}(T) \subseteq X \longrightarrow X$ be a continuous operator and let $\gamma(\cdot)$ be the Kuratowski measure of noncompactness in X. Let $k \geq 0$.*

T is said to be k-set-contraction if, for any bounded subset A of $\mathscr{D}(T)$, $T(A)$ is a bounded subset of X and

$$\gamma(T(A)) \leq k\gamma(A).$$

T is said to be condensing if, for any bounded subset A of $\mathscr{D}(T)$ such that $\gamma(A) > 0$, $T(A)$ is a bounded subset of X and

$$\gamma(T(A)) < \gamma(A).$$

(ii) Let $T : \mathscr{D}(T) \subseteq X \longrightarrow X$ be a continuous operator, $\overline{\gamma}(\cdot)$ being the Hausdorff measure of noncompactness in X, and $k \geq 0$. T is said to be k-ball-contraction if, for any bounded subset A of $\mathscr{D}(T)$, $T(A)$ is a bounded subset of X and

$$\overline{\gamma}(T(A)) \leq k\overline{\gamma}(A). \qquad \diamondsuit$$

Remark 2.6.1 *It is well known that*

(i) If $k < 1$, then every k-set-contraction operator is condensing.

(ii) Every condensing operator is 1-set-contraction. $\diamondsuit$

Let $T \in \mathscr{L}(X)$. We define Kuratowski measure of noncompactness, $\gamma(T)$, by

$$\gamma(T) := \inf\{k \text{ such that } T \text{ is } k\text{-set-contraction}\}, \tag{2.20}$$

and Hausdorff measure of noncompactness, $\overline{\gamma}(T)$, by

$$\overline{\gamma}(T) := \overline{\gamma}[T(B_X)] := \inf\{k \text{ such that } T \text{ is } k\text{-ball-contraction}\},$$

where B_X denotes the closed unit ball in X, that is, the set of all $x \in X$ satisfying $\|x\| \leq 1$. In the following lemma, we give some important properties of $\gamma(T)$ and $\overline{\gamma}(T)$.

Lemma 2.6.1 *([38, 66]) Let X be a Banach space and $T \in \mathscr{L}(X)$. We have the following*

(i) $\frac{1}{2}\gamma(T) \leq \overline{\gamma}(T) \leq 2\gamma(T)$.

(ii) $\gamma(T) = 0$ if, and only if, $\overline{\gamma}(T) = 0$ if, and only if, T is compact.

(iii) If $T, S \in \mathscr{L}(X)$, *then* $\gamma(ST) \leq \gamma(S)\gamma(T)$ *and* $\overline{\gamma}(ST) \leq \overline{\gamma}(S)\overline{\gamma}(T)$.

(iv) If $K \in \mathscr{K}(X)$, *then* $\gamma(T+K) = \gamma(T)$ *and* $\overline{\gamma}(T+K) = \overline{\gamma}(T)$.

(v) $\gamma(T^*) \leq \overline{\gamma}(T)$ *and* $\gamma(T) \leq \overline{\gamma}(T^*)$, *where* T^* *denotes the dual operator of* T.

(vi) If B *is a bounded subset of* X, *then* $\gamma(T(B)) \leq \gamma(T)\gamma(B)$.

(vii) $\overline{\gamma}(T) \leq \|T\|$. ♢

Theorem 2.6.1 *[6, Theorem 3.1] (see also [1]) Let* $A \in \mathscr{L}(X)$, *and* P, Q *be two complex polynomials satisfying* Q *divides* $P-1$.

(i) If $\gamma\big(P(A)\big) < 1$, *then*

$$Q(A) \in \Phi_+^b(X).$$

(ii) If $\gamma\big(P(A)\big) < \frac{1}{2}$, *then*

$$Q(A) \in \Phi^b(X).$$ ♢

Proposition 2.6.2 *(B. Abdelmoumen, A. Dehici, A. Jeribi, and M. Mnif [6, Corollary 3.4]) Let* X *be a Banach space and* $A \in \mathscr{L}(X)$. *If* $\gamma(A^n) < 1$, *for some* $n > 0$, *then* $I - A$ *is a Fredholm operator with* $i(I-A) = 0$. ♢

2.6.3 Measure of Non-Strict-Singularity

We recall the following definition of the measure of non-strict-singularity which is introduced by M. Schechter in [139].

Definition 2.6.6 *For* $A \in \mathscr{L}(X,Y)$, *set*

$$f_M(A) = \inf_{N \subset M} \overline{\gamma}(Ai_N) \tag{2.21}$$

and

$$f(A) = \sup_{M \subset X} f_M(A), \tag{2.22}$$

where M, N *represent infinite dimensional subspaces of* X *and* Ai_N *denotes the restriction of* A *to* N. ♢

The semi-norm $f(\cdot)$ is a measure of non-strict-singularity, it was introduced by M. Schechter in [142].

Proposition 2.6.3 *(V. R. Rakočević [130]) Let $A \in \mathscr{L}(X,Y)$. Then,*

(i) $A \in \mathscr{S}(X,Y)$ if, and only if,

$$f(A) = 0.$$

(ii) $A \in \mathscr{S}(X,Y)$ if, and only if,

$$f(A+T) = f(T)$$

for all $T \in \mathscr{L}(X,Y)$.

(iii) If Z is a Banach space and $B \in \mathscr{L}(Y,Z)$, then

$$f(BA) \leq f(B)f(A). \qquad \diamondsuit$$

Proposition 2.6.4 *(N. Moalla [122, Proposition 2.3]) Let $A \in \mathscr{L}(X)$. If $f(A^n) < 1$ for some integer $n \geq 1$, then $I - A \in \Phi^b(X)$ with $i(I-A) = 0$.* $\diamondsuit$

Proposition 2.6.5 *(N. Moalla [122]) For $A \in \mathscr{L}(X,Y)$, we have*

$$f(A) \leq \overline{\gamma}(A). \qquad \diamondsuit$$

Lemma 2.6.2 *(N. Moalla [122]) For all bounded operator*

$$T = \begin{pmatrix} T_1 & T_2 \\ T_3 & T_4 \end{pmatrix}$$

on $X \times Y$, we consider

$$g(T) = \max\Big\{f(T_1)+f(T_2),\ f(T_3)+f(T_4)\Big\}, \tag{2.23}$$

where $f(\cdot)$ is a measure of non-strict-singularity given in (2.22). Then, $g(\cdot)$ defines a measure of non-strict-singularity on the space $\mathscr{L}(X \times Y)$. $\diamondsuit$

2.6.4 γ-Relatively Bounded

Let us notice that, throughout the book, we are working on two spaces, for example X and Y with their respective measures $\gamma_X(\cdot)$ and $\gamma_Y(\cdot)$. However, and in order to simplify our reasoning, $\gamma_X(\cdot)$ and $\gamma_Y(\cdot)$ will be simply called $\gamma(\cdot)$. Of course, the reader will be able to link $\gamma(\cdot)$ either to $\gamma_X(\cdot)$ or to $\gamma_Y(\cdot)$.

Definition 2.6.7 *Let X and Y be Banach spaces, let $\gamma(\cdot)$ be a Kuratowski measure of noncompactness, and let S, A be two linear operators from X into Y bounded on its domains such that $\mathscr{D}(A) \subset \mathscr{D}(S)$. The operator S is called γ-relatively bounded with respect to A (or A-γ-bounded), if there exist constants $a_S \geq 0$ and $b_S \geq 0$, such that*

$$\gamma(S(\mathfrak{D})) \leq a_S\gamma(\mathfrak{D}) + b_S\gamma(A(\mathfrak{D})), \tag{2.24}$$

where $\mathfrak{D}$ is a bounded subset of $\mathscr{D}(A)$. The infimum of the constants b_S which satisfy (2.24) for some $a_S \geq 0$ is called the A-γ-bound of S. ♢

Theorem 2.6.2 *(A. Jeribi, B. Krichen, and M. Zarai Dhahri [92, Theorem 2.1]) If S is A-γ-bounded with a bound < 1 and if S and A are closed, then $A + S$ is closed.* ♢

Lemma 2.6.3 *(A. Jeribi, B. Krichen, and M. Zarai Dhahri [92, Remark 2.1]) If S is T-γ-bounded with T-γ-bound $\delta < 1$, then S is $(T+S)$-γ-bounded with $(T+S)$-γ-bound $\leq \frac{\delta}{1-\delta}$.* ♢

2.6.5 Perturbation Result

We start our investigation with the following lemma which is fundamental for our purpose.

Lemma 2.6.4 *Let A, $K \in \mathscr{L}(X)$ such that K commutes with A. If $f(K^n) < f_M(A^n)$ for some $n \geq 1$, then $A + K \in \Phi_+^b(X)$, $A \in \Phi_+^b(X)$ and*

$$i(A+K) = i(A),$$

where $f_M(\cdot)$ (resp. $f(\cdot)$) is given in (2.21) (resp. (2.22)). ♢

Proof. We will begin by the case $n = 1$. For this, we argue by contradiction. Suppose that $A+K \notin \Phi_+^b(X)$, then by Lemma 2.2.1, there is $S \in \mathscr{K}(X)$ such that

$$\alpha(A+K-S) = \infty.$$

We set $M = N(A+K-S)$. Then, the restriction of $A+K$ to M is compact, that is

$$(A+K)i_M = Si_M.$$

Let N be a subspace of M such that

$$\dim N = \infty.$$

So,

$$Ai_Mi_N = -Ki_Mi_N + Si_Mi_N.$$

From Proposition 2.6.3 (*ii*), it follows that

$$f(Ai_N) = f(Ki_N).$$

Thereby,

$$\begin{aligned} f_M(A) &= \inf_{N \subset M}(f(Ai_N)) \\ &\leq f(Ai_N) \\ &= f(Ki_N). \end{aligned}$$

Then,

$$f_M(A) \leq f_M(K) \leq f(K)$$

which is absurd. Consequently, $A+K \in \Phi_+^b(X)$. Let $\lambda \in [0,1]$, then

$$f(\lambda K) = \lambda f(K) < f_M(A),$$

and so $A+\lambda K \in \Phi_+^b(X)$. Thereby, $A \in \Phi_+^b(X)$. Since $A+\lambda K \in \Phi_+^b(X)$ for all $\lambda \in [0,1]$ and the use of Lemma 2.2.2 leads to

$$i(A+K) = i(A).$$

For $n > 1$, we have

$$A^n - K^n = (A^{n-1} + A^{n-2}K + \cdots + AK^{n-2} + K^{n-1})(A-K). \qquad (2.25)$$

Since

$$f(-K^n) = f(K^n) < f_M(A^n),$$

then

$$A^n - K^n \in \Phi_+^b(X). \tag{2.26}$$

In view of Theorem 2.2.10 and both (2.25) and (2.26), we get $A - K \in \Phi_+(X)$. Let $\lambda \in [-1,0]$. Thus,

$$\begin{aligned} f(-(\lambda K)^n) &= f((\lambda K)^n) \\ &= |\lambda^n| f(K^n) \\ &< f_M(A^n) \end{aligned}$$

and then, by what we have just showed,

$$A - \lambda K \in \Phi_+^b(X).$$

Therefore, $A + K \in \Phi_+^b(X)$ and $A \in \Phi_+^b(X)$. It remains to show that

$$i(A+K) = i(A).$$

To do this, we reason in the same way as the case $n = 1$. Q.E.D.

2.7 γ-DIAGONALLY DOMINANT

We denote by $\mathscr{L}$ the block matrix linear operator, acting on the Banach space $X \times Y$, of the form

$$\mathscr{L} = \begin{pmatrix} A & B \\ C & D \end{pmatrix}, \tag{2.27}$$

where the operator A acts on X and has domain $\mathscr{D}(A)$, D is defined on $\mathscr{D}(D)$ and acts on the Banach space Y, and the intertwining operator B (resp. C) is defined on the domain $\mathscr{D}(B)$ (resp. $\mathscr{D}(C)$) and acts on X (resp. Y). One of the problems in the study of such operators is that in general $\mathscr{L}$ is not closed or even closable, even if its entries are closed.

Definition 2.7.1 *Let $\gamma(\cdot)$ be a measure of noncompactness. The block operator matrix $\mathcal{L}$, given in (2.27), is called*

(i) γ-diagonally dominant if C is A-γ-bounded and B is D-γ-bounded.

(ii) off-γ-diagonally dominant if A is C-γ-bounded and D is B-γ-bounded.

Definition 2.7.2 *The block operator matrix $\mathcal{L}$ is called*

(i) γ-diagonally dominant with bound δ, if C is A-γ-bounded with A-γ-bound δ_C, and B is D-γ-bounded with D-γ-bound δ_B, and $\delta = \max\{\delta_C, \delta_B\}$.

(ii) off-γ-diagonally dominant with bound δ if A is C-γ-bounded with C-γ-bound δ_A, and D is B-γ-bounded with B-γ-bound δ_D, and $\delta = \max\{\delta_A, \delta_D\}$. $\diamondsuit$

2.8 GAP TOPOLOGY

2.8.1 Gap Between Two Subsets

Definition 2.8.1 *The gap between two subsets M and N of a normed space X is defined by the following formula:*

$$\delta(M,N) = \sup_{x \in M,\ \|x\|=1} dist(x,N),$$

whenever $M \neq \{0\}$. Otherwise we define $\delta(\{0\},N) = 0$ for every subset N. We can also define

$$\widehat{\delta}(M,N) = \max\Big\{\delta(M,N), \delta(N,M)\Big\}.$$

Sometimes the latter is called the symmetric or maximal gap between M and N in order to distinguish it from the former. The gap $\delta(M,N)$ can be characterized as the smallest number δ such that

$$dist(x,N) \leq \delta\|x\|, \ \textit{for all } x \in M.$$ $\diamondsuit$

Remark 2.8.1 *(i) The gap measures the distance between two subsets and it follows easily from the Definition 2.8.1,*

(a) $\delta(M,N) = \delta(\overline{M},\overline{N})$ and $\widehat{\delta}(M,N) = \widehat{\delta}(\overline{M},\overline{N})$,

(b) $\delta(M,N) = 0$ if, and only if, $\overline{M} \subset \overline{N}$,

(c) $\widehat{\delta}(M,N) = 0$ if, and only if, $\overline{M} = \overline{N}$,

where $\overline{M}$ (resp. $\overline{N}$) is the closure of M (resp. N).

(ii) Let us notice that $\widehat{\delta}(\cdot,\cdot)$ is a metric on the set $\mathscr{U}(X)$ of all linear, closed subspaces of X and the convergence $M_n \to N$ in $\mathscr{U}(X)$ is obviously defined by $\widehat{\delta}(M_n,N) \to 0$. Moreover, $(\mathscr{U}(X), \widehat{\delta})$ is a complete metric space.

2.8.2 Gap Between Two Operators

Definition 2.8.2 *Let X and Y be two normed spaces and let T, S be two closed linear operators acting from X into Y. Let us define*

$$\delta(T,S) = \delta\big(G(T),G(S)\big) \text{ and } \widehat{\delta}(T,S) = \widehat{\delta}\big(G(T),G(S)\big).$$

$\widehat{\delta}\big(T,S\big)$ is called the gap between S and T.

More explicitly,

$$\delta\big(G(T),G(S)\big) = \sup_{\substack{x \in \mathscr{D}(T) \\ \|x\|_X^2 + \|Tx\|_Y^2 = 1}} \left(\inf_{y \in \mathscr{D}(S)} \left(\|x-y\|_X^2 + \|Tx - Sy\|_Y^2 \right)^{\frac{1}{2}} \right).$$

The function $\widehat{\delta}(\cdot,\cdot)$ defines a metric on $\mathscr{C}(X,Y)$ called the gap metric and the topology induced by this metric is called the gap topology. The next theorem contains some basic properties of the gap between two closed linear operator.

Theorem 2.8.1 *(T. Kato [108, Chapter IV Section 2]) Let T and S be two closed densely defined linear operators. Then,*

(i) $\delta(T,S) = \delta(S^*,T^*)$ and $\widehat{\delta}(T,S) = \widehat{\delta}(S^*,T^*)$.

(ii) If S and T are one-to-one, then $\delta(S,T) = \delta(S^{-1},T^{-1})$ and $\widehat{\delta}(S,T) = \widehat{\delta}(S^{-1},T^{-1})$.

(iii) Let $A \in \mathscr{L}(X,Y)$, then $\widehat{\delta}(A+S,A+T) \leq 2(1+\|A\|^2)\widehat{\delta}(S,T)$.

(iv) Let T be a Fredholm (resp. semi-Fredholm) operator. If $\widehat{\delta}(T,S) < \gamma(T)(1+[\gamma(T)]^2)^{\frac{-1}{2}}$, then S is a Fredholm (resp. semi-Fredholm) operator, $\alpha(S) \leq \alpha(T)$ and $\beta(S) \leq \beta(T)$. Furthermore, there exists $b > 0$ such that $\widehat{\delta}(T,S) < b$, which implies $i(S) = i(T)$.

(v) Let $T \in \mathscr{L}(X,Y)$. If $S \in \mathscr{C}(X,Y)$ and $\widehat{\delta}(T,S) \leq \left[1+\|T\|^2\right]^{-\frac{1}{2}}$, then S is a bounded operator (so that $\mathscr{D}(S)$ is closed). ◇

A complete discussion and properties concerning the gap may be found in T. Kato [108]. For the case of closable linear operators, the authors A. Ammar and A. Jeribi have introduced in [32] the following definition.

Definition 2.8.3 *Let S and T be two closable operators. We define the gap between T and S by $\delta(T,S) = \delta(\overline{T},\overline{S})$ and $\widehat{\delta}(T,S) = \widehat{\delta}(\overline{T},\overline{S})$.* ◇

2.8.3 Convergence in the Generalized Sense

Definition 2.8.4 *Let $(T_n)_n$ be a sequence of closable linear operators from X into Y and let T be a closable linear operator from X into Y. The sequence $(T_n)_n$ is said to be converge in the generalized sense to T, written $T_n \xrightarrow{g} T$, if $\widehat{\delta}(T_n,T)$ converges to 0 when $n \to \infty$.* ◇

It should be remarked that the notion of generalized convergence introduced above for closed and closable operators can be thought as a generalization of convergence in norm for linear operators that may be unbounded. Moreover, an important passageway between these two notions is developed in the following theorem:

Theorem 2.8.2 *(A. Ammar [32, Theorem 2.3]) Let $(T_n)_n$ be a sequence of closable linear operators from X into Y and let T be a closable linear operator from X into Y. Then,*

(i) The sequence $(T_n)_n$ converges in the generalized sense to T, ($T_n \xrightarrow{g} T$), if, and only if, $(T_n+S)_n$ converges in the generalized sense to $T+S$, for all $S \in \mathscr{L}(X,Y)$.

(ii) Let $T \in \mathscr{L}(X,Y)$. The sequence $(T_n)_n$ converges in the generalized sense to T if, and only if, $T_n \in \mathscr{L}(X,Y)$ for sufficiently larger n and $(T_n)_n$ converges to T.

(iii) If $(T_n)_n$ converges in the generalized sense to T, then T^{-1} exists and $T^{-1} \in \mathscr{L}(Y,X)$, if, and only if, T_n^{-1} exists and $T_n^{-1} \in \mathscr{L}(Y,X)$ for sufficiently larger n and $(T_n^{-1})_n$ converges to T^{-1}. ♢

2.9 QUASI-INVERSE OPERATOR

Let A be a closed operator on a Banach space X, with the property that $\Phi_A \neq \emptyset$. If $f(\lambda)$ is a complex valued analytic function of a complex variable, we denote by $\Delta(f)$ the domain of analyticity of f.

Definition 2.9.1 *By $\mathscr{R}'_\infty(A)$ we mean the family of all analytic functions $f(\lambda)$ with the following properties:*

(i) $\mathbb{C}\backslash\Phi_A \subset \Delta(f)$, and

(ii) $\Delta(f)$ contains a neighborhood of ∞ and f is analytic at ∞. ♢

Definition 2.9.2 A bounded operator B is called a quasi-inverse of the closed operator A *if* $R(B) \subset \mathscr{D}(A)$,

$$AB = I + K_1,$$

and

$$BA = I + K_2,$$

where K_1, $K_2 \in \mathscr{K}(X)$. 

If A is a closed operator such that Φ_A is not empty, then by using Proposition 2.2.1 (i), Φ_A is open. Hence, it is the union of a disjoint collection of connected open sets. Each of them, $\Phi_i(A)$, will be called a component of Φ_A. In each $\Phi_i(A)$, a fixed point λ_i, is chosen in a prescribed manner. Since $\alpha(\lambda_i - A) < \infty$, $R(\lambda_i - A)$ is closed and $\beta(\lambda_i - A) < \infty$, then there exist a closed subspace X_i and a subspace Y_i, such that $\dim Y_i = \beta(\lambda_i - A)$ satisfying

$$X = N(\lambda_i - A) \bigoplus X_i$$

and

$$X = Y_i \bigoplus R(\lambda_i - A).$$

Now, let P_{1i} be the projection of X onto $N(\lambda_i - A)$ along X_i and let P_{2i} be the projection of X onto Y_i along $R(\lambda_i - A)$. P_{1i} and P_{2i} are bounded finite rank operators. It is shown in [138] that $(\lambda_i - A)|_{\mathcal{D}(A) \cap X_i}$ has a bounded inverse A_i, where

$$A_i : R(\lambda_i - A) \longrightarrow \mathcal{D}(A) \bigcap X_i.$$

Let T_i be the bounded operator defined by

$$T_i x := A_i(I - P_{2i})x \tag{2.28}$$

satisfying

$$T_i(\lambda_i - A) = I - P_{1i} \text{ on } \mathcal{D}(A)$$

and

$$(\lambda_i - A)T_i = I - P_{2i} \text{ on } X.$$

Hence, T_i is a quasi-inverse of $\lambda_i - A$. Moreover, when $\lambda \in \Phi_i(A)$ and $\frac{-1}{\lambda - \lambda_i} \in \rho(T_i)$, the operator

$$R'_\lambda(A) := T_i\left[(\lambda - \lambda_i)T_i + I\right]^{-1}$$

is shown in [71] to be a quasi-inverse of $\lambda - A$. In fact, $R'_\lambda(A)$ is defined and analytic for all $\lambda \in \Phi_A$ except for, at most, an isolated set, $\Phi^0(A)$, having no accumulation point in Φ_A.

Now, we are ready to declare this result:

Lemma 2.9.1 *(F. Abdmouleh, S. Charfi, and A. Jeribi [12, Lemma 3.2]) Let* $A \in \mathscr{C}(X)$, $B \in \mathscr{L}(X)$, $\lambda \in \Phi_A \backslash \Phi^0(A)$ *and* $\mu \in \Phi_B \backslash \Phi^0(B)$. *If there exist a positive integer* n *and a Fredholm perturbation* F_1, *such that*

$$B : \mathscr{D}(A^n) \longrightarrow \mathscr{D}(A)$$

and

$$ABx = BAx + F_1 x,$$

for all $x \in \mathscr{D}(A^n)$. *Then, there exists a Fredholm perturbation* F *depending analytically on* λ *and* μ *such that*

$$R'_\lambda(A)R'_\mu(B) = R'_\mu(B)R'_\lambda(A) + F. \qquad \diamondsuit$$

Definition 2.9.3 *A set D in the complex plane is called a Cauchy domain, if the following conditions are satisfied:*

(i) D is open,

(ii) D has a finite number of components, of which the closures of any two are disjoint, and

(iii) the boundary of D is composed of a finite positive number of closed rectifiable Jordan curves, of which any two are unable to intersect.

Theorem 2.9.1 *(A. E. Taylor [148, Theorem 3.3]) Let* F *and* Δ *be point sets in the plane. Let* F *be closed,* Δ *be open and* $F \subset \Delta$. *Suppose that the boundary* $\partial\Delta$ *of* Δ *is nonempty and bounded. Then, there exists a Cauchy domain D, such that*

(i) $F \subset D$,

(ii) $\overline{D} \subset \Delta$,

(iii) the curves forming ∂D *are polygons, and*

(iv) D is unbounded if Δ *is unbounded.*

Definition 2.9.4 *Let $f \in \mathscr{R}'_\infty(A)$. The class of operators $\mathscr{F}(A)$ will be defined as follows: $B \in \mathscr{F}(A)$ if*

$$B = f(\infty) + \frac{1}{2\pi i}\int_{+\partial\mathscr{D}} f(\lambda)R'_\lambda(A)d\lambda,$$

where

$$f(\infty) := \lim_{\lambda\to\infty} f(\lambda)$$

and D is an unbounded Cauchy domain such that $\mathbb{C}\backslash\Phi_A \subset D$, $\overline{D} \subset \Delta(f)$, and the boundary of D, $\partial\mathscr{D}$, does not contain any points of $\Phi^0(A)$. ◇

Theorem 2.9.2 *(R. M. Gethner and J. H. Shapiro [71, Theorem 7]) Let $B_1, B_2 \in \mathscr{F}(A)$. Then,*

$$B_1 - B_2 = K$$

where $K \in \mathscr{K}(X)$. ◇

Definition 2.9.5 *Let $f \in \mathscr{R}'_\infty(A)$. By $f(A)$ we mean an arbitrary operator in the set $\mathscr{F}(A)$.* ◇

Theorem 2.9.3 *(R. M. Gethner and J. H. Shapiro [71, Theorem 9]) Let $f(\lambda)$ and $g(\lambda)$ be in $\mathscr{R}'_\infty(A)$. Then,*

$$f(A).g(A) = (f.g)(A) + K,$$

where $K \in \mathscr{K}(X)$. ◇

Definition 2.9.6 *Let $A \in \mathscr{L}(X)$. By $\mathscr{R}'(A)$ we mean the family of all analytic functions, $f(\lambda)$, such that*

$$\mathbb{C}\backslash\Phi_A \subset \Delta(f).$$ ◇

Definition 2.9.7 *Let $f \in \mathscr{R}'(A)$. The class of operators $\mathscr{F}^*(A)$ will be defined as follows: $B \in \mathscr{F}^*(A)$ if*

$$B = \frac{1}{2\pi i}\int_{+\partial\mathscr{D}} f(\lambda)R'_\lambda(A)d\lambda,$$

where D is a bounded Cauchy domain such that $\mathbb{C}\backslash\Phi_A \subset D$, $\overline{D} \subset \Delta(f)$, and $\partial\mathscr{D}$ does not contain any points of $\Phi^0(A)$. ◇

Definition 2.9.8 *Let $f \in \mathscr{R}'(A)$. By $f^*(A)$ we mean an arbitrary operator in the set $\mathscr{F}^*(A)$.* ♢

Theorem 2.9.4 *(R. M. Gethner and J. H. Shapiro [71, Theorem 12]) Let $B_1, B_2 \in \mathscr{F}^*(A)$. Then,*

$$B_1 - B_2 = K,\ K \in \mathscr{K}(X). \qquad \diamondsuit$$

Theorem 2.9.5 *(R. M. Gethner and J. H. Shapiro [71, Theorem 13]) Let $A \in \mathscr{L}(X)$, and let $f(\lambda) = 1$. Then,*

$$f^*(A) = I + K,\ K \in \mathscr{K}(X). \qquad \diamondsuit$$

Theorem 2.9.6 *(R. M. Gethner and J. H. Shapiro [71, Theorem 14]) Let $A \in \mathscr{L}(X)$, and let $f(\lambda) = \lambda$. Then,*

$$f^*(A) = A + K,\ K \in \mathscr{K}(X). \qquad \diamondsuit$$

Lemma 2.9.2 *(R. M. Gethner and J. H. Shapiro [71, Lemma 7.4]) Let $\mu_i \in \Phi_i^0(A)$ ($\Phi_i^0(A)$ being the set of all $\lambda \in \Phi_i(A)$ such that $\frac{-1}{\lambda-\lambda_i} \in \sigma(T_i)$ and T_i is defined in Eq. (2.28)). Let D be a bounded Cauchy domain with the following $\overline{D} \subset \Phi_i(A)$, $\mu_i \in D$, and no other points of $\Phi^0(A)$ are contained in $\overline{D}$. Then,*

$$\frac{1}{2\pi i}\int_{+\partial\mathscr{D}} R'_\lambda(A)d\lambda = K \in \mathscr{K}(X). \qquad \diamondsuit$$

Lemma 2.9.3 *(J. Shapiro and M. Snow [144, Lemma 1.1]) Let $A \in \mathscr{C}(X)$, such that Φ_A is not empty, and let n be a positive integer. Then, for each $\lambda \in \Phi_A \backslash \Phi^0(A)$, there exists a subspace V_λ dense in X and depending on λ such that, for all $x \in V_\lambda$, we have*

$$R'_\lambda(A)x \in \mathscr{D}(A^n). \qquad \diamondsuit$$

2.10 LIMIT INFERIOR AND SUPERIOR

Let $\mathscr{S}$ be the collection of all non-empty compact subsets of $\mathbb{C}$. It is well known that the convergence of a sequence in $\mathscr{S}$ with respect to the Hausdorff metric can be characterized through the concepts of limit inferior and superior. Let $\{E_n\}_n$ be a sequence of arbitrary subsets of $\mathbb{C}$ and, define the limits inferior and superior of $\{E_n\}_n$, denoted respectively by $\liminf E_n$ and $\limsup E_n$, as follows:

$$\liminf E_n = \Big\{\lambda \in \mathbb{C}, \text{ for every } \varepsilon > 0, \text{ there exists } N \in \mathbb{N} \text{ such that } \mathbb{B}(\lambda,\varepsilon)\bigcap E_n \neq \emptyset \text{ for all } n \geq N\Big\},$$

and

$$\limsup E_n = \Big\{\lambda \in \mathbb{C}, \text{ for every } \varepsilon > 0, \text{ there exists } J \subset \mathbb{N} \text{ infinite such that } \mathbb{B}(\lambda,\varepsilon)\bigcap E_n \neq \emptyset \text{ for all } n \in J\Big\}.$$

We recall the following properties of limit inferior and superior.

Theorem 2.10.1 *Let $\{E_n\}_n$ be a sequence of non-empty subsets of $\mathbb{C}$. The following properties of limit inferior and superior are known:*

(i) $\liminf E_n$ and $\limsup E_n$ are closed subsets of $\mathbb{C}$.

(ii) $\lambda \in \limsup E_n$ if, and only if, there exists an increasing sequence of natural numbers $n_1 < n_2 < n_3 < \cdots$ and points $\lambda_{n_k} \in E_{n_k}$, for all $k \in \mathbb{N}^*$, such that

$$\lim \lambda_{n_k} = \lambda.$$

(iii) $\lambda \in \liminf E_n$ if, and only if, there exists a sequence $\{\lambda_n\}_n$ such that $\lambda_n \in E_n$ for all $n \in \mathbb{N}$, and

$$\lim \lambda_n = \lambda.$$

(iv) Suppose E, $E_n \in \mathscr{S}$ for all $n \in \mathbb{N}$, and there exists $K \in \mathscr{S}$ such that $E_n \subset K$, for all $n \in \mathbb{N}$. Then,

$$E_n \to E$$

in the Hausdorff metric if, and only if,

$$\limsup E_n \subset E$$

and

$$E \subset \liminf E_n. \qquad \diamondsuit$$

Chapter 3

Spectra

In this chapter, we investigate the essential spectra of the closed, densely defined linear operators on Banach space.

3.1 ESSENTIAL SPECTRA

3.1.1 Definitions

It is well known that if A is a self-adjoint operator on a Hilbert space, then the essential spectrum of A is the set of limit points of the spectrum of A (with eigenvalues counted according to their multiplicities), i.e., all points of the spectrum except isolated eigenvalues of finite algebraic multiplicity (see, for instance, Refs. [1, 136]). Irrespective of whether A is bounded or not on a Banach space X, there are several definitions of the essential spectrum, most of which constitute an enlargement of the continuous spectrum. Let us define the following sets:

$$
\begin{aligned}
\sigma_j(A) &:= \bigcap_{K \in \mathscr{W}^*(X)} \sigma(A+K),\\
\sigma_{e1}(A) &:= \Big\{\lambda \in \mathbb{C} \ \text{ such that } \ \lambda - A \notin \Phi_+(X)\Big\} := \mathbb{C}\backslash\Phi_+(X),\\
\sigma_{e2}(A) &:= \Big\{\lambda \in \mathbb{C} \ \text{ such that } \ \lambda - A \notin \Phi_-(X)\Big\} := \mathbb{C}\backslash\Phi_-(X),\\
\sigma_{e3}(A) &:= \sigma_{e1}(A) \bigcap \sigma_{e2}(A),\\
\sigma_{el}(A) &:= \Big\{\lambda \in \mathbb{C} \text{ such that } \lambda - A \notin \Phi_l(X)\Big\} := \mathbb{C}\backslash\Phi_{lA},\\
\sigma_{er}(A) &:= \Big\{\lambda \in \mathbb{C} \text{ such that } \lambda - A \notin \Phi_r(X)\Big\} := \mathbb{C}\backslash\Phi_{rA},\\
\sigma_{e4}(A) &:= \Big\{\lambda \in \mathbb{C} \text{ such that } \lambda - A \notin \Phi(X)\Big\} := \mathbb{C}\backslash\Phi_{A},\\
\sigma_{e5}(A) &:= \bigcap_{K \in \mathscr{K}(X)} \sigma(A+K),\\
\sigma_{e6}(A) &:= \mathbb{C}\backslash\Big\{\lambda \in \rho_5(A) \text{ such that all scalars near } \lambda \text{ are in } \rho(A)\Big\}\\
&:= \mathbb{C}\backslash\rho_6(A),\\
\sigma_{e7}(A) &:= \bigcap_{K \in \mathscr{K}(X)} \sigma_{ap}(A+K),\\
\sigma_{ewl}(A) &:= \bigcap_{K \in \mathscr{K}(X)} \sigma_{l}(A+K),\\
\sigma_{ewr}(A) &:= \bigcap_{K \in \mathscr{K}(X)} \sigma_{r}(A+K),\\
\sigma_{e8}(A) &:= \bigcap_{K \in \mathscr{K}(X)} \sigma_{\delta}(A+K),
\end{aligned}
$$

where $\mathscr{W}^*(X)$ stands for each one of the sets $\mathscr{W}(X)$ or $\mathscr{S}(X)$,

$\rho_5(A) = \{\lambda \in \mathbb{C} \text{ such that } \lambda - A \text{ is a Fredholm operator and } i(\lambda - A) = 0\}$,

$$\sigma_{ap}(A) = \Big\{\lambda \in \mathbb{C} \text{ such that } \inf_{x \in \mathscr{D}(A)\ \|x\|=1} \|(\lambda - A)x\| = 0\Big\},$$

$$\sigma_l(A) := \Big\{\lambda \in \mathbb{C} \text{ such that } \lambda - A \text{ is not left invertible}\Big\},$$

$$\sigma_r(A) := \Big\{\lambda \in \mathbb{C} \text{ such that } \lambda - A \text{ is not right invertible}\Big\},$$

and

$$\sigma_\delta(A) = \Big\{\lambda \in \mathbb{C} \text{ such that } \lambda - A \ \text{ is not surjective}\Big\}.$$

The subset $\sigma_j(\cdot)$ is the Jeribi's essential spectrum [1]. $\sigma_{e1}(\cdot)$ and $\sigma_{e2}(\cdot)$ are the Gustafson and Weidman's essential spectra, respectively

[79]. $\sigma_{e3}(\cdot)$ is the Kato's essential spectrum [108]. $\sigma_{e4}(\cdot)$ is the Wolf's essential spectrum [79, 139, 160]. $\sigma_{e5}(\cdot)$ is the Schechter's essential spectrum [79, 86–88, 139] and $\sigma_{e6}(\cdot)$ denotes the Browder's essential spectrum [79, 96, 139]. $\sigma_{e7}(\cdot)$ was introduced by V. Rakočević in [131] and designated the essential approximate point spectrum, $\sigma_{e8}(\cdot)$ is the essential defect spectrum and was introduced by V. Schmoeger [143], and $\sigma_{ap}(\cdot)$ is the approximate point spectrum. Let us notice that all these sets are closed and, in general, we have

$$\sigma_{e3}(A) := \sigma_{e1}(A) \bigcap \sigma_{e2}(A) \subseteq \sigma_{e4}(A) \subseteq \sigma_{e5}(A) \subseteq \sigma_{e6}(A),$$

$$\sigma_{e5}(A) = \sigma_{e7}(A) \bigcup \sigma_{e8}(A),$$

$$\sigma_{e1}(A) \subset \sigma_{e7}(A),$$

and

$$\sigma_{e2}(A) \subset \sigma_{e8}(A).$$

It is proved in [1] that

$$\sigma_{j}(A) \subset \sigma_{e5}(A).$$

However, if X is a Hilbert space and A is self-adjoint, then all these sets coincide.

Remark 3.1.1 *(i) If $\lambda \in \sigma_c(A)$ (the continuous spectrum of A), then $R(\lambda - A)$ is not closed (otherwise $\lambda \in \rho(A)$ see [137, Lemma 5.1 p. 179]). Therefore, $\lambda \in \sigma_{ei}(A)$, $i = 1, \cdots, 6$. Consequently, we have*

$$\sigma_c(A) \subset \bigcap_{i=1}^{6} \sigma_{ei}(A).$$

If the spectrum of A is purely continuous, then

$$\sigma(A) = \sigma_c(A) = \sigma_{ei}(A) \quad i = 1, \cdots, 6.$$

(ii) $\sigma_{e5}(A+K) = \sigma_{e5}(A)$ for all $K \in \mathscr{K}(X)$.

(iii) Let E_0 be a core of A, i.e., a linear subspace of $\mathscr{D}(A)$ such that the closure $\overline{A_{|E_0}}$ of its restriction $A_{|E_0}$ equals A. Then,

$$\sigma_{ap}(A) := \left\{\lambda \in \mathbb{C} \text{ such that } \inf_{\|x\|=1,\ x\in E_0} \|(\lambda - A)x\| = 0\right\}.$$

(iv) Set $\widetilde{\alpha}(A) := \inf\{\|Ax\|$ such that $x \in \mathscr{D}(A)$ and $\|x\| = 1\}$. Whenever E_0 is a core of A, then

$$\widetilde{\alpha}(A) := \inf\{\|Ax\| \text{ such that } x \in E_0 \text{ and } \|x\| = 1\}$$

holds. Using this quantity, we obtain

$$\sigma_{ap}(A) = \{\lambda \in \mathbb{C} \text{ such that } \widetilde{\alpha}(\lambda - A) = 0\}.$$

◇

Lemma 3.1.1 *Let $A \in \mathscr{C}(X)$ such that $0 \in \rho(A)$. Then, $\lambda \in \sigma_{ei}(A)$ if, and only if, $\frac{1}{\lambda} \in \sigma_{ei}(A^{-1})$ $i = 7,\ 8$.* ◇

3.1.2 Characterization of Essential Spectra

The following proposition gives a characterization of the Schechter essential spectrum by means of Fredholm operators.

Proposition 3.1.1 *(M. Schechter [137, Theorem 5.4, p. 180]) Let X be a Banach space and let $A \in \mathscr{C}(X)$. Then, $\lambda \notin \sigma_{e5}(A)$ if, and only if, $\lambda \in \Phi_A^0$, where $\Phi_A^0 = \{\lambda \in \Phi_A$ such that $i(\lambda - A) = 0\}$.* ◇

For $n \in \mathbb{N}^*$, let

$$\mathscr{I}_n(X) = \{K \in \mathscr{L}(X) \text{ satisfying } f((KB)^n) < 1 \text{ for all } B \in \mathscr{L}(X)\},$$

where $f(\cdot)$ is a measure of non-strict-singularity given in (2.22).

Theorem 3.1.1 *(N. Moalla [122]) Let $A \in \Phi(X)$, then for all $K \in \mathscr{I}_n(X)$, we have*

(i) $K + A \in \Phi(X)$ and $i(K + A) = i(A)$,
(ii) $\sigma_{e5}(A + K) = \sigma_{e5}(A)$. ◇

Theorem 3.1.2 *(N. Moalla [122]) Let $A,B \in \mathscr{C}(X)$ such that $\rho(A)\bigcap\rho(B) \neq \emptyset$. If, for some $\lambda \in \rho(A)\bigcap\rho(B)$, the operator $(\lambda - A)^{-1} - (\lambda - B)^{-1} \in \mathscr{I}_n(X)$, then*

$$\sigma_{e5}(A) = \sigma_{e5}(B). \qquad \diamondsuit$$

We recall some properties of $\sigma_{e7}(\cdot)$ (resp. $\sigma_{e8}(\cdot)$) due to V. Rakočević (resp. C. Schmoeger), given in [143].

Proposition 3.1.2 *(C. Schmoeger [143]) Let $A \in \mathscr{L}(X)$, then*

(i) $\sigma_{e7}(A) \neq \emptyset$,

(ii) $\sigma_{e8}(A) \neq \emptyset$, and

(iii) $\sigma_{e7}(A)$ and $\sigma_{e8}(A)$ are compact. $\diamondsuit$

The following proposition gives a characterization of the essential approximate point spectrum and the essential defect spectrum by means of upper semi-Fredholm and lower semi-Fredholm operators, respectively.

Proposition 3.1.3 *(A. Jeribi and N. Moalla [98, Proposition 3.1]) Let $A \in \mathscr{C}(X)$. Then,*

(i) $\lambda \notin \sigma_{e7}(A)$ if, and only if, $\lambda - A \in \Phi_+(X)$ and $i(\lambda - A) \leq 0$,

(ii) $\lambda \notin \sigma_{e8}(A)$ if, and only if, $\lambda - A \in \Phi_-(X)$ and $i(\lambda - A) \geq 0$, and

(iii) if A is a bounded linear operator, then $\sigma_{e8}(A) = \sigma_{e7}(A^*)$, where A^* stands for the adjoint operator of A.

This is equivalent that

$$\sigma_{e7}(A) = \sigma_{e1}(A)\bigcup\Big\{\lambda \in \mathbb{C} \text{ such that } i(A-\lambda) > 0\Big\},$$

and

$$\sigma_{e8}(A) = \sigma_{e4}(A)\bigcup\Big\{\lambda \in \mathbb{C} \text{ such that } i(A-\lambda) < 0\Big\}.$$

If, in addition, Φ_A is connected and $\rho(A) \neq \emptyset$, then

$$\sigma_{e1}(A) = \sigma_{e7}(A),$$

and

$$\sigma_{e4}(A) = \sigma_{e8}(A). \qquad \diamondsuit$$

Theorem 3.1.3 *(M. A. Kaashoek and D. C. Lay [105]) Let A be a closed linear operator on a Banach space X. Then, $\sigma_{e6}(A)$ is the largest subset of the spectrum of A which remains invariant under perturbations of A by Riesz operators which commute with A.* ◇

Theorem 3.1.4 *(F. Abdmouleh and A. Jeribi [14]) Let A and B be two bounded linear operators on a Banach space X. Then,*

(i) If $AB \in \mathcal{F}^b(X)$, then

$$\sigma_{ei}(A+B)\backslash\{0\} \subset \left[\sigma_{ei}(A)\bigcup\sigma_{ei}(B)\right]\backslash\{0\}, \ i=4, 5.$$

Furthermore, if $BA \in \mathcal{F}^b(X)$, then

$$\sigma_{e4}(A+B)\backslash\{0\} = \left[\sigma_{e4}(A)\bigcup\sigma_{e4}(B)\right]\backslash\{0\}.$$

Moreover, if $\mathbb{C}\backslash\sigma_{e4}(A)$ is connected, then

$$\sigma_{e5}(A+B)\backslash\{0\} = \left[\sigma_{e5}(A)\bigcup\sigma_{e5}(B)\right]\backslash\{0\}.$$

(ii) If the hypotheses of (i) are satisfied, and if $\mathbb{C}\backslash\sigma_{e5}(A+B)$, $\mathbb{C}\backslash\sigma_{e5}(A)$ and $\mathbb{C}\backslash\sigma_{e5}(B)$ are connected, then

$$\sigma_{e6}(A+B)\backslash\{0\} = [\sigma_{e6}(A)\bigcup\sigma_{e6}(B)]\backslash\{0\}.$$ ◇

Theorem 3.1.5 *Let $A, B \in \mathcal{C}(X)$ and let $\lambda \in \rho(A)\bigcap\rho(B)$. If $(\lambda-A)^{-1} - (\lambda-B)^{-1} \in \mathcal{F}^b(X)$, then*

$$\sigma_{ei}(A) = \sigma_{ei}(B), \ i=4,5.$$ ◇

Theorem 3.1.6 *[14, 96] Let $A \in \mathcal{C}(X)$ such that $\rho(A)$ is not empty. Then,*

(i) If $\mathbb{C}\backslash\sigma_{e4}(A)$ is connected, then

$$\sigma_{e4}(A) = \sigma_{e5}(A).$$

(ii) If $\mathbb{C}\backslash\sigma_{e5}(A)$ is connected, then

$$\sigma_{e5}(A) = \sigma_{e6}(A).$$ ◇

3.2 THE LEFT AND RIGHT JERIBI ESSENTIAL SPECTRA

In this section, we will give fine description of the definition of the left and right Jeribi essential spectra of a closed densely defined linear operators. For this, let X be a Banach space and let $A \in \mathscr{C}(X)$, we denote the left Jeribi essential spectrum by

$$\sigma_j^l(A) = \bigcap_{K \in \mathscr{W}^*(X)} \sigma_l(A+K)$$

and, the right Jeribi essential spectrum by

$$\sigma_j^r(A) = \bigcap_{K \in \mathscr{W}^*(X)} \sigma_r(A+K).$$

For $A \in \mathscr{C}(X)$, we observe that $\mathscr{K}(X) \subset \mathscr{W}^*(X)$. Then,

$$\sigma_j^l(A) \subseteq \sigma_{ewl}(A)$$

and

$$\sigma_j^r(A) \subseteq \sigma_{ewr}(A).$$

One of the central questions in the study of the left and right essential spectra of closed densely defined linear operators on Banach space X consists of showing what are the conditions that we must impose on space X such that for $A \in \mathscr{C}(X)$,

$$\sigma_{ewl}(A) = \sigma_j^l(A)$$

and

$$\sigma_{ewr}(A) = \sigma_j^r(A).$$

If X is a reflexive Banach space, then $\mathscr{L}(X) = \mathscr{W}(X)$. So, the left or right Jeribi essential spectrum is the smallest essential spectrum in the sense of inclusion of the other essential spectra.

The set of Browder essential spectrum [1, 70] of A is characterized by

$$\sigma_{e6}(A) := \mathbb{C} \backslash \left\{ \lambda \in \mathbb{C} \text{ such that } \lambda - A \in \mathscr{B}(X) \right\}.$$

3.3 S-RESOLVENT SET, S-SPECTRA, AND S-ESSENTIAL SPECTRA

3.3.1 The S-Resolvent Set

Let X and Y be two Banach space. Let $S \in \mathscr{L}(X,Y)$, for $A \in \mathscr{C}(X,Y)$ such that $A \neq S$ and $S \neq 0$, we define the S-resolvent set of A by

$$\rho_S(A) = \left\{ \lambda \in \mathbb{C} \text{ such that } \lambda S - A \text{ has a bounded inverse} \right\}.$$

We denote by

$$R_S(\lambda, A) := (\lambda S - A)^{-1}$$

and call it the S-resolvent operator of A. The name "S-resolvent" is appropriate, since $R_S(\lambda, A)$ helps to solve the equation

$$(\lambda S - A)x = y.$$

Thus,

$$x = (\lambda S - A)^{-1} y = R_S(\lambda, A)y$$

provided $R_S(\lambda, A)$ exists.

If $\rho_S(A)$ is not empty, then A is closed. Indeed, let $(x_n)_n \in \mathscr{D}(A)$ such that $x_n \to x$ and $Ax_n \to y$. Since $\rho_S(A) \neq \emptyset$, then there exist $\lambda_0 \in \rho_S(A)$ such that $(A - \lambda_0 S)^{-1} \in \mathscr{L}(X)$. Since $S \in \mathscr{L}(X)$, then

$$(A - \lambda_0 S)x_n \to y - \lambda_0 Sx.$$

Thus,

$$x_n \to (A - \lambda_0 S)^{-1}(y - \lambda_0 Sx) = x,$$

we deduce that $Ax = y$ and $x \in \mathscr{D}(A)$.

Lemma 3.3.1 *For all λ, μ $\in \rho_S(A)$, we have*

$$R_S(\lambda, A) - R_S(\mu, A) = (\mu - \lambda) R_S(\lambda, A) S R_S(\mu, A). \qquad \diamondsuit$$

Proof. Let λ, $\mu \in \rho_S(A)$, we have

$$\begin{aligned} R_S(\lambda,A) - R_S(\mu,A) &= (\lambda S - A)^{-1}\big[(\mu S - A) - (\lambda S - A)\big](\mu S - A)^{-1}, \\ &= (\mu - \lambda) R_S(\lambda,A) S R_S(\mu,A), \end{aligned}$$

which completes the proof of lemma. Q.E.D.

Proposition 3.3.1 *Let $A \in \mathscr{C}(X,Y)$, $S \in \mathscr{L}(X,Y)$ such that $S \neq 0$ and $S \neq A$, we have*

(i) the S-resolvent set $\rho_S(A)$ is open, and

(ii) the function

$$\varphi : \lambda \longrightarrow R_S(\lambda,A)$$

is holomorphic at any point of $\rho_S(A)$. 

Proof. (i) If $\rho_S(A) = \emptyset$, then $\rho_S(A)$ is open. If $\rho_S(A) \neq \emptyset$, then $\lambda_0 \in \rho_S(A)$. It sufficient to find $\varepsilon > 0$ such that

$$\mathbb{B}(\lambda_0, \varepsilon) \subset \rho_S(A),$$

where $\mathbb{B}(\lambda_0, \varepsilon)$ designates the open ball centered at λ_0 with radius ε. We can write for any $\lambda \in \mathbb{C}$

$$\begin{aligned} \lambda S - A &= \lambda_0 S - A + (\lambda - \lambda_0) S, \\ &= (\lambda_0 S - A)\left[I + (\lambda - \lambda_0) R_S(\lambda_0,A) S\right]. \end{aligned}$$

If $\|(\lambda - \lambda_0) R_S(\lambda_0,A) S\| < 1$, then

$$I + (\lambda - \lambda_0) R_S(\lambda_0,A) S$$

is invertible. This is equivalent to say that if

$$|\lambda - \lambda_0| < \frac{1}{\|R_S(\lambda_0,A) S\|},$$

then $\lambda \in \rho_S(A)$. Consequently, it suffices to take $\varepsilon = \frac{1}{\|R_S(\lambda_0,A)S\|}$.

(ii) Let λ, $\mu \in \rho_S(A)$, we have

$$\begin{aligned}\frac{\varphi(\mu)-\varphi(\lambda)}{\mu-\lambda} &= \frac{R_S(\mu,A)-R_S(\lambda,A)}{\mu-\lambda},\\ &= \frac{R_S(\mu,A)(\lambda-\mu)S\,R_S(\lambda,A)}{\mu-\lambda},\\ &= -R_S(\mu,A)SR_S(\lambda,A).\end{aligned}$$

Hence,

$$\lim_{\mu\longrightarrow\lambda}\frac{\varphi(\mu)-\varphi(\lambda)}{\mu-\lambda} = -R_S(\lambda,A)SR_S(\lambda,A)\in\mathscr{L}(Y,X),$$

which completes the proof of proposition. Q.E.D.

Remark 3.3.1 *If $A \in \mathscr{L}(X)$ and S is an invertible bounded operator, then*

$$\rho_S(A) = \rho(S^{-1}A)\bigcap\rho(AS^{-1}).$$

Because

$$\lambda S - A = S(\lambda - S^{-1}A),$$

and

$$\lambda S - A = (\lambda - AS^{-1})S,$$

it follows that $\lambda \in \rho_S(A)$ if, and only if, $\lambda \in \rho(S^{-1}A)\bigcap\rho(AS^{-1})$. ♢

Theorem 3.3.1 *Let $A \in \mathscr{C}(X,Y)$, $S \in \mathscr{L}(X,Y)$ and $\lambda \in \mathbb{C}$. If $\lambda S - A$ is one-to-one and onto, then $(\lambda S - A)^{-1}$ is a bounded linear operator.* ♢

Proof. Let $\lambda \in \mathbb{C}$, we shall first prove that $(\lambda S - A)^{-1}$ is a closed linear operator. For this, we take the sequence $(y_n)_n \subset Y$ such that

$$\begin{cases} y_n \to y \text{ in } Y,\\ (\lambda S - A)^{-1}y_n \to x \text{ in } X.\end{cases}$$

In the following, we set $x_n = (\lambda S - A)^{-1}y_n$, then $x_n \in \mathscr{D}(A)$ and

$$y_n = (\lambda S - A)x_n \in Y.$$

Since $(\lambda S - A)x_n \to y$, $x_n \to x$ and $\lambda S - A$ is a closed linear operator which implies that $x \in \mathscr{D}(A)$ and

$$y = (\lambda S - A)x.$$

So, $y \in Y$ and

$$x = (\lambda S - A)^{-1}y.$$

This proves that $(\lambda S - A)^{-1}$ is a closed operator. Furthermore, $(\lambda S - A)^{-1}$ is a closed operator defined on all Y. Using the closed graph theorem (see Theorem 2.1.1), we obtain that

$$(\lambda S - A)^{-1}$$

is a bounded linear operator. Q.E.D.

Proposition 3.3.2 *Let A and $S \in \mathscr{L}(X)$ such that $S \neq A$ and $S \neq 0$. Then,*

(i) If S is an operator which commutes with A, then for any $\lambda \in \rho_S(A)$, we have

$$\begin{aligned} AR_S(\lambda, A) &= R_S(\lambda, A)A, \\ R_S(\lambda, A)S &= SR_S(\lambda, A). \end{aligned}$$

(ii) If S commutes with A, then for every λ, $\mu \in \rho_S(A)$, we have

$$R_S(\lambda, A)R_S(\mu, A) = R_S(\mu, A)R_S(\lambda, A).$$ ◇

Proof. (i) Since, by hypothesis, S commutes with A, then the last two operators commute with the operator $\lambda S - A$. In addition, since $\lambda \in \rho_S(A)$, we obtain that S and A commute with $R_S(\lambda, A)$.

(ii) Let λ, $\mu \in \rho_S(A)$, we have

$$\begin{aligned} R_S(\lambda, A)R_S(\mu, A) &= \Big[(\mu S - A)(\lambda S - A)\Big]^{-1}, \\ &= \Big[\lambda S(\mu S - A) - A(\mu S - A)\Big]^{-1}, \\ &= \Big[(\lambda S - A)(\mu S - A)\Big]^{-1}, \\ &= R_S(\mu, A)R_S(\lambda, A), \end{aligned}$$

which completes the proof of proposition. Q.E.D.

Theorem 3.3.2 *Let A and $S \in \mathscr{L}(X)$ such that $S \neq A$ and $S \neq 0$. If S commutes with A, then for any λ and $\lambda_0 \in \rho_S(A)$ with $|\lambda - \lambda_0| < \frac{1}{\|R_S(\lambda_0,A)S\|}$, we have*

$$R_S(\lambda,A) = \sum_{n\geq 0} (\lambda_0 - \lambda)^n S^n R_S(\lambda_0,A)^{n+1}. \qquad \diamondsuit$$

Proof. Let λ, $\lambda_0 \in \rho_S(A)$ such that $|\lambda - \lambda_0| < \frac{1}{\|R_S(\lambda_0,A)S\|}$. Then,

$$\begin{aligned} R_S(\lambda,A) &= (\lambda_0 S - A - \lambda_0 S + \lambda S)^{-1}, \\ &= \Big[(\lambda_0 S - A)\big(I - (\lambda_0 - \lambda)R_S(\lambda_0,A)S\big)\Big]^{-1}, \\ &= \Big[I - (\lambda_0 - \lambda)R_S(\lambda_0,A)S\Big]^{-1} R_S(\lambda_0,A), \\ &= \sum_{n\geq 0} (\lambda_0 - \lambda)^n (R_S(\lambda_0,A)S)^n R_S(\lambda_0,A). \end{aligned}$$

According to the Proposition 3.3.2 (i),

$$R_S(\lambda_0,A)S = SR_S(\lambda_0,A). \tag{3.1}$$

So, Eq. (3.1) implies that

$$(R_S(\lambda_0,A)S)^n = S^n R_S(\lambda_0,A)^n.$$

Consequently,

$$R_S(\lambda,A) = \sum_{n\geq 0} (\lambda_0 - \lambda)^n S^n R_S(\lambda_0,A)^{n+1},$$

which completes the proof of theorem. Q.E.D.

Theorem 3.3.3 *Let A and $S \in \mathscr{L}(X)$ such that $S \neq A$ and $S \neq 0$. If S commutes with A, then*

$$\frac{d^n}{d\lambda^n} R_S(\lambda,A) = (-1)^n\, n!\, S^n R_S(\lambda,A)^{n+1} \;\; \forall\, \lambda \in \rho_S(A). \qquad \diamondsuit$$

Proof. We argue by recurrence. We know that the function

$$\varphi : \lambda \longrightarrow R_S(\lambda, A)$$

is holomorphic at any point λ of $\rho_S(A)$. Then, in view of Proposition 3.3.2 (i), we have

$$\begin{aligned} \frac{d}{d\lambda} R_S(\lambda,A) &= \lim_{\mu \to \lambda} \frac{R_S(\mu,A) - R_S(\lambda,A)}{\mu - \lambda}, \\ &= \lim_{\mu \to \lambda} -R_S(\mu,A) S R_S(\lambda,A), \\ &= -S R_S(\lambda,A)^2. \end{aligned}$$

For $n = 1$, the equality is true. We assume that is true until the order n and we prove that it remains true to the order $n+1$. We will show that for any $\lambda \in \rho_S(A)$,

$$\frac{d^{n+1}}{d\lambda^{n+1}} R_S(\lambda,A) = (-1)^{n+1}\,(n+1)!\,S^{n+1}\,R_S(\lambda,A)^{n+2}.$$

In fact,

$$\begin{aligned} \frac{d^{n+1}}{d\lambda^{n+1}} R_S(\lambda,A) &= \frac{d}{d\lambda}\Big(\frac{d^n}{d\lambda^n} R_S(\lambda,A)\Big), \\ &= \lim_{\mu \to \lambda} \frac{\frac{d^n}{d\mu^n} R_S(\mu,A) - \frac{d^n}{d\lambda^n} R_S(\lambda,A)}{\mu - \lambda}, \\ &= \lim_{\mu \to \lambda} (-1)^n\, n!\, S^n \frac{R_S(\mu,A)^{n+1} - R_S(\lambda,A)^{n+1}}{\mu - \lambda}. \end{aligned}$$

According to the Proposition 3.3.2 (ii), we have

$$R_S(\lambda,A) R_S(\mu,A) = R_S(\mu,A) R_S(\lambda,A) \quad \forall\ \lambda,\ \mu \in \rho_S(A).$$

So, we can write

$$\begin{aligned} R_S(\mu,A)^{n+1} - R_S(\lambda,A)^{n+1} &= \Big(R_S(\mu,A) - R_S(\lambda,A)\Big)\Big(R_S(\mu,A)^n \\ &\quad + R_S(\mu,A)^{n-1} R_S(\lambda,A) + R_S(\mu,A)^{n-2} R_S(\lambda,A)^2 + \cdots + R_S(\lambda,A)^n\Big). \end{aligned}$$

This implies that

$$\begin{aligned}
&\frac{d^{n+1}}{d\lambda^{n+1}}R_S(\lambda,A)\\
&= \lim_{\mu\to\lambda}(-1)^n\, n!\, S^n \frac{\Big(R_S(\mu,A)-R_S(\lambda,A)\Big)\Big(R_S(\mu,A)^n+\cdots+R_S(\lambda,A)^n\Big)}{\mu-\lambda},\\
&= (-1)^n\, n!\, S^n\Big(-S\,R_S(\lambda,A)^2\Big)(n+1)\,R_S(\lambda,A)^n,\\
&= (-1)^{n+1}\,(n+1)!\,S^{n+1}\,R_S(\lambda,A)^{n+2},
\end{aligned}$$

which completes the proof of theorem. Q.E.D.

3.3.2 S-Spectra

Let $A\in\mathscr{C}(X)$ and $S\in\mathscr{L}(X)$ such that $S\neq A$ and $S\neq 0$. We define the S-spectrum of A by

$$\sigma_S(A):=\mathbb{C}\backslash\rho_S(A).$$

We see the following examples, where $\sigma_S(A)$ can be discrete or the whole complex plane:

(i) Let $A=\begin{pmatrix}2&2\\0&3\end{pmatrix}$ and $S=\begin{pmatrix}1&0\\0&0\end{pmatrix}$, then $\sigma_S(A)=\{2\}$.

(ii) Let $A=\begin{pmatrix}1&2\\0&0\end{pmatrix}$ and $S=\begin{pmatrix}1&0\\0&0\end{pmatrix}$, then $\sigma_S(A)=\mathbb{C}$.

(iii) Let $A=\begin{pmatrix}1&2\\0&3\end{pmatrix}$ and $S=\begin{pmatrix}0&0\\0&1\end{pmatrix}$, then $\sigma_S(A)=\emptyset$.

The S-left spectrum of A is defined by

$$\sigma_{l,S}(A)=\Big\{\lambda\in\mathbb{C}\text{ such that }\lambda S-A\notin\mathscr{G}_l(X)\Big\}$$

and the S-right spectrum of A is defined by

$$\sigma_{r,S}(A)=\Big\{\lambda\in\mathbb{C}\text{ such that }\lambda S-A\notin\mathscr{G}_r(X)\Big\}.$$

It is clear that

$$\sigma_S(A)=\sigma_{l,S}(A)\bigcup\sigma_{r,S}(A).$$

Corollary 3.3.1 *Let A be a linear operator and S be a non null bounded linear operator from X into Y. If A is a non closed operator, then*

$$\sigma_S(A) = \mathbb{C}. \qquad \diamondsuit$$

Proof. Let A be a linear operator which is not closed. We argue by contradiction. Suppose that $\rho_S(A)$ is not empty, then there exists $\lambda \in \mathbb{C}$ such that $\lambda \in \rho_S(A)$ and consequently, $(\lambda S - A)^{-1}$ is a bounded operator. Hence, $\lambda S - A$ is a closed operator. In addition, we can write

$$A = A - \lambda S + \lambda S.$$

We conclude that A is a closed operator, which is a contradiction. Q.E.D.

Definition 3.3.1 *Let $A \in \mathscr{C}(X)$ and $S \in \mathscr{L}(X)$ such that $S \neq A$ and $S \neq 0$. We define the following sets:*

(i) The S-point spectrum of A and it is denoted by

$$\sigma_{p,S}(A) = \Big\{\lambda \in \mathbb{C} \text{ such that } \lambda S - A \text{ is not one-to-one}\Big\}.$$

(ii) The S-continuous spectrum of A and it is defined by

$$\sigma_{c,S}(A) = \Big\{\lambda \in \mathbb{C} : \lambda S - A \text{ is one-to-one}, \ \overline{(\lambda S - A)(\mathscr{D}(A))} = X, \text{ and } (\lambda S - A)^{-1} \text{ is unbounded}\Big\}.$$

(iii) The S-residual spectrum of A and it is denoted by

$$\sigma_{R,S}(A) = \Big\{\lambda \in \mathbb{C} : \lambda S - A \text{ is one-to-one}, \ \overline{(\lambda S - A)(\mathscr{D}(A))} \neq X\Big\}.$$

(iv) The S-approximate point spectrum and it is denoted by

$$\sigma_{ap,S}(A) = \Big\{\lambda \in \mathbb{C} : \exists\, x_n \in \mathscr{D}(A), \ \|x_n\| = 1 \text{ and } \lim_{n \to +\infty} \|(\lambda S - A)x_n\| = 0\Big\}. \qquad \diamondsuit$$

Remark 3.3.2 *Let $A \in \mathscr{C}(X)$ and $S \in \mathscr{L}(X)$ such that $S \neq A$ and $S \neq 0$, then the S-spectrum, $\sigma_S(A)$, is partitioned into three disjoint sets as follows*

$$\sigma_S(A) = \sigma_{p,S}(A) \bigcup \sigma_{c,S}(A) \bigcup \sigma_{R,S}(A). \qquad \diamondsuit$$

3.3.3 S-Browder's Resolvent

Definition 3.3.2 *Let $A \in \mathscr{C}(X)$ and λ_0 be an isolated point of $\sigma_S(A)$. For an admissible contour Γ_{λ_0},*

$$P_{\lambda_0,S} = -\frac{S}{2\pi i}\oint_{\Gamma_{\lambda_0}} (A - \lambda S)^{-1}\, d\lambda,$$

is called the S-Riesz integral for A, S and λ_0 with range and kernel denote by $R_{\lambda,S}$ and $K_{\lambda,S}$, respectively. ◇

The S-discrete spectrum of A, denoted $\sigma_{d_S}(A)$, is just the set of isolated points $\lambda \in \mathbb{C}$ of the spectrum such that the corresponding S-Riesz projectors $P_{\lambda,S}$ are finite-dimensional. Another part of the spectrum, which is generally larger than $\sigma_{e6,S}(A)$, is $\sigma_S(A)\backslash\sigma_{d_S}(A)$. We will also use this terminology here and the notation

$$\sigma_{e6,S}(A) = \sigma_S(A)\backslash\sigma_{d_S}(A)$$

and

$$\rho_{6,S}(A) = \mathbb{C}\backslash\sigma_{e6,S}(A).$$

The largest open set on which the resolvent is finitely meromorphic is precisely

$$\rho_{6,S}(A) = \sigma_{d_S}(A)\bigcup\rho_S(A).$$

For $\lambda \in \rho_{6,S}(A)$, let $P_{\lambda,S}(A)$ (or $P_{\lambda,S}$) denotes the corresponding (finite rank) S-Riesz projector with a range and a kernel denoted by $R_{\lambda,S}$ and $K_{\lambda,S}$, respectively. Since $P_{\lambda,S}$ is invariant, we may define the operator

$$A_{\lambda,S} = (A - \lambda S)(I - P_{\lambda,S}) + P_{\lambda,S}, \tag{3.2}$$

with respect to the decomposition

$$X = K_{\lambda,S}\bigoplus R_{\lambda,S}$$

and

$$A_{\lambda,S} = ((A - \lambda S)_{|K_{\lambda,S}})\bigoplus I.$$

We have just cut off the finite-dimensional part of $A-\lambda S$ in the S-Riesz decomposition. Since

$$\sigma\big((A-\lambda S)_{|K_{\lambda,S}}\big)=\sigma(A-\lambda S)\backslash\{0\},$$

$A_{\lambda,S}$ has a bounded inverse, denoted by $R_{b,S}(A,\lambda)$ and called the "**S**-Browder's resolvent", i.e.,

$$R_{b,S}(A,\lambda)=((A-\lambda)_{|K_{\lambda,S}})^{-1}\bigoplus I$$

with respect to

$$X=K_{\lambda,S}\bigoplus R_{\lambda,S}.$$

Or, alternatively

$$R_{b,S}(A,\lambda)=\big((A-\lambda S)_{|K_{\lambda,S}}\big)^{-1}(I-P_{\lambda,S})+P_{\lambda,S}$$

for $\lambda\in\rho_{6,S}(A)$. This clearly extends the usual resolvent $R_S(A,\lambda)$ from $\rho_S(A)$ to $\rho_{6,S}(A)$ and retains many of its important properties.

Proposition 3.3.3 *Let $A\in\mathscr{C}(X)$ and $S\in\mathscr{L}(X)$ such that $S\neq A$ and $S\neq 0$. Then, for any μ, $\lambda\in\rho_{6,S}(A)$, we have*

$$R_{b,S}(A,\lambda)-R_{b,S}(A,\mu)=(\lambda-\mu)R_{b,S}(A,\lambda)SR_{b,S}(A,\mu)+\mathscr{M}(\lambda,\mu), \tag{3.3}$$

where $\mathscr{M}(\lambda,\mu)$ is a finite rank operator with the following expression

$$\mathscr{M}(\lambda,\mu)=R_{b,S}(A,\lambda)\big[(A-(\lambda S+I))P_{\lambda,S}-(A-(\mu S+I))P_{\mu,S}\big]R_{b,S}(A,\mu) \tag{3.4}$$

is a finite rank operator with $rank(\mathscr{M}(\lambda,\mu))=rank(P_{\lambda,S})+rank(P_{\mu,S})$ in case $\lambda\neq\mu$. ◇

Proof. We have

$$R_{b,S}(A,\lambda)-R_{b,S}(A,\mu)=R_{b,S}(A,\lambda)[A_{\mu,S}-A_{\lambda,S}]R_{b,S}(A,\mu),$$

where

$$A_{\mu,S}:=(A-\mu S)(I-P_{\mu,S})+P_{\mu,S}.$$

So,

$$\begin{array}{rcl} A_{\mu,S}-A_{\lambda,S} & = & [(A-\mu S)(I-P_{\mu,S})+P_{\mu,S}]-[(A-\lambda S)(I-P_{\lambda,S})+P_{\lambda,S}] \\ & = & [(A-(\lambda S+I))P_{\lambda,S}-(A-(\mu S+I))P_{\mu,S}]+(\lambda-\mu)S. \end{array}$$

Therefore,

$$R_{b,S}(A,\lambda)-R_{b,S}(A,\mu)=(\lambda-\mu)R_{b,S}(A,\lambda)SR_{b,S}(A,\mu)+\mathscr{M}(\lambda,\mu).$$

Q.E.D.

Proposition 3.3.4 *Let X and Y be two complex Banach spaces. Let $A\in\mathscr{C}(X)$, $S\in\mathscr{L}(X)$, $B\in\mathscr{L}(Y,X)$, and $C:X\longrightarrow Y$ be two linear operators. Then,*

(i) $R_{b,S}(A,\mu)B$ is continuous for some $\mu\in\rho_{6,S}(A)$ if, and only if, it is continuous for all $\mu\in\rho_{6,S}(A)$.

(ii) $C(A-\mu S)^{-1}$ is bounded for some $\lambda\in\rho_S(A)$ if, and only if, $CR_{b,S}(A,\mu)$ is bounded for some (hence for every) $\mu\in\rho_{6,S}(A)$.

(iii) If B and C satisfy the conditions (i) and (ii), respectively, and B is densely defined, then $C\mathscr{M}(\lambda,\mu)$, $\overline{\mathscr{M}(\lambda,\mu)B}$, and $\overline{C\mathscr{M}(\lambda,\mu)B}$ are operators of finite rank for any μ, $\lambda\in\rho_{6,S}(A)$. ◇

Proof. From the resolvent identity, we have for any μ, $\lambda\in\rho_S(A)$,

$$(A-\lambda S)^{-1}C=(A-\mu S)^{-1}C+(\lambda-\mu)(A-\lambda S)^{-1}S(A-\mu S)^{-1}C, \tag{3.5}$$

and for any μ, $\lambda\in\rho_{6,S}(A)$,

$$R_{b,S}(A,\lambda)B=R_{b,S}(A,\mu)B+(\lambda-\mu)R_{b,S}(A,\lambda)SR_{b,S}(A,\mu)B+\mathscr{M}(\lambda,\mu)B, \tag{3.6}$$

and

$$CR_{b,S}(A,\lambda) = CR_{b,S}(A,\mu) + (\lambda - \mu)CR_{b,S}(A,\lambda)SR_{b,S}(A,\mu) + C\mathscr{M}(\lambda,\mu). \tag{3.7}$$

(i) Since S is bounded, then

$$R_{b,S}(A,\lambda)SR_{b,S}(A,\mu)B$$

is bounded. According to Proposition 3.3.3 that the operator

$$(A - (\lambda S + I))P_{\lambda,S} - (A - (\mu S + I))P_{\mu,S}$$

is bounded. Thus,

$$\mathscr{M}(\lambda,\mu)B$$

has finite dimensional range, and in view of Eq. (3.6), we have

$$R_{b,S}(A,\lambda)B - R_{b,S}(A,\mu)B$$

is bounded. Hence, $R_{b,S}(A,\mu)B$ is continuous for some $\mu \in \rho_{6,S}(A)$ if, and only if, it is continuous for all $\mu \in \rho_{6,S}(A)$.

(ii) If $CR_{b,S}(A,\lambda)$ is bounded for some $\lambda \in \rho_{6,S}(A)$, then clearly $CR_{b,S}(A,\mu)$ is also bounded for any μ and, it follows from the Eq. (3.7) that

$$CR_{b,S}(A,\mu)$$

is bounded for any μ. By using Eq. (3.5) we have $C(A - \mu S)^{-1}$ is bounded for every $\mu \in \rho_S(A)$.

(iii) According to Proposition 3.3.3 that the operator $\mathscr{M}(\lambda,\mu)$ is a finite rank operator. So, $C\mathscr{M}(\lambda,\mu)$ and $\mathscr{M}(\lambda,\mu)B$ are a finite rank operator. Hence, it is clear that

$$\overline{\mathscr{M}(\lambda,\mu)B}$$

is of finite rank if B is densely defined. Since

$$\begin{aligned} C\mathscr{M}(\lambda,\mu)B &= CR_{b,S}(A,\lambda)\big[(A - (\lambda S + I))P_{\lambda,S} - \\ &\qquad (A - (\mu S + I))P_{\mu,S}\big]R_{b,S}(A,\mu)B \\ &= CR_{b,S}(A,\lambda)[(A - \lambda S)(I - P_{\lambda,S}) + P_{\lambda,S}]R_{b,S}(A,\mu)B \end{aligned}$$

and if B and C satisfy the conditions (i) and (ii), respectively, then

$$\overline{C\mathcal{M}(\lambda,\mu)B}$$

will again be continuous and densely defined with finite-dimensional range. Q.E.D.

3.3.4 S-Essential Spectra

Let $A \in \mathcal{C}(X)$ and $S \in \mathcal{L}(X)$ such that $S \neq 0$ and $S \neq A$. We are concerned with the following S-essential spectra (see [1, 10, 101]) defined by the following sets

$$\begin{aligned}
\sigma_{e1,S}(A) &:= \{\lambda \in \mathbb{C} \text{ such that } \lambda S - A \notin \Phi_+(X)\} := \mathbb{C}\backslash\Phi_{+A,S},\\
\sigma_{e2,S}(A) &:= \{\lambda \in \mathbb{C} \text{ such that } \lambda S - A \notin \Phi_-(X)\} := \mathbb{C}\backslash\Phi_{-A,S},\\
\sigma_{e3,S}(A) &:= \{\lambda \in \mathbb{C} \text{ such that } \lambda S - A \notin \Phi_\pm(X)\},\\
\sigma_{e4,S}(A) &:= \{\lambda \in \mathbb{C} \text{ such that } \lambda S - A \notin \Phi(X)\},\\
\sigma_{e5,S}(A) &:= \bigcap_{K \in \mathcal{K}(X)} \sigma_S(A+K),\\
\sigma_{e6,S}(A) &:= \{\lambda \in \mathbb{C} \text{ such that } \lambda S - A \notin \mathcal{B}(X)\},\\
\sigma_{el,S}(A) &:= \{\lambda \in \mathbb{C} \text{ such that } \lambda S - A \notin \Phi_l(X)\},\\
\sigma_{er,S}(A) &:= \{\lambda \in \mathbb{C} \text{ such that } \lambda S - A \notin \Phi_r(X)\},\\
\sigma_{ewl,S}(A) &:= \{\lambda \in \mathbb{C} \text{ such that } \lambda S - A \notin \mathcal{W}_l(X)\},\\
\sigma_{ewr,S}(A) &:= \{\lambda \in \mathbb{C} \text{ such that } \lambda S - A \notin \mathcal{W}_r(X)\},\\
\sigma_{bl,S}(A) &:= \{\lambda \in \mathbb{C} \text{ such that } \lambda S - A \notin \mathcal{B}^l(X)\},\\
\sigma_{br,S}(A) &:= \{\lambda \in \mathbb{C} \text{ such that } \lambda S - A \notin \mathcal{B}^r(X)\},\\
\sigma_{eap,S}(A) &:= \Big\{\lambda \in \mathbb{C} \text{ such that } \lambda S - A \notin \Phi_+(X) \text{ and}\\
&\qquad\qquad i(\lambda S - A) \leq 0\Big\} := \mathbb{C}\backslash\rho_{eap,S}(A),\\
\sigma_{e\delta,S}(A) &:= \Big\{\lambda \in \mathbb{C} \text{ such that } \lambda S - A \notin \Phi_-(X) \text{ and}\\
&\qquad\qquad i(\lambda S - A) \geq 0\Big\} := \mathbb{C}\backslash\rho_{e\delta,S}(A).
\end{aligned}$$

They can be ordered as

$$\sigma_{e3,S}(A) = \sigma_{e1,S}(A)\bigcap\sigma_{e2,S}(A) \subseteq \sigma_{e4,S}(A) \subseteq \sigma_{e5,S}(A) \subseteq \sigma_{e6,S}(A),$$

$$\sigma_{el,S}(A) \subseteq \sigma_{ewl,S}(A) \subseteq \sigma_{bl,S}(A),$$

and

$$\sigma_{er,S}(A) \subseteq \sigma_{ewr,S}(A) \subseteq \sigma_{br,S}(A).$$

In [29], the authors proved that

$$\sigma_{e4,S}(A) = \sigma_{el,S}(A) \bigcup \sigma_{er,S}(A),$$

$$\sigma_{e1,S}(A) \subset \sigma_{el,S}(A),$$

and

$$\sigma_{e2,S}(A) \subset \sigma_{er,S}(A).$$

Remark 3.3.3 *(i) If $S = I$, we recover the usual definition of the essential spectra of a closed densely defined linear operator A.*

(ii) If $A \in \mathscr{L}(X)$ and S is invertible, then

$$\sigma_{ei,S}(A) = \sigma_{ei}(S^{-1}A) \bigcup \sigma_{ei}(AS^{-1}), \ \ i \in \{1,2,3,4,5,6,ap,\delta\}. \quad \diamondsuit$$

Theorem 3.3.4 *(F. Abdmouleh, A. Ammar and A. Jeribi [10]) Let S and A be two bounded linear operators on a Banach space X. Then,*

$$\sigma_{e5,S}(A) = \sigma_{e4,S}(A) \bigcup \{\lambda \in \mathbb{C} \text{ such that } i(A - \lambda S) \neq 0\}. \quad \diamondsuit$$

Proof. Let $\lambda \notin \sigma_{e4,S}(A) \bigcup \{\lambda \in \mathbb{C}$ such that $i(A - \lambda S) \neq 0\}$. Then,

$$A - \lambda S \in \Phi^b(X)$$

and

$$i(A - \lambda S) = 0.$$

By applying Theorem 2.2.11, there exists $K \in \mathscr{K}(X)$ such that $A - \lambda S + K$ is invertible. Hence,

$$\lambda \in \rho_S(A + K).$$

This shows that

$$\lambda \notin \bigcap_{K \in \mathscr{K}(X)} \sigma_S(A + K).$$

Then, we have

$$\sigma_{e5,S}(A) \subset \sigma_{e4,S}(A) \bigcup \{\lambda \in \mathbb{C} \ \text{ such that } i(A-\lambda S) \neq 0\}. \tag{3.8}$$

To prove the inverse inclusion of Eq. (3.8). Suppose $\lambda \notin \sigma_{e5,S}(A)$, then there exists $K \in \mathscr{K}(X)$ such that $\lambda \in \rho_S(A+K)$. Hence, $A - \lambda S + K \in \Phi^b(X)$ and

$$i(A-\lambda S+K) = 0.$$

Now, the operator $A - \lambda S$ can be written in the form

$$A - \lambda S = A + K - \lambda S - K.$$

Since $K \in \mathscr{K}(X)$, using Lemma 2.3.1, we get $A - \lambda S \in \Phi^b(X)$ and

$$i(A-\lambda S) = i(A+K-\lambda S) = 0.$$

We conclude that

$$\lambda \notin \sigma_{e4,S}(A) \bigcup \{\lambda \in \mathbb{C} \text{ such that } i(A-\lambda S) \neq 0\}.$$

Hence,

$$\sigma_{e4,S}(A) \bigcup \{\lambda \in \mathbb{C} \text{ such that } i(A-\lambda S) \neq 0\} \subset \sigma_{e5,S}(A).$$

Therefore,

$$\sigma_{e5,S}(A) = \sigma_{e4,S}(A) \bigcup \{\lambda \in \mathbb{C} \ \text{ such that } i(A-\lambda S) \neq 0\}.$$

This completes the proof. Q.E.D.

Proposition 3.3.5 *Let A and $S \in \mathscr{L}(X)$ such that $S \neq A$. If $\mathbb{C} \backslash \sigma_{e4,S}(A)$ is connected and $\rho_S(A)$ is not empty, then*

$$\sigma_{e5,S}(A) = \sigma_{e4,S}(A). \qquad \diamondsuit$$

Proof. (i) By using Theorem 3.3.4, we have

$$\sigma_{e5,S}(A) = \sigma_{e4,S}(A) \bigcup \{\lambda \in \mathbb{C} \text{ such that } i(A - \lambda S) \neq 0\}.$$

Hence,

$$\sigma_{e4,S}(A) \subset \sigma_{e5,S}(A).$$

Conversely, let $\lambda \in \sigma_{e5,S}(A)$. It suffices to show that

$$\{\lambda \in \mathbb{C} \text{ such that } i(A - \lambda S) \neq 0\} \subset \sigma_{e4,S}(A)$$

which is equivalent to

$$[\mathbb{C} \backslash \sigma_{e4,S}(A)] \bigcap \{\lambda \in \mathbb{C} \text{ such that } i(A - \lambda S) \neq 0\} = \emptyset.$$

Suppose that

$$[\mathbb{C} \backslash \sigma_{e4,S}(A)] \bigcap \{\lambda \in \mathbb{C} \text{ such that } i(A - \lambda S) \neq 0\} \neq \emptyset$$

and let

$$\lambda_0 \in [\mathbb{C} \backslash \sigma_{e4,S}(A)] \bigcap \{\lambda \in \mathbb{C} \text{ such that } i(A - \lambda S) \neq 0\}.$$

Since $\rho_S(A) \neq \emptyset$, then there exists $\lambda_1 \in \mathbb{C}$ such that $\lambda_1 \in \rho_S(A)$ and consequently, $\lambda_1 S - A \in \Phi^b(X)$ and

$$i(\lambda_1 S - A) = 0.$$

On the other side, $\mathbb{C} \backslash \sigma_{e4,S}(A)$ is connected, it follows from Proposition 2.2.1 (ii) that $i(\lambda S - A)$ is constant on any component of $\Phi_{A,S}$. Therefore,

$$i(\lambda_1 S - A) = i(\lambda_0 S - A) = 0,$$

which is a contradiction. Then,

$$[\mathbb{C} \backslash \sigma_{e4,S}(A)] \bigcap \{\lambda \in \mathbb{C} \text{ such that } i(A - \lambda S) \neq 0\} = \emptyset.$$

Hence,

$$\sigma_{e5,S}(A) \subset \sigma_{e4,S}(A). \qquad \text{Q.E.D.}$$

Lemma 3.3.2 *(A. Jeribi, N. Moalla, and S. Yengui [101, Lemma 2.1]) Let $A \in \mathscr{C}(X)$ and $S \in \mathscr{L}(X)$, such that $\rho_S(A)$ is not empty.*

(i) If $\mathbb{C} \backslash \sigma_{e4,S}(A)$ is connected, then

$$\sigma_{e4,S}(A) = \sigma_{e5,S}(A).$$

(ii) If $\mathbb{C} \backslash \sigma_{e5,S}(A)$ is connected, then

$$\sigma_{e5,S}(A) = \sigma_{e6,S}(A). \qquad \diamondsuit$$

3.4 INVARIANCE OF THE *S*-ESSENTIAL SPECTRUM

The following result gives a characterization of the S-essential spectrum by means of a Fredholm operators.

Proposition 3.4.1 *Let $S \in \mathscr{L}(X)$ and $A \in \mathscr{C}(X)$. Then,*

$$\lambda \notin \sigma_{e5,S}(A) \text{ if, and only if, } A - \lambda S \in \Phi(X) \text{ and } i(A - \lambda S) = 0. \qquad \diamondsuit$$

Proof. Let $\lambda \notin \sigma_{e5,S}(A)$. Then, there exists a compact operator K on X such that $\lambda \in \rho_S(A+K)$. So,

$$A + K - \lambda S \in \Phi(X)$$

and

$$i(A + K - \lambda S) = 0.$$

Now, the operator $A - \lambda S$ can be written in the form

$$A - \lambda S = A + K - \lambda S - K.$$

By Theorem 2.2.14, we have

$$A - \lambda S \in \Phi(X)$$

and

$$i(A - \lambda S) = 0.$$

Conversely, we suppose that $A - \lambda S \in \Phi(X)$ and $i(A - \lambda S) = 0$. Let $n = \alpha(A - \lambda S) = \beta(A - \lambda S)$, $\{x_1, \cdots, x_n\}$ being the basis for the $N((A - \lambda S))$

and $\{y'_1, \cdots, y'_n\}$ being the basis for annihilator $R(A-\lambda S)^{\perp}$. By Lemma 2.1.1, there are functionals $x'_1, \cdots, x'_n$ in X^* (the adjoint space of X) and elements $y_1, \cdots, y_n$ such that

$$x'_j(x_k) = \delta_{jk} \text{ and } y'_j(y_k) = \delta_{jk}, \ \ 1 \leq j, \ k \leq n,$$

where $\delta_{jk} = 0$ if $j \neq k$ and $\delta_{jk} = 1$ if $j = k$. Consider the operator K defined by

$$K : X \ni x \longrightarrow Kx := \sum_{i=1}^{n} x'_i(x) y_i \in X.$$

Clearly, K is a linear operator defined everywhere on X. It is bounded, since

$$\|Kx\| \leq \Big(\sum_{k=1}^{n} \|x'_k\| \|y_k\| \Big) \|x\|.$$

Moreover, the range of K is contained in a finite dimensional subspace of X. Then, K is a finite rank operator in X (see Lemma 2.1.4). By Lemma 2.1.5, K is a compact operator in X. We may prove that

$$N(A-\lambda S) \bigcap N(K) = \{0\} \text{ and } R(A-\lambda S) \bigcap R(K) = \{0\}. \tag{3.9}$$

Indeed, let $x \in N(A-\lambda S)$, then

$$x = \sum_{k=1}^{n} \alpha_k x_k.$$

Therefore,

$$x'_j(x) = \alpha_j, \ 1 \leq j \leq n.$$

On the one hand, if $x \in N(K)$, then

$$x'_j(x) = 0, \ 1 \leq j \leq n.$$

This proves the first relation in Eq. (3.9). The second inclusion is similar. In fact, if $y \in R(K)$, then

$$y = \sum_{k=1}^{n} \alpha_k y_k,$$

and hence,

$$y'_j(y) = \alpha_j, \ 1 \leq j \leq n.$$

But, if $y \in R(A-\lambda S)$, then

$$y'_j(y) = 0,\ 1 \le j \le n.$$

This gives the second relation in Eq. (3.9). On the other hand, K is a compact operator, hence we deduce, from Theorem 2.2.14, that $\lambda S - A - K \in \Phi(X)$ and

$$i(A-\lambda S-K) = 0.$$

If $x \in N(A-\lambda S-K)$, then $(A-\lambda S)x$ is in $R(A-\lambda S)\bigcap R(K)$. This implies that $x \in N(A-\lambda S)\bigcap N(K)$ and so, $x = 0$. Thus,

$$\alpha(A+K-\lambda S) = 0.$$

In the same way, one proves that

$$R(A+K-\lambda S) = X.$$

Using Theorem 2.1.2, we get $\lambda \in \rho_S(A+K)$. Also,

$$\lambda \notin \bigcap_{K \in \mathscr{K}(X)} \sigma_S(A+K).$$

So, $\lambda \notin \sigma_{e5,S}(A)$. Q.E.D.

Corollary 3.4.1 *Let $A \in \mathscr{C}(X)$ and $S \in \mathscr{L}(X)$. If $\Phi_{A,S}$ is connected and $\rho_S(A)$ is not empty, then*

$$\sigma_{e5,S}(A) = \left\{\lambda \in \mathbb{C} \text{ such that } A-\lambda S \notin \Phi(X)\right\} = \mathbb{C}\backslash\Phi_{A,S}. \qquad \diamondsuit$$

Theorem 3.4.1 *Let $S \in \mathscr{L}(X)$ and $\lambda \in \rho_S(A)\bigcap\rho_S(A+B)$. If*

$$\|x_n\| + \|Ax_n\| + \|Bx_n\| \le c,$$

for all $x_n \in \mathscr{D}(A)$ implies that $(A-\lambda S)^{-1}Bx_n$ has a convergent subsequence, then

$$\sigma_{e5,S}(A+B) = \sigma_{e5,S}(A). \tag{3.10}$$

Proof. We employ the identities

$$(A+B-\mu S)-(A-\mu S)(A-\lambda S)^{-1}(A+B-\lambda S)=(\mu-\lambda)S(A-\lambda S)^{-1}B. \tag{3.11}$$

Since, $\rho_S(A)$ and $\rho_S(A+B)$ are not empty, then A and $A+B$ are closed. Hence, $A+B$ is A-bounded. This shows that the hypotheses imply that $(A-\lambda S)^{-1}B$ is A-compact and $(A+B)$-compact. Let $\mu \notin \sigma_{e5,S}(A+B)$, then from Proposition 3.4.1, we get

$$A+B-\mu S \in \Phi(X)$$

and

$$i(A+B-\mu S)=0.$$

By Eq. (3.11), we have

$$(A-\mu S)(A-\lambda S)^{-1}(A+B-\lambda S) \in \Phi(X)$$

and

$$i\Big((A-\mu S)(A-\lambda S)^{-1}(A+B-\lambda S)\Big)=0.$$

Since $\lambda \in \rho_S(A+B)$, then

$$A+B-\lambda S \in \Phi(X)$$

and

$$i(A+B-\lambda S)=0.$$

Using Theorem 2.2.21, we get

$$(A-\mu S)(A-\lambda S)^{-1} \in \Phi(X)$$

and

$$i\Big((A-\mu S)(A-\lambda S)^{-1}\Big)=0.$$

From this and the identity

$$A-\mu S=(A-\mu S)(A-\lambda S)^{-1}(A-\lambda S),$$

we obtain

$$A - \mu S \in \Phi(X)$$

and

$$i(A - \mu S) = 0.$$

Thus, in view of Proposition 3.4.1, $\mu \notin \sigma_{e5,S}(A)$ and, we conclude that

$$\sigma_{e5,S}(A) \subset \sigma_{e5,S}(A+B).$$

Conversely, if $\mu \notin \sigma_{e5,S}(A)$, then

$$A - \mu S \in \Phi(X) \text{ and } i(A - \mu S) = 0.$$

Since $\lambda \in \rho_S(A+B)$, then by Proposition 3.4.1, we have

$$A + B - \lambda S \in \Phi(X)$$

and

$$i(A + B - \lambda S) = 0.$$

Thus,

$$(A - \mu S)(A - \lambda S)^{-1}(A + B - \lambda S) \in \Phi(X)$$

and

$$i\Big((A - \mu S)(A - \lambda S)^{-1}(A + B - \lambda S)\Big) = 0.$$

By Eq. (3.11), we have

$$A + B - \mu S \in \Phi(X)$$

and

$$i(A + B - \mu S) = 0.$$

Then, in view of Proposition 3.4.1, $\mu \notin \sigma_{e5,S}(A+B)$, and we obtain

$$\sigma_{e5,S}(A+B) \subset \sigma_{e5,S}(A).$$

This proves (3.10) and completes the proof. Q.E.D.

Remark 3.4.1 *If A and B are bounded operators. Theorem* 3.4.1 *remains true if we replace* $(A-\lambda S)^{-1}B$ *by* $B(A-\lambda S)^{-1}$. *Indeed, it suffices to replace Eq.* (3.11) *by*

$$(A+B-\mu S)-(A+B-\lambda S)(A-\lambda S)^{-1}(A-\mu S) = (\mu-\lambda)B(A-\lambda S)^{-1}S. \quad \diamondsuit$$

3.5 PSEUDOSPECTRA

3.5.1 Pseudospectrum

Let us start by giving the definition of the pseudospectrum of densely closed linear operator A for every $\varepsilon > 0$,

$$\sigma_\varepsilon(A) := \sigma(A)\bigcup\left\{\lambda \in \mathbb{C} \text{ such that } \|(\lambda - A)^{-1}\| > \frac{1}{\varepsilon}\right\},$$

or by

$$\Sigma_\varepsilon(A) := \sigma(A)\bigcup\left\{\lambda \in \mathbb{C} \text{ such that } \|(\lambda - A)^{-1}\| \geq \frac{1}{\varepsilon}\right\},$$

with the convention $\|(\lambda - A)^{-1}\| = \infty$ if, and only if, $\lambda \in \sigma(A)$. The pseudospectrum, $\sigma_\varepsilon(A)$, is the open subset of the complex plane bounded by the ε^{-1} level curve of the norm of the resolvent. For $\varepsilon > 0$, it can be shown that $\sigma_\varepsilon(A)$ is a larger set and is never empty. The pseudospectra of A are a family of strictly nested closed sets, which grow to fill the whole complex plane as $\varepsilon \to \infty$ (see [80, 151, 152]). From these definitions, it follows that the pseudospectra associated with various ε are nested sets. Then, for all $0 < \varepsilon_1 < \varepsilon_2$, we have

$$\sigma(A) \subset \sigma_{\varepsilon_1}(A) \subset \sigma_{\varepsilon_2}(A)$$

and

$$\sigma(A) \subset \Sigma_{\varepsilon_1}(A) \subset \Sigma_{\varepsilon_2}(A),$$

and that the intersections of all the pseudospectra are the spectrum

$$\bigcap_{\varepsilon>0} \sigma_\varepsilon(A) = \sigma(A) = \bigcap_{\varepsilon>0} \Sigma_\varepsilon(A).$$

Theorem 3.5.1 *Let $A \in \mathscr{C}(X)$. The following three conditions are equivalent:*

(i) $\lambda \in \sigma_\varepsilon(A)$.

(ii) There exists a bounded operator D such that $\|D\| < \varepsilon$ and

$$\lambda \in \sigma(A+D).$$

(iii) Either $\lambda \in \sigma(A)$ or $\|(\lambda - A)^{-1}\| < \varepsilon^{-1}$. ◇

Proof. $(i) \Rightarrow (ii)$ If $\lambda \in \sigma(A)$, we may put $D = 0$. Otherwise, let $f \in \mathscr{D}(A-\lambda)$, $\|f\| = 1$ and

$$\|(A-\lambda)f\| < \varepsilon.$$

Let $\phi \in X^*$ satisfy $\|\phi\| = 1$ and $\phi(f) = 1$. Then, let us define the rank one operator $D : X \longrightarrow X$ by

$$Dg := -\phi(g)(A-\lambda)f.$$

We see immediately that $\|D\| < \varepsilon$ and

$$(A - \lambda + D)f = 0.$$

$(ii) \Rightarrow (iii)$ We derive a contradiction from the assumption that $\lambda \notin \sigma(A)$ and

$$\|(A-\lambda)^{-1}\| \leq \varepsilon^{-1}.$$

Let $B : X \longrightarrow X$ be the bounded operator defined by the norm convergent series

$$\begin{aligned} B &:= \sum_{n=0}^{\infty} (A-\lambda)^{-1}\left(-D(A-\lambda)^{-1}\right)^n \\ &= (A-\lambda)^{-1}(I + D(A-\lambda)^{-1})^{-1}. \end{aligned}$$

It is immediate from these formulae that B is one-to-one with a range equal to $\mathscr{D}(A-\lambda)$. We also see that

$$B(I+D(A-\lambda)^{-1})f=(A-\lambda)^{-1}f,$$

for all $f\in X$. Putting

$$g=(A-\lambda)^{-1}f,$$

we conclude that

$$B(A-\lambda+D)g=g$$

for all $g\in\mathscr{D}(A-\lambda)$. The proof that

$$(A-\lambda+D)Bh=h,$$

for all $h\in X$ is similar. Hence, $A-\lambda+D$ is invertible, with an inverse B.

$(iii)\Rightarrow(i)$ We assume for non-triviality that $\lambda\notin\sigma(A)$. There exists $g\in X$ such that

$$\|(A-\lambda)^{-1}g\|>\varepsilon^{-1}\|g\|.$$

Putting $f:=(A-\lambda)^{-1}g$, we see that

$$\|(A-\lambda)f\|<\varepsilon\|f\|. \qquad \text{Q.E.D.}$$

Remark 3.5.1 From Theorem 3.5.1, it follows immediately that

$$\sigma_\varepsilon(A)=\bigcup_{\|D\|<\varepsilon}\sigma(A+D). \qquad \diamondsuit$$

3.5.2 S-Pseudospectrum

Let $A\in\mathscr{C}(X)$ and $S\in\mathscr{L}(X)$. Consider the equation defined by

$$(\lambda S-A)x=b, \tag{3.12}$$

where $\lambda\in\mathbb{C}$, $b\in X$ and $x\in\mathscr{D}(A)$. For $\lambda\in\rho_S(A)$, the existence of a solution of Eq. (3.12) and it is uniqueness are guaranteed. So,

$$x=R_S(\lambda,A)b. \tag{3.13}$$

Under the perturbation of b by adding $r \in X$ to b such that $0 < \|r\| < \varepsilon$. We obtain the equation

$$(\lambda S - A)\overline{x} = b + r.$$

This implies that

$$\overline{x} = R_S(\lambda, A)(b + r). \tag{3.14}$$

We want the solution of the Eq. (3.12) remains stable despite the perturbation. It means that $\overline{x}$ is very close to x and so,

$$\|\overline{x} - x\| < 1.$$

Since the two Eqs. (3.13) and (3.14) provide

$$\begin{aligned} \|\overline{x} - x\| &= \|R_S(\lambda, A)r\| \\ &< \|R_S(\lambda, A)\|\varepsilon. \end{aligned}$$

For this, it suffices to take

$$\|R_S(\lambda, A)\| \leq \frac{1}{\varepsilon}.$$

In this case, we define

$$\rho_{S,\varepsilon}(A) = \rho_S(A) \bigcap \left\{ \lambda \in \mathbb{C} \text{ such that } \|R_S(\lambda, A)\| \leq \frac{1}{\varepsilon} \right\}.$$

Therefore,

$$\sigma_{S,\varepsilon}(A) = \mathbb{C} \backslash \rho_{S,\varepsilon}(A).$$

Definition 3.5.1 *Let $A \in \mathscr{C}(X)$, $S \in \mathscr{L}(X)$ and $\varepsilon > 0$ such that $S \neq A$ and $S \neq 0$. We define the S-pseudospectrum of A by*

$$\sigma_{S,\varepsilon}(A) = \sigma_S(A) \bigcup \left\{ \lambda \in \mathbb{C} \textit{ such that } \|R_S(\lambda, A)\| > \frac{1}{\varepsilon} \right\}.$$

Convention: $\|R_S(\lambda, A)\| = +\infty$ *if, and only if,* $\lambda \in \sigma_S(A)$. $\diamondsuit$

If we replace the bounded operator S by the identity operator I, we obtain the definition of pseudospectrum $\sigma_\varepsilon(A)$ (see [58, 153]).

Proposition 3.5.1 *Let $A \in \mathscr{C}(X)$, $S \in \mathscr{L}(X)$ and $\varepsilon > 0$, then the S-pseudospectra verifies the following properties:*

(i) $\sigma_{S,\varepsilon}(A) \neq \emptyset$,

(ii) the S-pseudospectra $\left\{\sigma_{S,\varepsilon}(A)\right\}_{\varepsilon>0}$ *are increasing sets in the sense of inclusion relative to the parameter strictly positive* ε*, and*

(iii) $\bigcap\limits_{\varepsilon>0} \sigma_{S,\varepsilon}(A) = \sigma_S(A)$. ◇

Proof. *(i)* We argue by contradiction. Suppose that $\sigma_{S,\varepsilon}(A) = \emptyset$, then

$$\rho_{S,\varepsilon}(A) = \mathbb{C}.$$

It means exactly that

$$\rho_S(A) \bigcap \left\{\lambda \in \mathbb{C} \text{ such that } \|R_S(\lambda,A)\| \leq \frac{1}{\varepsilon}\right\} = \mathbb{C},$$

and so

$$\rho_S(A) = \mathbb{C}.$$

Hence,

$$\left\{\lambda \in \mathbb{C} \text{ such that } \|R_S(\lambda,A)\| \leq \frac{1}{\varepsilon}\right\} = \mathbb{C}.$$

Consider the function

$$\begin{aligned} \varphi : \ \mathbb{C} &\longrightarrow \mathscr{L}(X) \\ \lambda &\longrightarrow R_S(\lambda,A). \end{aligned}$$

Since φ is analytic on $\mathbb{C}$ and for every $\lambda \in \mathbb{C}$, we have

$$\|\varphi(\lambda)\| \leq \frac{1}{\varepsilon},$$

then φ is an entire bounded function. Therefore, using Liouville theorem (see Theorem 2.1.5), we obtain that φ is constant. It follows that $R_S(\lambda,A)$ is null, which is a contradiction.

(ii) Let $\varepsilon_1,\ \varepsilon_2 > 0$ such that $\varepsilon_1 < \varepsilon_2$. If $\lambda \in \sigma_{S,\,\varepsilon_1}(A)$, then

$$\|R_S(\lambda,A)\| > \frac{1}{\varepsilon_1} > \frac{1}{\varepsilon_2}.$$

Therefore,

$$\lambda \in \sigma_{S,\,\varepsilon_2}(A).$$

(iii) We have

$$\begin{aligned}\bigcap_{\varepsilon>0} \sigma_{S,\,\varepsilon}(A) &= \bigcap_{\varepsilon>0} \left(\sigma_S(A) \bigcup \left\{\lambda \in \mathbb{C} \text{ such that } \|R_S(\lambda,A)\| > \frac{1}{\varepsilon}\right\}\right), \\ &= \sigma_S(A) \bigcup \left(\bigcap_{\varepsilon>0} \left\{\lambda \in \mathbb{C} \text{ such that } \|R_S(\lambda,A)\| > \frac{1}{\varepsilon}\right\}\right).\end{aligned}$$

It suffices to prove that

$$\bigcap_{\varepsilon>0} \left\{\lambda \in \mathbb{C} \text{ such that } \|R_S(\lambda,A)\| > \frac{1}{\varepsilon}\right\} = \{\lambda \in \mathbb{C} \text{ such that } \|R_S(\lambda,A)\| = +\infty\}.$$

For the inclusion in the direct sense: let $\lambda \in \mathbb{C}$ such that for all $\varepsilon > 0$

$$\|R_S(\lambda,A)\| > \frac{1}{\varepsilon}.$$

Consequently,

$$\lim_{\varepsilon \to 0^+} \|R_S(\lambda,A)\| = \|R_S(\lambda,A)\| = +\infty.$$

For the other inclusion: let $\lambda \in \mathbb{C}$ such that

$$\|R_S(\lambda,A)\| = +\infty > \frac{1}{\varepsilon}$$

for all $\varepsilon > 0$. Hence,

$$\begin{aligned}\bigcap_{\varepsilon>0} \sigma_{S,\,\varepsilon}(A) &= \sigma_S(A) \bigcup \left\{\lambda \in \mathbb{C} \text{ such that } \|R_S(\lambda,A)\| = +\infty\right\}, \\ &= \sigma_S(A),\end{aligned}$$

which completes the proof of proposition. Q.E.D.

3.5.3 Ammar-Jeribi Essential Pseudospectrum

In Ref. [29], A. Ammar and A. Jeribi introduced the definition of Ammar-Jeribi essential pseudospectrum of closed densely defined linear operators by

$$\sigma_{e5,\varepsilon}(A) = \bigcap_{K \in \mathcal{K}(X)} \sigma_{\varepsilon}(A+K).$$

Theorem 3.5.2 *Let X be a Banach space, $\varepsilon > 0$ and $A \in \mathcal{C}(X)$. Then, $\lambda \notin \sigma_{e5,\varepsilon}(A)$ if, and only if, for all $D \in \mathcal{L}(X)$ such that $\|D\| < \varepsilon$, we have $A+D-\lambda \in \Phi(X)$ and*

$$i(A+D-\lambda) = 0. \qquad \diamondsuit$$

Proof. Let $\lambda \notin \sigma_{e5,\varepsilon}(A)$. By using Theorem 3.5.1, we infer that there exists a compact operator K on X, such that

$$\lambda \notin \bigcup_{\|D\|<\varepsilon} \sigma(A+K+D).$$

So, for all $D \in \mathcal{L}(X)$ such that $\|D\| < \varepsilon$, we have

$$\lambda \in \rho(A+D+K).$$

Therefore, $A+D+K-\lambda \in \Phi(X)$ and

$$i(A+D+K-\lambda) = 0,$$

for all $D \in \mathcal{L}(X)$ such that $\|D\| < \varepsilon$. From Theorem 2.2.14, we deduce that for all $D \in \mathcal{L}(X)$ such that $\|D\| < \varepsilon$, we have

$$A+D-\lambda \in \Phi(X)$$

and

$$i(A+D-\lambda) = 0.$$

Conversely, we suppose that, for all $D \in \mathcal{L}(X)$ such that $\|D\| < \varepsilon$, we have

$$A+D-\lambda \in \Phi(X)$$

and

$$i(A+D-\lambda)=0.$$

Without loss of generality, we may assume that $\lambda=0$. Let $n=\alpha(A+D)=\beta(A+D)$, $\{x_1,\cdots,x_n\}$ being the basis for $N(A+D)$ and $\{y'_1,\cdots,y'_n\}$ being the basis for the $N((A+D)^*)$. By using Lemma 2.1.1, there are functionals $x'_1,\cdots,x'_n$ in X^* (the adjoint space of X) and elements $y_1,\cdots,y_n$, such that

$$x'_j(x_k)=\delta_{jk}$$

and

$$y_j(y_k)=\delta_{jk},$$

$1\leq j,\ k\leq n$, where $\delta_{jk}=0$ if $j\neq k$ and $\delta_{jk}=1$ if $j=k$. Let K be the operator defined by

$$Kx=\sum_{k=1}^{n}x'_k(x)y_k,$$

$x\in X$. Clearly, K is a linear operator defined everywhere on X. It is bounded, since

$$\|Kx\|\leq\|x\|\Big(\sum_{k=1}^{n}\|x'_k\|\|y_k\|\Big).$$

Moreover, the range of K is contained in a finite-dimensional subspace of X. Then, K is a finite rank operator in X. So, K is a compact operator in X. Now, we may prove that

$$N(A+D)\bigcap N(K)=\{0\}\ \text{ and } R(A+D)\bigcap R(K)=\{0\}, \tag{3.15}$$

for all $D\in\mathscr{L}(X)$ such that $\|D\|<\varepsilon$. Indeed, let $x\in N(A+D)$, then

$$x=\sum_{k=1}^{n}\alpha_k x_k,$$

and therefore,

$$x_j(x)=\alpha_j,\ 1\leq j\leq n.$$

Moreover, if $x\in N(K)$, then

$$x'_j(x)=0,$$

with $1 \leq j \leq n$. This proves the first relation in Eq. (3.15). The second inclusion is similar. In fact, if $y \in R(K)$, then

$$y = \sum_{k=1}^{n} \alpha_k y_k,$$

and hence,

$$y_j(y) = \alpha_j, \text{ with } 1 \leq j \leq n.$$

However, if $y \in R(A+D)$, then

$$y'_j(y) = 0,$$

with $1 \leq j \leq n$. This gives the second relation in Eq. (3.15). Besides, K is a compact operator. We deduce, from Theorem 2.2.14, that $0 \in \Phi_{A+K+D}$ and

$$i(A+D+K) = 0.$$

If $x \in N(A+D+K)$, then

$$(A+D)x \in R(A+D) \bigcap R(K).$$

This implies that $x \in N(A+D) \bigcap N(K)$ and $x = 0$. Thus,

$$\alpha(A+D+K) = 0.$$

In the same way, we can prove that

$$R(A+D+K) = X.$$

Hence, $0 \in \rho(A+D+K)$. This implies that, for all $D \in \mathscr{L}(X)$ such that $\|D\| < \varepsilon$, we have

$$0 \notin \sigma(A+D+K).$$

Also,

$$0 \notin \bigcap_{K \in \mathscr{K}(X)} \sigma_{\varepsilon}(A+K).$$

So,

$$0 \notin \sigma_{e5,\varepsilon}(A). \qquad \text{Q.E.D.}$$

Remark 3.5.2 *Let X be a Banach space, $\varepsilon > 0$ and $A \in \mathscr{C}(X)$.*

(i) Using Theorem 3.5.2, we have

$$\sigma_{e5,\varepsilon}(A) := \bigcup_{\|D\|<\varepsilon} \sigma_{e5}(A+D).$$

(ii) We have

$$\sigma_{e5,\varepsilon}(A) := \bigcap_{D\in\mathscr{F}(X)} \sigma_{\varepsilon}(A+D). \qquad \diamondsuit$$

3.5.4 Essential Pseudospectra

Let $A \in \mathscr{C}(X)$ and $\varepsilon > 0$. We are concerned with the following essential pseudospectra

$$\begin{aligned}
\sigma_{e1,\varepsilon}(A) &:= \{\lambda \in \mathbb{C} \text{ such that } \lambda - A \notin \Phi_{+}^{\varepsilon}(X)\} := \mathbb{C}\backslash\Phi_{+A}^{\varepsilon},\\
\sigma_{e2,\varepsilon}(A) &:= \{\lambda \in \mathbb{C} \text{ such that } \lambda - A \notin \Phi_{-}^{\varepsilon}(X)\} := \mathbb{C}\backslash\Phi_{-A}^{\varepsilon},\\
\sigma_{e3,\varepsilon}(A) &:= \{\lambda \in \mathbb{C} \text{ such that } \lambda - A \notin \Phi_{\pm}^{\varepsilon}(X)\} := \mathbb{C}\backslash\Phi_{\pm A}^{\varepsilon},\\
\sigma_{e4,\varepsilon}(A) &:= \{\lambda \in \mathbb{C} \text{ such that } \lambda - A \notin \Phi^{\varepsilon}(X)\} := \mathbb{C}\backslash\Phi_{A}^{\varepsilon},\\
\sigma_{e5,\varepsilon}(A) &:= \bigcap_{K\in\mathscr{K}(X)} \sigma_{\varepsilon}(A+K),\\
\sigma_{eap,\varepsilon}(A) &:= \bigcap_{K\in\mathscr{K}(X)} \sigma_{ap,\varepsilon}(A+K),\\
\Sigma_{eap,\varepsilon}(A) &:= \bigcap_{K\in\mathscr{K}(X)} \Sigma_{ap,\varepsilon}(A+K),\\
\sigma_{e\delta,\varepsilon}(A) &:= \sigma_{e2,\varepsilon}(A) \bigcup\\
&\qquad \{\lambda \in \mathbb{C} \text{ such that } i(\lambda - A - D) < 0, \text{ for all } \|D\| < \varepsilon\},
\end{aligned}$$

where

$$\sigma_{ap,\varepsilon}(A) := \left\{\lambda \in \mathbb{C} \text{ such that } \inf_{x\in\mathscr{D}(A),\ \|x\|=1} \|(\lambda - A)x\| < \varepsilon\right\},$$

and

$$\Sigma_{ap,\varepsilon}(A) := \left\{\lambda \in \mathbb{C} \text{ such that } \inf_{x\in\mathscr{D}(A),\ \|x\|=1} \|(\lambda - A)x\| \le \varepsilon\right\}.$$

Note that if ε tends to 0, we recover the usual definition of the essential spectra of a closed linear operator A.

Remark 3.5.3 *Let $A \in \mathscr{C}(X)$ and consider $i \in \{1,..,5,ap,\delta\}$. Then,*

(i) $\sigma_{ei,\varepsilon}(A) \subset \sigma_{\varepsilon}(A)$,

(ii) $\bigcap_{\varepsilon>0} \sigma_{ei,\varepsilon}(A) = \sigma_{ei}(A)$,

(iii) if $\varepsilon_1 < \varepsilon_2$, *then* $\sigma_{ei,\varepsilon_1}(A) \subset \sigma_{ei,\varepsilon_2}(A)$, *and*

(iv) they can be ordered as

$$\sigma_{e3,\varepsilon}(A) = \sigma_{e1,\varepsilon}(A) \bigcap \sigma_{e2,\varepsilon}(A) \subseteq \sigma_{e4,\varepsilon}(A) \subseteq \sigma_{e5,\varepsilon}(A) = \sigma_{eap,\varepsilon}(A) \bigcap \sigma_{e\delta,\varepsilon}(A). \quad \diamondsuit$$

3.5.5 Conditional Pseudospectrum

In this section we define the conditional pseudospectrum of a linear operators in infinite dimensional Banach spaces and consider some basic properties in order to put this definition in its due place. We begin with the following definition.

Definition 3.5.2 *Let $A \in \mathscr{L}(X)$ and $0 < \varepsilon < 1$. The condition pseudospectrum of A is denoted by $\Sigma^{\varepsilon}(A)$ and is defined as,*

$$\Sigma^{\varepsilon}(A) := \sigma(A) \bigcup \left\{ \lambda \in \mathbb{C} : \ \|\lambda - A\| \|(\lambda - A)^{-1}\| > \frac{1}{\varepsilon} \right\},$$

with the convention that $\|\lambda - A\| \|(\lambda - A)^{-1}\| = \infty$, if $\lambda - A$ is not invertible. The condition pseudoresolvent of A is denoted by $\rho^{\varepsilon}(A)$ and is defined as,

$$\rho^{\varepsilon}(A) := \rho(A) \bigcap \left\{ \lambda \in \mathbb{C} : \ \|\lambda - A\| \|(\lambda - A)^{-1}\| \leq \frac{1}{\varepsilon} \right\}. \qquad \diamondsuit$$

For $0 < \varepsilon < 1$, it can be shown that $\rho^{\varepsilon}(A)$ is a larger set and is never empty. Here also $\rho^{\varepsilon}(A) \subseteq \rho(A)$ for $0 < \varepsilon < 1$. Recall that the usual condition pseudospectral radius $r_{\varepsilon}(A)$ of $A \in \mathscr{L}(X)$ by

$$r_{\varepsilon}(A) := \sup \left\{ |\lambda| : \lambda \in \Sigma^{\varepsilon}(A) \right\},$$

and the spectral radius of $A \in \mathscr{L}(X)$ by

$$r_{\sigma}(A) = \sup \left\{ |\lambda| : \lambda \in \sigma(A) \right\}.$$

3.6 STRUCTURED PSEUDOSPECTRA

3.6.1 Structured Pseudospectrum

We refer to E. B. Davies [58] which defined the structured pseudospectrum, or spectral value sets of a closed densely defined linear operator A on X by

$$\sigma(A,B,C,\varepsilon) = \bigcup_{\|D\|<\varepsilon} \sigma(A+CDB),$$

where $B \in \mathscr{L}(X,Y)$ and $C \in \mathscr{L}(Z,X)$. It is clear that

$$\sigma(A,I,I,\varepsilon) = \sigma_\varepsilon(A).$$

We first prove the following proposition which is basic for our purpose.

Proposition 3.6.1 *Let* $A \in \mathscr{L}(X)$, $B \in \mathscr{L}(X,Y)$, $C \in \mathscr{L}(Z,X)$, $G(s) := B(s-A)^{-1}C$ *and* $\varepsilon > 0$. *Then,* $\sigma(A,B,C,\varepsilon)\backslash\sigma(A)$ *is a bounded open subset of* $\mathbb{C}$ *such that*

$$\sigma(A,B,C,\varepsilon) = \sigma(A)\bigcup\left\{s \in \rho(A) \text{ such that } \|G(s)\| > \frac{1}{\varepsilon}\right\}. \qquad \diamondsuit$$

Proof. We will discuss these two cases:

<u>1st case</u> : If B or $C = 0$, then $\sigma(A,B,C,\varepsilon) = \sigma(A,\varepsilon)$. Hence,

$$\sigma(A,B,C,\varepsilon)\backslash\sigma(A) = \emptyset$$

is bounded.

<u>2nd case</u> : If B and $C \neq 0$. We consider the following function

$$\left\{\begin{array}{lll} \varphi : & \rho(A) & \longrightarrow \mathbb{R} \\ & s & \longrightarrow \|G(s)\|. \end{array}\right.$$

φ is continuous. It is clear that

$$\sigma(A,B,C,\varepsilon)\backslash\sigma(A) = \varphi^{-1}\left(\left]\frac{1}{\varepsilon},+\infty\right[\right).$$

So, we can deduce that $\sigma(A,B,C,\varepsilon)\backslash\sigma(A)$ is open. Furthermore, we argue by contradiction so we suppose that there exists $s \in \sigma(A,B,C,\varepsilon)\backslash\sigma(A)$ such that $|s| \to +\infty$. Since

$$\lim_{|s|\to+\infty} \|G(s)\| = 0,$$

then $0 \geq \frac{1}{\varepsilon}$ which is impossible. As a consequence we obtain that

$$\sigma(A,B,C,\varepsilon)\backslash\sigma(A)$$

is a bounded set. Q.E.D.

Remark 3.6.1 *As a consequence of Proposition 3.6.1, that the structured pseudospectra, $\sigma(A,B,C,\varepsilon)$, of a bounded operator A is a bounded subset of $\mathbb{C}$.* ◇

The following identity was established in [58, Theorem 9.2.18].

Theorem 3.6.1 *Let $A \in \mathscr{C}(X)$, $B \in \mathscr{L}(X,Y)$, $C \in \mathscr{L}(Z,X)$ and $\varepsilon > 0$. Then,*

$$\sigma(A,B,C,\varepsilon) = \sigma(A) \bigcup \left\{ s \in \mathbb{C} \text{ such that } \|B(s-A)^{-1}C\| > \frac{1}{\varepsilon} \right\}. \quad \diamondsuit$$

3.6.2 The Structured Essential Pseudospectra

In this section, we are interested in studying the following structured essential pseudospectra of a closed, densely defined linear operator A introduced in [67] with respect to the perturbation structure (B,C), where $B \in \mathscr{L}(X,Y)$, $C \in \mathscr{L}(Z,X)$ and uncertainty level $\varepsilon > 0$. We note that the structured essential pseudospectrum consists of large sets of all complex numbers to which at least one essential spectrum can be shifted by disturbance operator D of norm $\|D\| < \varepsilon$.

Definition 3.6.1 *Let $A \in \mathscr{C}(X)$, $B \in \mathscr{L}(X,Y)$, $C \in \mathscr{L}(Z,X)$, and $\varepsilon > 0$.*

(i) The structured Jeribi essential pseudospectrum is defined by

$$\sigma_j(A,B,C,\varepsilon) := \bigcap_{K \in \mathscr{W}^*(X)} \sigma(A+K,B,C,\varepsilon).$$

(ii) The structured Wolf essential pseudospectrum is defined by

$$\sigma_{e4}(A,B,C,\varepsilon) := \bigcup_{\|D\|<\varepsilon} \sigma_{e4}(A+CDB).$$

(iii) The structured Ammar-Jeribi essential pseudospectrum is defined by

$$\sigma_{e5}(A,B,C,\varepsilon) := \bigcup_{\|D\|<\varepsilon} \sigma_{e5}(A+CDB).$$

(iv) The structured Browder essential pseudospectrum is defined by

$$\sigma_{e6}(A,B,C,\varepsilon) := \bigcup_{\|D\|<\varepsilon} \sigma_{e6}(A+CDB). \qquad \diamondsuit$$

Remark 3.6.2 *All these sets satisfy the following inclusions:*

$$\sigma_{e4}(A,B,C,\varepsilon) \subseteq \sigma_{e5}(A,B,C,\varepsilon) \subseteq \sigma_{e6}(A,B,C,\varepsilon). \qquad \diamondsuit$$

By Theorem 3.1.6, we have the following.

Lemma 3.6.1 *Let $A \in \mathscr{C}(X)$, $B \in \mathscr{L}(X,Y)$, $C \in \mathscr{L}(Z,X)$ and $\varepsilon > 0$. Suppose that for every $\|D\| < \varepsilon$, we have $\rho(A+CDB) \neq \emptyset$. Then,*

(i) If for all $\|D\| < \varepsilon$, we have $\mathbb{C}\backslash\sigma_{e4}(A+CDB)$ is connected, then

$$\sigma_{e4}(A,B,C,\varepsilon) = \sigma_{e5}(A,B,C,\varepsilon).$$

(ii) If $\mathbb{C}\backslash\sigma_{e5}(A+CDB)$ is connected for every $\|D\| < \varepsilon$, then

$$\sigma_{e5}(A,B,C,\varepsilon) = \sigma_{e6}(A,B,C,\varepsilon). \qquad \diamondsuit$$

3.6.3 The Structured S-Pseudospectra

Definition 3.6.2 *Let $A \in \mathscr{C}(X)$, $S \in \mathscr{L}(X)$, $B \in \mathscr{L}(X,Y)$, $C \in \mathscr{L}(Z,X)$ and $\varepsilon > 0$. We define the structured S-pseudospectrum by*

$$\sigma_S(A,B,C,\varepsilon) := \bigcup_{\|D\|<\varepsilon} \sigma_S(A+CDB). \qquad \diamondsuit$$

Remark 3.6.3 *If S is an invertible operator, then*

$$\sigma_S(A,B,C,\varepsilon) = \sigma(S^{-1}A,B,S^{-1}C,\varepsilon). \qquad \diamondsuit$$

Definition 3.6.3 *Let $A \in \mathscr{C}(X)$, $S \in \mathscr{L}(X)$, $B \in \mathscr{L}(X,Y)$, $C \in \mathscr{L}(Z,X)$ and $\varepsilon > 0$.*

(i) The structured S-point pseudospectrum is defined by

$$\sigma_{p,S}(A,B,C,\varepsilon) := \bigcup_{\|D\|<\varepsilon} \sigma_{p,S}(A+CDB).$$

(ii) The structured S-continuous pseudospectrum is defined by

$$\sigma_{c,S}(A,B,C,\varepsilon) := \bigcup_{\|D\|<\varepsilon} \sigma_{c,S}(A+CDB).$$

(iii) The structured S-residual pseudospectrum is defined by

$$\sigma_{R,S}(A,B,C,\varepsilon) := \bigcup_{\|D\|<\varepsilon} \sigma_{R,S}(A+CDB).$$

(iv) The structured S-approximate point pseudospectrum is defined by

$$\sigma_{ap,S}(A,B,C,\varepsilon) := \bigcup_{\|D\|<\varepsilon} \sigma_{ap,S}(A+CDB). \qquad \diamondsuit$$

Remark 3.6.4 *Let $A \in \mathscr{C}(X)$, $S \in \mathscr{L}(X)$, $B \in \mathscr{L}(X,Y)$, $C \in \mathscr{L}(Z,X)$ and $\varepsilon > 0$, then the structured S-pseudospectra is partitioned into three disjoint sets as follows:*

$$\sigma_S(A,B,C,\varepsilon) = \sigma_{p,S}(A,B,C,\varepsilon) \bigcup \sigma_{c,S}(A,B,C,\varepsilon) \bigcup \sigma_{R,S}(A,B,C,\varepsilon). \quad \diamondsuit$$

Lemma 3.6.2 *Let $A \in \mathscr{C}(X)$, $S \in \mathscr{L}(X)$, $B \in \mathscr{L}(X,Y)$, $C \in \mathscr{L}(Z,X)$ and $\varepsilon > 0$, then*

$$\sigma_{p,S}(A,B,C,\varepsilon) \subset \sigma_{ap,S}(A,B,C,\varepsilon). \qquad \diamondsuit$$

Proof. Let $\lambda \in \sigma_{p,S}(A,B,C,\varepsilon)$. Then, there exists $D \in \mathscr{L}(Y,Z)$ such that $\|D\| < \varepsilon$ and

$$\lambda \in \sigma_{p,S}(A+CDB).$$

Therefore, there exists $x \in \mathscr{D}(A) \backslash \{0\}$ with

$$(\lambda S - A - CDB)x = 0.$$

In this case, it is sufficient to take $x_n = \frac{x}{\|x\|}$. Consequently,

$$\lambda \in \sigma_{ap,S}(A,B,C,\varepsilon).$$ Q.E.D.

3.6.4 The Structured S-Essential Pseudospectra

Definition 3.6.4 *Let $A \in \mathscr{C}(X)$, $S \in \mathscr{L}(X)$, $B \in \mathscr{L}(X,Y)$, $C \in \mathscr{L}(Z,X)$ and $\varepsilon > 0$, we define the structured S-essential pseudospectra in the following way*

$$\begin{aligned}
\sigma_{e1,S}(A,B,C,\varepsilon) &:= \bigcup_{\|D\|<\varepsilon} \sigma_{e1,S}(A+CDB),\\
\sigma_{e2,S}(A,B,C,\varepsilon) &:= \bigcup_{\|D\|<\varepsilon} \sigma_{e2,S}(A+CDB),\\
\sigma_{e3,S}(A,B,C,\varepsilon) &:= \bigcup_{\|D\|<\varepsilon} \sigma_{e3,S}(A+CDB),\\
\sigma_{e4,S}(A,B,C,\varepsilon) &:= \bigcup_{\|D\|<\varepsilon} \sigma_{e4,S}(A+CDB),
\end{aligned}$$

$$\sigma_{e5,S}(A,B,C,\varepsilon) := \bigcup_{\|D\|<\varepsilon} \sigma_{e5,S}(A+CDB).$$ ◇

Remark 3.6.5 *Let A, $S \in \mathscr{L}(X)$, $B \in \mathscr{L}(X,Y)$, $C \in \mathscr{L}(Z,X)$, $\lambda \in \mathbb{C}$ and $\varepsilon > 0$. If $SA = AS$, then we can write*

$$(\lambda S - A)(\lambda S - CDB) = ACDB + \lambda S(\lambda S - A - CDB) \tag{3.16}$$

and

$$(\lambda S - CDB)(\lambda S - A) = CDBA + \lambda(\lambda S - A - CDB)S. \tag{3.17}$$

Chapter 4

Perturbation of Unbounded Linear Operators by γ-Relative Boundedness

In Ref. [61], T. Diagana studied some sufficient conditions such that if S, T and K are three unbounded linear operators with S being a closed operator, then their algebraic sum $S+T+K$ is also a closed operator. The main focus of this chapter is to extend these results to the closable operator by adding a new concept of the gap and the γ-relative boundedness inspired by the work of A. Jeribi, B. Krichen and M. Zarai Dhahri in Ref. [92]. After that, we apply the obtained results to study the specific properties of some block operator matrices. All results of this chapter are given in Ref. [22].

4.1 SUM OF CLOSABLE OPERATORS

4.1.1 Norm Operators

The goal of this part consists in establishing some properties of closable operators in Banach spaces.

Theorem 4.1.1 *Let X and Y be two Banach spaces and let S, T and K be three operators from X into Y such that $\mathscr{D}(T) \subset \mathscr{D}(S) \subset \mathscr{D}(K)$. If*

(i) *there exist two constants a, $b > 0$ such that*

$$\|Sx\| \leq a\|x\| + b\|Tx\|, \;\; x \in \mathscr{D}(T),$$

(ii) *there exist two constants e, $d > 0$ such that $\mu = b(1+d) < 1$ and*

$$\|Kx\| \leq e\|x\| + d\|Sx\|, \;\; x \in \mathscr{D}(S).$$

Then,

$$S+T+K \text{ is closable if, and only if, } T \text{ is also closable.}$$

In this case, $\mathscr{D}(\overline{S+T+K}) = \mathscr{D}(\overline{T})$ and

$$\widehat{\delta}(S+T+K,T) \leq (1-\mu)^{-1}(\nu^2+\mu^2)^{\frac{1}{2}}, \tag{4.1}$$

where $\nu = (a(1+d)+e)$ $(\nu > 0)$ and $\widehat{\delta}(\cdot,\cdot)$ is the gap between two operators. ◇

Proof. Let $x \in \mathscr{D}(S+T+K) = \mathscr{D}(T)$. Then,

$$\|(S+T+K)x\| \leq (1+b)\|Tx\| + (a+e)\|x\| + d\big(a\|x\| + b\|Tx\|\big),$$

and hence,

$$\|(S+T+K)x\| \leq (a+e+da)\|x\| + (1+b+bd)\|Tx\|. \tag{4.2}$$

Similarly, we also have, for all $x \in \mathscr{D}(T)$,

$$\|(S+K)x\| \leq b\|Tx\| + (a+e)\|x\| + d(a\|x\| + b\|Tx\|)$$

and hence,

$$\|(S+K)x\| \leq (a+e+da)\|x\| + b(1+d)\|Tx\|. \tag{4.3}$$

Combining (4.2) and (4.3), it follows that for all $x \in \mathscr{D}(T)$, we have

$$\|(S+T+K)x\| \geq -(a+e+da)\|x\| + [1-b(1+d)]\|Tx\|,$$

which yields

$$\left[1-b(1+d)\right]\|Tx\| \leq \|(S+T+K)x\|+(a+e+da)\|x\|.$$

Setting $\mu = b(1+d)$ $(0<\mu<1)$ and $\nu = a+e+da$, we deduce that

$$\|Tx\| \leq (1-\mu)^{-1}\Big(\|(S+T+K)x\|+\nu\|x\|\Big). \tag{4.4}$$

Let $(x_n)_n$ be a sequence in $\mathscr{D}(S+T+K)=\mathscr{D}(T)$ such that $x_n \to 0$ and

$$(S+T+K)x_n \to \chi.$$

By using (4.4), it follows that $(Tx_n)_n$ is a Cauchy sequence in the Banach space Y and therefore, there exists $\psi_1 \in Y$ such that

$$Tx_n \to \psi_1.$$

Since T is closable, then

$$\psi_1 = 0.$$

It follows from (4.2) that

$$(S+T+K)x_n \to 0.$$

Then, $\chi = 0$ and $S+T+K$ is closable. Similarly, T is closable if $S+T+K$ is also closable.

Now, we may prove that

$$\mathscr{D}(\overline{S+T+K}) = \mathscr{D}(\overline{T}).$$

Let $x \in \mathscr{D}(\overline{S+T+K})$, then there exists a sequence $(x_n)_n$ such that $(x_n)_n$ is $(S+T+K)$-convergent to x. By using (4.4), we deduce that $(x_n)_n$ is T-convergent to x, then $x \in \mathscr{D}(\overline{T})$ so that

$$\mathscr{D}(\overline{S+T+K}) \subset \mathscr{D}(\overline{T}).$$

The opposite inclusion is proved similarly. In order to prove (4.1). Let $\varphi = (u,(\overline{S+T+K})u) \in G(\overline{S+T+K})$ with $\|\varphi\| = 1$ so that, we have

$$\|u\|^2+\|(\overline{S+T+K})u\|^2 = \|\varphi\|^2 = 1. \tag{4.5}$$

Since $G(\overline{S+T+K}) = \overline{G(S+T+K)}$, then there exists a sequence $(\varphi_n)_n$ in $G(S+T+K)$, such that

$$\varphi_n \to \varphi.$$

Hence,

$$\varphi_n = (u_n, (S+T+K)u_n),$$

where $u_n \in \mathscr{D}(S+T+K)$, $u_n \to u$ and

$$(S+T+K)u_n \to (\overline{S+T+K})u.$$

Now, we take $\psi_n = (u_n, Tu_n) \in G(T)$. By using (4.4), we have $(u_n)_n$ is T-convergent to u. Then,

$$\psi_n \to \psi_0 \in G(\overline{T}).$$

Combining both (4.3) and (4.4), we have

$$\begin{array}{rcl} \|\varphi_n - \psi_n\| & = & \|(S+K)u_n\| \\ & \leq & (1-\mu)^{-1}\Big(\nu\|u_n\| + \mu\|(S+T+K)u_n\|\Big). \end{array}$$

If $n \to \infty$, then we obtain

$$\|\varphi - \psi_0\| \leq (1-\mu)^{-1}\Big(\nu\|u\| + \mu\|(\overline{S+T+K})u\|\Big).$$

From Schwarz inequality and Eq. (4.5), it follows that

$$\|\varphi - \psi_0\| \leq (1-\mu)^{-1}(\nu^2+\mu^2)^{\frac{1}{2}}.$$

Hence,

$$\text{dist}(\varphi, G(\overline{T})) \leq (1-\mu)^{-1}(\nu^2+\mu^2)^{\frac{1}{2}}.$$

Now, since φ is an arbitrary element of the unit sphere of $G(\overline{S+T+K})$, then

$$\delta(S+T+K, T) \leq (1-\mu)^{-1}(\nu^2+\mu^2)^{\frac{1}{2}}.$$

Similarly, we can estimate the quantity $\delta(T, S+T+K)$. As a result, we obtain (4.1). Q.E.D.

Remark 4.1.1 *If the hypotheses of Theorem* 4.1.1 *are satisfied, then*

(i) *$S+T+K$ is closed if, and only if, T is also closed.*

(ii) *If $S+K \in \mathscr{L}(X,Y)$, then $\widehat{\delta}(S+T+K,T) \leq \|S+K\|$.* $\diamondsuit$

Theorem 4.1.2 *Let S, T and K be three operators from X into Y such that $\mathscr{D}(T) \subset \mathscr{D}(S) \subset \mathscr{D}(K)$ and*

(a) *there exist two constants a, $b > 0$*

$$\|Sx\| \leq a\|x\| + b\|Tx\|, \;\; x \in \mathscr{D}(T),$$

(b) *there exist two constants e, $d > 0$ such that*

$$\|Kx\| \leq e\|x\| + d\|Sx\|, \;\; x \in \mathscr{D}(S),$$

and

$$\widetilde{\gamma}(T)(1-\nu) > \mu, \tag{4.6}$$

where $\mu = b(1+d) < 1$, $\nu = a(1+d)+e$ $(\nu > 0)$ and $\widetilde{\gamma}(T)$ is given in (2.13). If T is a semi-Fredholm operator, then $S+T+K$ is semi-Fredholm and it satisfies the following properties:

(i) *$\alpha(S+T+K) \leq \alpha(T)$,*

(ii) *$\beta(S+T+K) \leq \beta(T)$, and*

(iii) *$i(S+T+K) = i(T)$.* $\diamondsuit$

Proof. Referring to (4.6), we have $\nu < 1$. Then, by virtue of Remark 4.1.1 (i), we have $S+T+K$ is closed. Let us introduce the new norm in $\mathscr{D}(T)$ by

$$[\| x\|] = (\nu+\varepsilon)\|x\| + (\mu+\varepsilon)\|Tx\|, \;\; x \in \mathscr{D}(T),$$

where ε is fixed and

$$0 < \varepsilon < \frac{\widetilde{\gamma}(T)(1-\nu)-\mu}{1+\widetilde{\gamma}(T)}. \tag{4.7}$$

It is clear that $[\|\cdot\|]$ and $\|\cdot\|_T$ (see Eq. (2.5)) are equivalent and, if we denote by $\widehat{X} = (\mathscr{D}(T), \; [\|\cdot\|])$, then $\widehat{X}$ is a Banach space and we have the following inequalities

$$\|\widehat{T}\| \leq (\mu+\nu)^{-1}$$

and

$$\|\widehat{S}+\widehat{K}\| \leq 1.$$

Since

$$\begin{cases} \alpha(\widehat{T}) = \alpha(T), \ \beta(\widehat{T}) = \beta(T), \ R(\widehat{T}) = R(T), \\ \beta(S\widehat{+T+}K) = \beta(S+T+K) \ \text{and} \ R(S\widehat{+T+}K) = R(S+T+K), \end{cases}$$

then we deduce that $\widehat{T} \in \Phi_{\pm}(X,Y)$ and, it is sufficient to show that $S\widehat{+T+}K \in \Phi_{\pm}(X,Y)$. It is clear that

$$\widetilde{\gamma}(\widehat{T}) = \frac{\widetilde{\gamma}(T)}{\nu+\varepsilon+(\mu+\varepsilon)\widetilde{\gamma}(T)}. \tag{4.8}$$

By using (4.7) and Eq.(4.8), we have $\gamma(\widehat{T}) > 1$. Thus,

$$\|\widehat{S}+\widehat{K}\| < \widetilde{\gamma}(\widehat{T}).$$

Accordingly, we have

$$S\widehat{+T+}K \in \Phi_{\pm}(X,Y),$$

$$\alpha(S\widehat{+T+}K) \leq \alpha(\widehat{T}),$$

and

$$\beta(S\widehat{+T+}K) \leq \beta(\widehat{T}),$$

which completes the proof of theorem. Q.E.D.

Corollary 4.1.1 *Let $\varepsilon > 0$ such that*

$$\|S+K\| \leq \frac{\widetilde{\gamma}(T)}{\sqrt{1+(\varepsilon\widetilde{\gamma}(T))^2}},$$

where $\widetilde{\gamma}(T)$ is given in (2.13). If T is a semi-Fredholm operator, then $S+T+K$ is semi-Fredholm and it satisfies the following properties:

(i) $\alpha(S+T+K) \leq \alpha(T)$,

(ii) $\beta(S+T+K) \leq \beta(T)$, and

(iii) $i(S+T+K) = i(T)$. ◊

4.1.2 Kuratowski Measure of Noncompactness

In this part, we provide some sufficient conditions for three closed operators S, T and K to have their algebaric sum also closed.

Theorem 4.1.3 *Let $\gamma(\cdot)$ be the Kuratowski measure of noncompactness and let S, T and K be three closed operators such that $\mathscr{D}(S) \subset \mathscr{D}(T) \subset \mathscr{D}(K)$. If*

(i) there exist two constants $a_s, b_s > 0$ such that

$$\gamma(S(\mathfrak{D})) \leq a_s\gamma(\mathfrak{D}) + b_s\gamma(T(\mathfrak{D})), \ \ \mathfrak{D} \subset \mathscr{D}(T),$$

(ii) there exist two constants a_k, $b_k > 0$ such that $b_k < 1$, $b_s(1+b_k) < 1$ and

$$\gamma(K(\mathfrak{D})) \leq a_k\gamma(\mathfrak{D}) + b_k\gamma(S(\mathfrak{D})), \ \ \mathfrak{D} \subset \mathscr{D}(T).$$

Then, the operator $S+T+K$ is closed. ◊

Proof. Let $\mathfrak{D} \subset \mathscr{D}(T)$. The fact that

$$\gamma(T(\mathfrak{D})) = \gamma((T+S+K-S-K)(\mathfrak{D})) \tag{4.9}$$

allows us to

$$\gamma((T+S+K-S-K)(\mathfrak{D})) \leq \gamma((T+S+K)(\mathfrak{D})) + \gamma((S+K)(\mathfrak{D})). \tag{4.10}$$

By using Eq. (4.9) and the relation (4.10), we have

$$\gamma((T+S+K)(\mathfrak{D})) \geq \gamma(T(\mathfrak{D})) - \gamma((S+K)(\mathfrak{D})). \tag{4.11}$$

Moreover,

$$\begin{aligned}
&\gamma((S+K)(\mathfrak{D})) \\
&\quad \leq \ \gamma(S(\mathfrak{D})) + \gamma(K(\mathfrak{D})) \\
&\quad \leq \ a_s\gamma(\mathfrak{D}) + b_s\gamma(T(\mathfrak{D})) + a_k\gamma(\mathfrak{D}) + b_k(a_s\gamma(\mathfrak{D}) + b_s\gamma(T(\mathfrak{D}))) \\
&\quad \leq \ a_s\gamma(\mathfrak{D}) + b_s\gamma(T(\mathfrak{D})) + a_k\gamma(\mathfrak{D}) + b_k a_s\gamma(\mathfrak{D}) + b_k b_s\gamma(T(\mathfrak{D})) \\
&\quad \leq \ (a_s + a_k + b_k a_s)\gamma(\mathfrak{D}) + b_s(1+b_k)\gamma(T(\mathfrak{D})),
\end{aligned}$$

we infer, from the last equation and Eq. (4.11), that

$$\begin{aligned}
&\gamma((T+S+K)(\mathfrak{D})) \\
&\quad \geq \gamma(T(\mathfrak{D})) - \gamma((S+K)(\mathfrak{D})) \\
&\quad \geq \gamma(T(\mathfrak{D})) - (a_s + a_k + b_k a_s)\gamma(\mathfrak{D}) - b_s(1+b_k)\gamma(T(\mathfrak{D})) \\
&\quad \geq -(a_s + a_k + b_k a_s)\gamma(\mathfrak{D}) + (1 - b_s(1+b_k))\gamma(T(\mathfrak{D})).
\end{aligned}$$

Then,

$$\gamma((T+S+K)(\mathfrak{D})) \geq -(a_s + a_k + b_k a_s)\gamma(\mathfrak{D}) + (1 - b_s(1+b_k))\gamma(T(\mathfrak{D})) \tag{4.12}$$

provided that $0 < 1 - b_s(1+b_k) < 1$ and $a_s + a_k + b_k a_s > 0$. Let $(x_n)_n$ be a sequence in $\mathscr{D}(T+S+K) = \mathscr{D}(T)$ such that $x_n \to x$ and

$$(T+S+K)x_n \to y.$$

This implies that $(x_n)_n$ and $((T+S+K)x_n)_n$ are relatively compact. Thus,

$$\gamma((x_n)_n) = \gamma(((T+S+K)x_n)_n) = 0.$$

By using (4.12), we deduce

$$\gamma(Tx_n) = 0,$$

and there exist a subsequence $x_{n_k} \to x \in \mathscr{D}(T)$ such that

$$Tx_{n_k} \to \alpha.$$

Since T is closed, then

$$Tx = \alpha.$$

In view of Theorem 2.6.2, we have $S+K$ is closed. Since

$$(S+K)x_{n_k} \to y - \alpha,$$

then $x \in \mathscr{D}(T)$ and

$$(T+S+K)x = y.$$

We conclude that the algebraic sum $T+S+K$ is a closed operator. Q.E.D.

4.2 BLOCK OPERATOR MATRICES

4.2.1 2×2 Block Operator Matrices

We denote by $\mathscr{L}$ the linear block operator matrices, acting on the Banach space $X \times Y$, of the form

$$\mathscr{L} = \begin{pmatrix} A & B \\ C & D \end{pmatrix},$$

where the operator A acts on X and has domain $\mathscr{D}(A)$, D is defined on $\mathscr{D}(D)$ and acts on the Banach space Y, and the intertwining operator B (resp. C) is defined on the domain $\mathscr{D}(B)$ (resp. $\mathscr{D}(C)$) and acts on X (resp. Y). One of the problems in the study of such operators is that in general $\mathscr{L}$ is not closed or even closable, even if its entries are closed.

Proposition 4.2.1 *Consider the block operator matrices acting on the Banach space $X \times Y$ by*

$$\mathscr{T} = \begin{pmatrix} A & 0 \\ 0 & D \end{pmatrix},$$

$$\mathscr{S} = \begin{pmatrix} 0 & B \\ 0 & 0 \end{pmatrix},$$

and

$$\mathscr{K} = \begin{pmatrix} 0 & 0 \\ C & 0 \end{pmatrix}.$$

If the operator

$$\mathscr{T} + \mathscr{S} + \mathscr{K}$$

is γ-diagonally dominant with bound δ, then

(i) $\mathscr{S}$ is $\mathscr{T}$-γ-bounded with $\mathscr{T}$-γ-bound δ, and

(ii) $\mathscr{K}$ is $\mathscr{S}$-γ-bounded with $\mathscr{S}$-γ-bound δ. ♢

Proof. Let $\pi_i(\cdot)$, $i = 1,2$ be denote the natural projections on X and Y, respectively.

(*i*) For $\mathfrak{D} \subset \mathscr{D}(\mathscr{K})$, we get

$$\gamma\left[\begin{pmatrix} 0 & B \\ 0 & 0 \end{pmatrix}\begin{pmatrix} \pi_1(\mathfrak{D}) \\ \pi_2(\mathfrak{D}) \end{pmatrix}\right] = \gamma(B(\pi_2(\mathfrak{D}))).$$

According to these assumptions, there exist two constants $a_B, b_B \geq 0$ such that

$$\gamma(B(\pi_2(\mathfrak{D}))) \leq a_B\gamma(\pi_2(\mathfrak{D})) + b_B\gamma(D(\pi_2(\mathfrak{D}))).$$

Since

$$\gamma(\pi_2(\mathfrak{D})) \leq \gamma(\pi_2)\gamma(\mathfrak{D}),$$

then

$$\gamma(B(\pi_2(\mathfrak{D}))) \leq a_B\gamma(\pi_2)\gamma(\mathfrak{D}) + b_B\gamma(D(\pi_2(\mathfrak{D}))).$$

Hence,

$$\begin{aligned} \gamma\left[\begin{pmatrix} 0 & B \\ 0 & 0 \end{pmatrix}\begin{pmatrix} \pi_1(\mathfrak{D}) \\ \pi_2(\mathfrak{D}) \end{pmatrix}\right] & \\ \leq \; & a_B\gamma(\pi_2)\gamma(\mathfrak{D}) + b_B\gamma(D(\pi_2(\mathfrak{D}))) \\ \leq \; & a_B\gamma(\pi_2)\gamma(\mathfrak{D}) + b_B \max\gamma\left[\begin{pmatrix} A & 0 \\ 0 & D \end{pmatrix}\begin{pmatrix} \pi_1(\mathfrak{D}) \\ \pi_2(\mathfrak{D}) \end{pmatrix}\right]. \end{aligned}$$

(*ii*) Also,

$$\gamma\left[\begin{pmatrix} 0 & 0 \\ C & 0 \end{pmatrix}\begin{pmatrix} \pi_1(\mathfrak{D}) \\ \pi_2(\mathfrak{D}) \end{pmatrix}\right] = \gamma(C(\pi_1(\mathfrak{D}))).$$

According to these assumptions, there exist two constants $a_C, b_C \geq 0$ such that

$$\gamma(C(\pi_1(\mathfrak{D}))) \leq a_C\gamma(\pi_1(\mathfrak{D})) + b_C\gamma(A(\pi_1(\mathfrak{D}))).$$

Since

$$\gamma(\pi_1(\mathfrak{D})) \leq \gamma(\pi_1)\gamma(\mathfrak{D}),$$

then

$$\gamma(C(\pi_1(\mathfrak{D}))) \leq a_C\gamma(\pi_1)\gamma(\mathfrak{D}) + b_C\gamma(A(\pi_1(\mathfrak{D}))).$$

Hence, we get

$$\begin{aligned}
\gamma\left[\begin{pmatrix} 0 & 0 \\ C & 0 \end{pmatrix}\begin{pmatrix} \pi_1(\mathfrak{D}) \\ \pi_2(\mathfrak{D}) \end{pmatrix}\right] \\
&\leq a_C\gamma(\pi_1)\gamma(\mathfrak{D}) + b_C\gamma(A(\pi_1(\mathfrak{D}))) \\
&\leq a_C\gamma(\pi_1)\gamma(\mathfrak{D}) + b_C \max\gamma\left[\begin{pmatrix} A & 0 \\ 0 & D \end{pmatrix}\begin{pmatrix} \pi_1(\mathfrak{D}) \\ \pi_2(\mathfrak{D}) \end{pmatrix}\right].
\end{aligned}$$

This completes the proof. Q.E.D.

4.2.2 3×3 *Block Operator Matrices*

In what follows, we are concerned with the block operator matrices

$$\mathcal{M} = \begin{pmatrix} A & B & C \\ D & E & F \\ G & H & K \end{pmatrix},$$

which acts on the product of Banach spaces $X \times Y \times Z$, where the entire of the matrix are linear operators. The operator A, having the domain $\mathscr{D}(A)$, acts on X, the operator E, having the domain $\mathscr{D}(E)$, acts on Y, and the operator K, having the domain $\mathscr{D}(K)$, acts on Z. Similarly, the operator B, H, C, F, D, and G are defined, respectively, by the domains $\mathscr{D}(B) \subset Y$ into X, $\mathscr{D}(H) \subset Y$ into Z, $\mathscr{D}(C) \subset Z$ into X, $\mathscr{D}(F) \subset Z$ into Y, $\mathscr{D}(D) \subset X$ into Y, and $\mathscr{D}(G) \subset X$ to Z. Let $\mathcal{M}$ be the operator

$$\left\{\begin{array}{l}
G(\mathcal{M}) = \left\{\left[\begin{pmatrix} u_1 \\ u_2 \\ u_3 \end{pmatrix}, \begin{pmatrix} v_1 \\ v_2 \\ v_3 \end{pmatrix}\right] \in (X \times Y \times Z)^2 \text{ such that } \begin{array}{l} v_1 = Au_1 + Bu_2 + Cu_3 \\ v_2 = Du_1 + Eu_2 + Fu_3 \\ v_3 = Gu_1 + Hu_2 + Ku_3 \end{array}\right\} \\[2ex]
\mathscr{D}(\mathcal{M}) = (\mathscr{D}(A) \cap \mathscr{D}(D) \cap \mathscr{D}(G)) \times (\mathscr{D}(B) \cap \mathscr{D}(E) \cap \mathscr{D}(H)) \times \\
\qquad (\mathscr{D}(C) \cap \mathscr{D}(F) \cap \mathscr{D}(K)).
\end{array}\right.$$

We denote by

$$\mathscr{T} = \begin{pmatrix} A & 0 & 0 \\ 0 & E & 0 \\ 0 & 0 & K \end{pmatrix},$$

$$\mathscr{S} = \begin{pmatrix} 0 & B & 0 \\ 0 & 0 & F \\ G & 0 & 0 \end{pmatrix},$$

and

$$\mathscr{K} = \begin{pmatrix} 0 & 0 & C \\ D & 0 & 0 \\ 0 & H & 0 \end{pmatrix}.$$

Then, it is clear that

$$\mathscr{M} = \mathscr{T} + \mathscr{S} + \mathscr{K}.$$

Remark 4.2.1 *It is clear that $\mathscr{T}$ is closed if, and only if, A, E and K are closed.* ◇

Theorem 4.2.1 (a) *Let us consider the following conditions*

(i) *B is E-γ-bounded with E-γ-bound δ_1,*

(ii) *F is K-γ-bounded with K-γ-bound δ_2, and*

(iii) *G is A-γ-bounded with A-γ-bound δ_3.*

Then, $\mathscr{S}$ is $\mathscr{T}$-γ-bounded with $\mathscr{T}$-γ-bound

$$\delta = \max\{\delta_1, \delta_2, \delta_3\}.$$

(b) *Let us consider the following conditions*

(i) *C is F-γ-bounded with F-γ-bound δ_1,*

(ii) *D is G-γ-bounded with G-γ-bound δ_2, and*

(iii) *H is B-γ-bounded with B-γ-bound δ_3.*

Then, $\mathscr{K}$ is $\mathscr{S}$-γ-bounded with $\mathscr{S}$-γ-bound $\delta = \max\{\delta_1, \delta_2, \delta_3\}$. ◇

Proof. (a) Let $\pi_i(\cdot)$, $i = 1, 2, 3$ be the natural projections on X, Y and Z, respectively, and let $\varepsilon > 0$. Consider the constants, a_B, a_G, a_F, b_B, b_G, and b_F such that

$$\delta_1 \leq b_1 \leq \delta_1 + \varepsilon,$$

$$\delta_2 \leq b_2 \leq \delta_2 + \varepsilon,$$

$$\delta_3 \leq b_3 \leq \delta_3 + \varepsilon,$$

$$a_1 = \{a_G \gamma(\pi_1), a_B \gamma(\pi_2), a_F \gamma(\pi_3)\},$$

and

$$a_2 = \{b_G, b_B, b_F\}.$$

For $\mathfrak{D} \subset \mathscr{D}(\mathscr{S})$, we get

$$\gamma\left[\begin{pmatrix} 0 & B & 0 \\ 0 & 0 & F \\ G & 0 & 0 \end{pmatrix}\begin{pmatrix} \pi_1(\mathfrak{D}) \\ \pi_2(\mathfrak{D}) \\ \pi_3(\mathfrak{D}) \end{pmatrix}\right] =$$

$$\max\{\gamma(G(\pi_1(\mathfrak{D}))), \gamma(B(\pi_2(\mathfrak{D}))), \gamma(F(\pi_3(\mathfrak{D})))\}.$$

Hence,

$$\begin{cases} \gamma(G(\pi_1(\mathfrak{D}))) \leq a_G \gamma(\pi_1(\mathfrak{D})) + b_G \gamma(A(\pi_1(\mathfrak{D}))), \\ \gamma(B(\pi_2(\mathfrak{D}))) \leq a_B \gamma(\pi_2(\mathfrak{D})) + b_B \gamma(E(\pi_2(\mathfrak{D}))), \\ \gamma(F(\pi_3(\mathfrak{D}))) \leq a_F \gamma(\pi_3(\mathfrak{D})) + b_F \gamma(K(\pi_3(\mathfrak{D}))). \end{cases}$$

Since,

$$\gamma(\pi_i(\mathfrak{D})) \leq \gamma(\pi_i)\gamma(\mathfrak{D}),\ i = 1, 2, 3,$$

then

$$\begin{cases} \gamma(G(\pi_1(\mathfrak{D}))) \leq a_G \gamma(\pi_1)\gamma(\mathfrak{D}) + b_G \gamma(A(\pi_1(\mathfrak{D}))), \\ \gamma(B(\pi_2(\mathfrak{D}))) \leq a_B \gamma(\pi_2)\gamma(\mathfrak{D}) + b_B \gamma(E(\pi_2(\mathfrak{D}))), \\ \gamma(F(\pi_3(\mathfrak{D}))) \leq a_F \gamma(\pi_3)\gamma(\mathfrak{D}) + b_F \gamma(K(\pi_3(\mathfrak{D}))). \end{cases}$$

Hence, we get

$$\gamma\left[\begin{pmatrix} 0 & B & 0 \\ 0 & 0 & F \\ G & 0 & 0 \end{pmatrix}\begin{pmatrix} \pi_1(\mathfrak{D}) \\ \pi_2(\mathfrak{D}) \\ \pi_3(\mathfrak{D}) \end{pmatrix}\right]$$

$$\leq \max a_2 \max\{\gamma(A(\pi_1(\mathfrak{D}))), \gamma(E(\pi_2(\mathfrak{D}))), \gamma(K(\pi_3(\mathfrak{D})))\}+$$

$$\gamma(\mathfrak{D}) \max a_1$$

$$\leq \gamma(\mathfrak{D}) \max a_1 + \gamma \left[\begin{pmatrix} A & 0 & 0 \\ 0 & E & 0 \\ 0 & 0 & K \end{pmatrix} \begin{pmatrix} \pi_1(\mathfrak{D}) \\ \pi_2(\mathfrak{D}) \\ \pi_3(\mathfrak{D}) \end{pmatrix} \right] \max a_2.$$

Since

$$\max\{b_G, b_B, b_F\} = \{\delta_1 + \varepsilon, \delta_2 + \varepsilon, \delta_3 + \varepsilon\} = \delta + \varepsilon,$$

then $\mathscr{S}$ is $\mathscr{T}$-γ-bounded with $\mathscr{T}$-γ-bound $< \delta$.

The proof of (b) may be checked in the same way as in the proof of the item (a). Q.E.D.

Chapter 5

Essential Spectra

In this chapter, we characterize the essential spectra of the closed, densely defined linear operators on Banach space.

5.1 CHARACTERIZATION OF ESSENTIAL SPECTRA

The goal of this section consists in establishing some preliminary results which will be needed in the sequel.

5.1.1 Characterization of Left and Right Weyl Essential Spectra

Theorem 5.1.1 *Let* $A \in \mathscr{C}(X)$. *Then,*

(*i*) $\lambda \notin \sigma_{ewl}(A)$ *if, and only if,* $\lambda - A \in \Phi_l(X)$ *and* $i(\lambda - A) \leq 0$.

(*ii*) $\lambda \notin \sigma_{ewr}(A)$ *if, and only if,* $\lambda - A \in \Phi_r(X)$ *and* $i(\lambda - A) \geq 0$.

(*iii*) $\sigma_{e5}(A) = \sigma_{ewl}(A) \bigcup \sigma_{ewr}(A)$. ◊

Proof. (*i*) If $\lambda - A \in \Phi_l(X)$ and $i(\lambda - A) \leq 0$. Then, by Lemma 2.2.7, $\lambda - A$ can be expressed in the form

$$\lambda - A = U + K,$$

where $K \in \mathscr{K}(X)$ and $U \in \mathscr{C}(X)$ such that $R(U)$ is closed and U is injective. Furthermore, by Lemma 2.2.10, $R(\lambda - A - K)$ is complemented. So, $\lambda \notin \sigma_l(A+K)$ and hence,

$$\lambda \notin \sigma_{ewl}(A).$$

Conversely, if $\lambda \notin \sigma_{ewl}(A)$, then there exists $K \in \mathscr{K}(X)$ such that $\lambda - A - K$ is left invertible. Hence, $\lambda - A - K$ is left Fredholm and

$$\alpha(\lambda - A - K) = 0.$$

Using Lemma 2.2.10, we get $\lambda - A \in \Phi_l(X)$ and

$$i(\lambda - A) \leq 0.$$

(ii) Let $\lambda \in \Phi_{rA}$ such that

$$i(\lambda - A) \geq 0.$$

Using Theorem 2.3.3, there exist $K \in \mathscr{K}(X)$ and $V \in \mathscr{C}(X)$ such that

$$\lambda - A = V + K, \;\; V(X) = X.$$

Thus, $\lambda - A - K$ is surjective. Furthermore, by Lemma 2.2.10, $N(\lambda - A - K)$ is complemented. So, $\lambda \notin \sigma_r(A+K)$. This gives

$$\lambda \notin \sigma_{ewr}(A).$$

(iii) From Proposition 3.1.1, $\lambda \notin \sigma_{e5}(A)$ if, and only if, $\lambda - A \in \Phi(X)$ and

$$i(\lambda - A) = 0.$$

Then, the result follows from *(i)* and *(ii)*. Q.E.D.

Remark 5.1.1 *Let $A \in \mathscr{C}(X)$. Then,*

$$\sigma_c(A) \subset \sigma_{ewl}(A) \bigcap \sigma_{ewr}(A).$$

In fact, let $\lambda \in \sigma_c(A)$, then $R(\lambda - A)$ is not closed (otherwise $\lambda \in \rho(A)$ see [137, Lemma 5.1 p. 179]). Therefore,

$$\lambda \in \sigma_{ewl}(A) \bigcap \sigma_{ewr}(A). \qquad \diamondsuit$$

Lemma 5.1.1 *Let $A \in \mathscr{C}(X)$ such that $0 \in \rho(A)$. Then, for $\lambda \neq 0$, we have*

$$\lambda \in \sigma_{ei}(A) \text{ if, and only if, } \frac{1}{\lambda} \in \sigma_{ei}(A^{-1}),\ i = l,\ r,\ wl,\ wr. \qquad \diamondsuit$$

Proof. For $\lambda \neq 0$, assume that $\frac{1}{\lambda} \in \Phi_{A^{-1}}$, then

$$\lambda^{-1} - A^{-1} \in \Phi^b(X).$$

The operator $\lambda - A$ can be written in the form

$$\lambda - A = -\lambda(\lambda^{-1} - A^{-1})A. \tag{5.1}$$

Since A is one-to-one and onto, then by Eq. (5.1), we have $N(\lambda - A)$ and $N(\lambda^{-1} - A^{-1})$ are isomorphic and, $R(\lambda - A)$ and $R(\lambda^{-1} - A^{-1})$ are isomorphic. This shows that $\lambda \in \Phi_{+A}$ $\big($resp. $\Phi_{-A}\big)$ and $R(\lambda - A)$ $\big($resp. $N(\lambda - A)\big)$ is complemented if, and only if, $\lambda^{-1} \in \Phi_{+A^{-1}}$ $\big($resp. $\Phi_{-A^{-1}}\big)$ and $R(\lambda^{-1} - A^{-1})$ $\big($resp. $N(\lambda^{-1} - A^{-1})\big)$ is complemented. Therefore,

$$\lambda \in \Phi^l_A \ \big(\text{resp. } \Phi^r_A\big) \text{ if, and only if, } \frac{1}{\lambda} \in \Phi^l_{A^{-1}} \ \big(\text{resp. } \Phi^r_{A^{-1}}\big).$$

Since $0 \in \rho(A)$, then

$$i(A) = 0.$$

Using both Theorem 2.2.5 and Eq. (5.1), we conclude that

$$\begin{aligned} i(\lambda - A) &= i(A) + i(\lambda^{-1} - A^{-1}) \\ &= i(\lambda^{-1} - A^{-1}). \end{aligned}$$

So, $\lambda \in \Phi^l_A$ $\big($resp. $\Phi^r_A\big)$ and $i(\lambda - A) \leq 0$ $\big($resp. $i(\lambda - A) \geq 0\big)$ if, and only if, $\frac{1}{\lambda} \in \Phi^l_{A^{-1}}$ $\big($resp. $\Phi^r_{A^{-1}}\big)$ and $i(\lambda^{-1} - A^{-1}) \leq 0$ $\big($resp. $i(\lambda^{-1} - A^{-1}) \geq 0\big)$. Hence,

$$\lambda \in \sigma_{ei}(A), \text{ if, and only if, } \frac{1}{\lambda} \in \sigma_{ei}(A^{-1}),\ i = l,\ r,\ wl,\ wr,$$

which completes the proof. Q.E.D.

Now, we discuss the left and right Weyl essential spectra by means of the polynomially Riesz operators. Let $A \in \mathscr{C}(X)$, with a non-empty resolvent set, we will give a refinement of the definition of the left Weyl essential spectrum and the right Weyl essential spectrum of A, respectively, by

$$\sigma_e^l(A) = \bigcap_{R \in \mathfrak{G}(X)} \sigma_l(A+R)$$

and

$$\sigma_e^r(A) = \bigcap_{R \in \mathfrak{G}(X)} \sigma_r(A+R),$$

where

$$\mathfrak{G}(X) = \{R \in \mathscr{L}(X) \text{ such that } (\lambda - A - R)^{-1}R \in \mathscr{PR}(X) \text{ for all } \lambda \in \rho(A+R)\}.$$

For $A \in \mathscr{L}(X)$, we observe that

$$\mathscr{K}(X) \subset \mathfrak{G}(X).$$

Indeed, let $K \in \mathscr{K}(X)$. If we take $P(z) = z$, then for all $\lambda \in \rho(A+K)$,

$$P((\lambda - A - K)^{-1}K) \in \mathscr{K}(X),$$

so $P((\lambda - A - K)^{-1}K) \in \mathscr{R}(X)$. Evidently, the following inclusions hold

$$\sigma_e^l(A) \subseteq \sigma_{ewl}(A)$$

and

$$\sigma_e^r(A) \subseteq \sigma_{ewr}(A).$$

Note that if R and R' are two operators in $\mathfrak{G}(X)$, we have not necessarily $R + R' \in \mathfrak{G}(X)$. So, we can note deduce the stability of $\sigma_{ewl}(A)$ and $\sigma_{ewr}(A)$ by perturbations of operators in the class $\mathfrak{G}(X)$. This bring us to introduce the following subset of $\mathfrak{G}(X)$. Let

$$\widetilde{\mathscr{I}}(X) = \{R \in \mathscr{L}(X) \text{ satisfying } SR \in \mathscr{PR}(X) \text{ for all } S \in \mathscr{L}(X)\}.$$

By [165, Theorem 2.18], $\widetilde{\mathscr{I}}(X)$ is equivalent to

$$\{R \in \mathscr{L}(X) \text{ satisfying } RS \in \mathscr{PR}(X) \text{ for all } S \in \mathscr{L}(X)\}.$$

Proposition 5.1.1 *Let $A \in \mathscr{C}(X)$. Then,*

(i) If $A \in \Phi_l(X)$ and $R \in \widetilde{\mathscr{I}}(X)$, then $A+R \in \Phi_l(X)$ and

$$i(A+R) = i(A).$$

(ii) If $A \in \Phi_r(X)$ and $R \in \widetilde{\mathscr{I}}(X)$, then $A+R \in \Phi_r(X)$ and

$$i(A+R) = i(A). \qquad \diamondsuit$$

Proof. Let $A \in \Phi_l(X)$, then by Lemma 2.2.8 (i), there exist $A_l \in \mathscr{L}(X)$ and $K \in \mathscr{K}(X)$ such that

$$A_l(A+R) = I - K + A_l R \text{ on } \mathscr{D}(A).$$

Since $R \in \widetilde{\mathscr{I}}(X)$, then $A_l R \in \mathscr{PR}(X)$. By applying Corollary 2.4.2, we get $I + A_l R \in \Phi^b(X)$ and

$$i(I + A_l R) = 0.$$

Since K is compact, then

$$A_l(A+R) \in \Phi(X)$$

and

$$R(A_l) \bigcup R(A_l R - K) \subset \mathscr{D}(A).$$

Using Lemma 2.2.8 (i), we have

$$A + R \in \Phi_l(X).$$

Since $A+R \in \mathscr{C}(X)$, we can make $\mathscr{D}(A+R) = \mathscr{D}(A)$ into a Banach space by introducing the norm

$$\|x\|_A = \|x\| + \|Ax\|.$$

Let $X_A = (\mathscr{D}(A), \|\cdot\|_A)$ be the Banach space for the graph norm $\|\cdot\|_A$. We can regard A as operator from X_A into X. This will be denoted by $\widehat{A}$. Clearly that $\widehat{A}+\widehat{R}$ and $\widehat{R}$ are bounded operators from X_A into X, $\widehat{K} \in \mathscr{L}(X_A)$ and $\widehat{A_l} \in \mathscr{L}(X, X_A)$. Clearly,

$$\widehat{A_l}\widehat{A} = I_{X_A} - \widehat{K}$$

is a Fredholm operator satisfying

$$i(\widehat{A_l}\widehat{A}) = 0$$

and

$$i(\widehat{A_l}(\widehat{A}+\widehat{R})) = i(A_l(A+R)).$$

It is clear $\widehat{A_l}$ is Fredholm if, and only if, $\widehat{A}$ is Fredholm if, and only if, $\widehat{A}+\widehat{R}$ is Fredholm, and therefore, by (2.6)

$$i(A+R) = i(\widehat{A}+\widehat{R}) = i(\widehat{A}) = i(A).$$

(ii) A same reasoning allows us to reach the result of (ii). Q.E.D.

Theorem 5.1.2 *Let $A \in \mathscr{C}(X)$. If $R \in \widetilde{\mathscr{I}}(X)$, then*

$$\sigma_{ewl}(A) = \sigma_{ewl}(A+R) \tag{5.2}$$

and

$$\sigma_{ewr}(A) = \sigma_{ewr}(A+R). \tag{5.3}$$

◇

Proof. Let $\lambda \in \Phi_{lA}$ such that $i(\lambda - A) \leq 0$. By Proposition 5.1.1 (i), we have

$$\lambda - A - R \in \Phi_l(X)$$

and

$$i(\lambda - A - R) = i(\lambda - A) \leq 0.$$

The opposite inclusion follows by symmetry and, we obtain (5.2). The proof of (5.3) may be checked in a similar way to that (5.2). Q.E.D.

Theorem 5.1.3 *Let A and $B \in \mathscr{C}(X)$ such that $\rho(A)\bigcap\rho(B) \neq \emptyset$. If, for some $\lambda \in \rho(A)\bigcap\rho(B)$, the operator $(\lambda - A)^{-1} - (\lambda - B)^{-1} \in \widetilde{\mathscr{I}}(X)$, then*

$$\sigma_{ewl}(A) = \sigma_{ewl}(B)$$

and

$$\sigma_{ewr}(A) = \sigma_{ewr}(B). \qquad \diamondsuit$$

Proof. Without loss of generality, we may assume that $\lambda = 0$. Hence, $0 \in \rho(A)\bigcap\rho(B)$ and therefore, we can write for $\mu \neq 0$

$$\mu - A = -\mu(\mu^{-1} - A^{-1})A.$$

Since A is one-to-one and onto, then

$$\alpha(\mu - A) = \alpha(\mu^{-1} - A^{-1})$$

and

$$R(\mu - A) = R(\mu^{-1} - A^{-1}).$$

This shows that $\mu \in \Phi_{lA}$ (resp. Φ_{rA}) if, and only if, $\mu^{-1} \in \Phi_{lA^{-1}}$ (resp. $\Phi_{rA^{-1}}$), in this case we have

$$i(\mu - A) = i(\mu^{-1} - A^{-1}).$$

Therefore, it follows from Proposition 5.1.1 that $\Phi_{lA} = \Phi_{lB}$ (resp. $\Phi_{rA} = \Phi_{rB}$) and

$$i(\mu - A) = i(\mu - B)$$

for each $\mu \in \Phi_{lA}$ (resp. Φ_{rA}), since $A^{-1} - B^{-1} \in \widetilde{\mathscr{I}}(X)$. Hence, to use both Lemma 5.1.1 and Theorem 5.1.1 makes us conclude that $\sigma_{ewl}(A) = \sigma_{ewl}(B)$ (resp. $\sigma_{ewr}(A) = \sigma_{ewr}(B)$), which completes the proof. Q.E.D.

5.1.2 Characterization of Left and Right Jeribi Essential Spectra

We begin with the following theorem which gives a refinement of the definition of the left Jeribi essential spectrum on L_1-spaces.

Theorem 5.1.4 *Let (Ω,Σ,μ) be an arbitrary positive measure space. Let A be a closed, densely defined and linear operator on $L_1(\Omega,\mu)$. Then,*

$$\sigma_j^l(A) = \sigma_{ewl}(A). \qquad \diamondsuit$$

Proof. We first claim that

$$\sigma_{ewl}(A) \subseteq \sigma_j^l(A).$$

Indeed, if $\lambda \notin \sigma_j^l(A)$, then there exists $F \in \mathscr{W}^*(L_1(\Omega,\mu))$ such that $\lambda \notin \sigma_l(A+F)$. Then, $\lambda - A - F \in \Phi_l(X)$ and

$$i(\lambda - A - F) \leq 0. \tag{5.4}$$

Since $F \in \mathscr{W}^*(L_1(\Omega,\mu))$, we have

$$(\lambda - A - F)^{-1}F \in \mathscr{W}^*(L_1(\Omega,\mu)).$$

Hence, by applying both Eqs. (2.8) and (2.9), we get

$$[(\lambda - A - F)^{-1}F]^2 \in \mathscr{K}(L_1(\Omega,\mu)).$$

So, $I + (\lambda - A - F)^{-1}F \in \Phi_l(X)$ and

$$i(I + (\lambda - A - F)^{-1}F) = 0. \tag{5.5}$$

By using the equality

$$\lambda - A = (\lambda - A - F)(I + (\lambda - A - F)^{-1}F),$$

together with Lemma 2.2.9 and both Eqs. (5.4) and (5.5), we have $\lambda - A \in \Phi_l(X)$ and

$$i(\lambda - A) = i(\lambda - A - F) \leq 0.$$

Thus, $\lambda \notin \sigma_{ewl}(A)$. This proves the claim. Moreover, since

$$\mathscr{K}(L_1(\Omega,\mu)) \subset \mathscr{W}^*(L_1(\Omega,\mu)),$$

we infer that

$$\sigma_j^l(A) \subset \sigma_{ewl}(A),$$

which completes the proof of theorem. Q.E.D.

The following theorem will allow us to show that the equalities between the right Weyl essential spectrum and the right Jeribi essential spectrum.

Theorem 5.1.5 *Let* (Ω, Σ, μ) *be an arbitrary positive measure space. Let* A *be a closed, densely defined and linear operator on* $L_1(\Omega, \mu)$. *Then, we have*

$$\sigma_j^r(A) = \sigma_{ewr}(A). \qquad \diamondsuit$$

Proof. Since $\mathscr{K}(L_1(\Omega, \mu)) \subset \mathscr{W}^*(L_1(\Omega, \mu))$, then

$$\sigma_j^r(A) \subset \sigma_{ewr}(A).$$

It remains to show that

$$\sigma_{ewr}(A) \subseteq \sigma_j^r(A).$$

For this, we consider $\lambda \notin \sigma_j^r(A)$, then there exists $F \in \mathscr{W}^*(L_1(\Omega, \mu))$ such that

$$\lambda \notin \sigma_l(A+F).$$

Since $F \in \mathscr{W}^*(L_1(\Omega, \mu))$, we have

$$(\lambda - A - F)^{-1}F \in \mathscr{W}^*(L_1(\Omega, \mu)).$$

Hence, by applying Eqs. (2.8) and (2.9), we get

$$[(\lambda - A - F)^{-1}F]^2 \in \mathscr{K}(L_1(\Omega, \mu)).$$

Using Lemma 2.2.5, we get

$$I + (\lambda - A - F)^{-1}F \in \Phi(X)$$

and

$$i(I + (\lambda - A - F)^{-1}F) = 0.$$

Writing

$$\lambda - A = (\lambda - A - F)(I + (\lambda - A - F)^{-1}F).$$

Since $\lambda \notin \sigma_r(A+F)$, then $\lambda - A - F$ is right Fredholm and

$$i(\lambda - A) \geq 0.$$

Therefore, by Lemma 2.2.9 (ii),

$$(\lambda - A - F)(I + (\lambda - A - F)^{-1}F) \in \Phi_r(X)$$

and

$$i(\lambda - A) = i(\lambda - A - F) \geq 0,$$

which completes the proof. Q.E.D.

Remark 5.1.2 (i) *Theorems 5.1.4 and 5.1.5 provide a unified definition of the left and right Weyl spectrum on L_1-spaces.*

(ii) *We have seen in the previous that $\sigma_j^l(A)$ and $\sigma_{ewl}(A)$ (resp. $\sigma_j^r(A)$ and $\sigma_{ewr}(A)$) are not equal. As we observed in Theorem 5.1.4 that*

$$\sigma_j^l(A) = \sigma_{ewl}(A)$$

and in Theorem 5.1.5 that

$$\sigma_j^r(A) = \sigma_{ewr}(A)$$

on L_1-spaces. 

As a consequence of Theorems 5.1.4 and 5.1.5, we obtain the equalities between $\sigma_{ewl}(A)$ (resp. $\sigma_{ewr}(A)$) and $\sigma_j^l(A)$ (resp. $\sigma_j^r(A)$) for $A \in \mathscr{C}(X)$ with X satisfies the Dunford-Pettis property.

Corollary 5.1.1 *If X satisfies the Dunford-Pettis property and A is a closed densely defined and linear operator on X, then*

$$\sigma_j^l(A) = \sigma_{ewl}(A),$$

and

$$\sigma_j^r(A) = \sigma_{ewr}(A).$$ ◇

Proof. The proof is obtained in the same way of the proof of Theorems 5.1.4 and 5.1.5. Q.E.D.

Remark 5.1.3 *If X satisfies the Dunford-Pettis property and A is a closed densely defined and linear operator on X, then*

$$\sigma_j^l(A) = \sigma_j^l(A+K)$$

and

$$\sigma_j^r(A) = \sigma_j^r(A+K)$$

for all $K \in \mathscr{W}(X)$. 

Recall that the relationship between the left-right Jeribi essential spectra and the left-right Weyl essential spectra is the inclusion strict on a Banach space. This suggests the following theorem.

Theorem 5.1.6 *If A is a closed densely defined and linear operator on* $L_p(\Omega, d\mu)$ *with* $p \in [1, \infty)$. *If* $\mathscr{W}^*(L_p(\Omega, \mu)) = \mathscr{S}(L_p(\Omega, \mu))$, *then*

$$\sigma_j^l(A) = \sigma_{ewl}(A),$$

and

$$\sigma_j^r(A) = \sigma_{ewr}(A).$$ ◇

Proof. The fact that $\mathscr{K}(L_p(\Omega, \mu)) \subset \mathscr{S}(L_p(\Omega, \mu))$, then

$$\sigma_j^l(A) \subset \sigma_{ewl}(A).$$

Conversely, let $\lambda \notin \sigma_j^l(A)$, then there exists $F \in \mathscr{S}(L_p(\Omega, \mu))$ such that

$$\lambda \notin \sigma_l(A+F).$$

This implies that $\lambda \in \Phi_{lA+F}$ and

$$i(\lambda - A - F) \leq 0.$$

Since $F \in \mathscr{S}(L_p(\Omega,\mu))$ and $\mathscr{S}(L_p(\Omega,\mu))$ is a two-sided ideal of $L_p(\Omega,\mu)$, we have

$$(\lambda - A - F)^{-1}F \in \mathscr{S}(L_p(\Omega,\mu)).$$

Hence, by applying Lemma 2.1.6, we get

$$[(\lambda - A - F)^{-1}F]^2 \in \mathscr{K}(L_p(\Omega,\mu)).$$

Using Lemma 2.2.5, we get

$$I + (\lambda - A - F)^{-1}F \in \Phi^b(X)$$

and

$$i(I + (\lambda - A - F)^{-1}F) = 0.$$

Using the equality $\lambda - A$ as

$$\lambda - A = (\lambda - A - F)(I + (\lambda - A - F)^{-1}F),$$

together with Lemma 2.2.9 (i), we have

$$(\lambda - A - F)(I + (\lambda - A - F)^{-1}F) \in \Phi_l(X)$$

and

$$i(\lambda - A) = i(\lambda - A - F) \leq 0.$$

Finally, the use of Theorem 5.1.1 shows

$$\lambda \notin \sigma_{ewl}(A),$$

which completes the proof of theorem. Q.E.D.

Corollary 5.1.2 *Let $A \in \mathscr{C}(L_p(\Omega,d\mu))$ with $p \in [1,\infty)$. Then,*

(i) If $A \in \Phi_l(L_p(\Omega,d\mu))$ and $R \in \mathscr{S}(L_p(\Omega,d\mu))$, then $A + R \in \Phi_l(L_p(\Omega,d\mu))$ and

$$i(A+R) = i(A).$$

(ii) If $A \in \Phi_r(L_p(\Omega,d\mu))$ and $R \in \mathscr{S}(L_p(\Omega,d\mu))$, then $A + R \in \Phi_r(L_p(\Omega,d\mu))$ and

$$i(A+R) = i(A). \qquad \diamondsuit$$

Theorem 5.1.7 *Let A, $B \in \mathscr{C}(L_p(\Omega, d\mu))$ such that $\rho(A) \bigcap \rho(B) \neq \emptyset$ ($p \in [1, \infty)$). If for some $\lambda \in \rho(A) \bigcap \rho(B)$, the operator $(\lambda - A)^{-1} - (\lambda - B)^{-1} \in \mathscr{S}(L_p(\Omega, \mu))$, then*

$$\sigma_j^l(A) = \sigma_j^l(B),$$

and

$$\sigma_j^r(A) = \sigma_j^r(B). \qquad \diamondsuit$$

Proof. Without loss of generality, we suppose that $\lambda = 0$. Hence, $0 \in \rho(A) \bigcap \rho(B)$. Therefore, we can write for $\mu \neq 0$

$$\mu - A = -\mu(\mu^{-1} - A^{-1})A.$$

Since, A is one-to-one and onto, then

$$\alpha(\mu - A) = \alpha(\mu^{-1} - A^{-1})$$

and

$$R(\mu - A) = R(\mu^{-1} - A^{-1}).$$

This shows that $\mu \in \Phi_{lA}$ (resp. Φ_{rA}) if, and only if, $\mu^{-1} \in \Phi_{lA^{-1}}$ (resp. $\Phi_{rA^{-1}}$), in this case, we have

$$i(\mu - A) = i(\mu^{-1} - A^{-1}).$$

Therefore, it follows, from Corollary 5.1.2, that $\Phi_{lA} = \Phi_{lB}$ (resp. $\Phi_{rA} = \Phi_{rB}$) and

$$i(\mu - A) = i(\mu - B)$$

for each $\mu \in \Phi_{lA}$ (resp. Φ_{rA}), since $A^{-1} - B^{-1} \in \mathscr{S}(L_p(\Omega, \mu))$. Hence, to use from Theorems 5.1.1 and 5.1.6 makes us conclude that $\sigma_j^l(A) = \sigma_j^l(B)$ (resp. $\sigma_j^r(A) = \sigma_j^r(B)$). Q.E.D.

5.2 STABILITY OF ESSENTIAL APPROXIMATE POINT SPECTRUM AND ESSENTIAL DEFECT SPECTRUM OF LINEAR OPERATOR

5.2.1 Stability of Essential Spectra

The reader interested in the results of this section may also refer to [28], which constitutes the real basis of our work. The purpose of this section, is to present the following useful stability of essential spectra.

Theorem 5.2.1 *Let X be a Banach space, A and B be two operators in $\mathscr{L}(X)$.*

(i) Assume that for each $\lambda \in \Phi_{+A}$, there exists a left Fredholm inverse $A_{\lambda l}$ of $\lambda - A$ such that $\|BA_{\lambda l}\| < 1$. Then,

$$\sigma_{e7}(A+B) = \sigma_{e7}(A).$$

(ii) Assume that for each $\lambda \in \Phi_{-A}$, there exists a right Fredholm inverse $A_{\lambda r}$ of $\lambda - A$ such that $\|A_{\lambda r}B\| < 1$. Then,

$$\sigma_{e8}(A+B) = \sigma_{e8}(A). \qquad \diamondsuit$$

Proof. Let

$$\mathscr{P}_\gamma(X) = \{A \in \mathscr{L}(X) \text{ such that } \gamma(A^n) < 1, \text{ for some } n > 0\},$$

where $\gamma(\cdot)$ is the Kuratowski measure of noncompactness. Since $\|BA_{\lambda l}\| < 1$ (resp. $\|A_{\lambda r}B\| < 1$), then $\gamma(BA_{\lambda l}) < 1$ (resp. $\gamma(A_{\lambda r}B) < 1$). So, $BA_{\lambda l} \in \mathscr{P}_\gamma(X)$ (resp. $A_{\lambda r}B \in \mathscr{P}_\gamma(X)$). Applying Proposition 2.6.2, we have

$$I - BA_{\lambda l} \in \Phi^b(X) \text{ and } i(I - BA_{\lambda l}) = 0, \tag{5.6}$$

and

$$I - A_{\lambda r}B \in \Phi^b(X) \text{ and } i(I - A_{\lambda r}B) = 0. \tag{5.7}$$

(i) Let $\lambda \notin \sigma_{e7}(A)$, then by Proposition 3.1.3 (i), we get

$$\lambda - A \in \Phi^b_+(X)$$

and

$$i(\lambda - A) \leq 0.$$

As, $A_{\lambda l}$ is a left Fredholm inverse of $\lambda - A$, then there exists $F \in \mathscr{F}^b(X)$ such that

$$A_{\lambda l}(\lambda - A) = I - F \text{ on } X. \tag{5.8}$$

By Eq. (5.8), the operator $\lambda - A - B$ can be written in the form

$$\lambda - A - B = \lambda - A - B(A_{\lambda l}(\lambda - A) + F) = (I - BA_{\lambda l})(\lambda - A) - BF. \tag{5.9}$$

According to the Eq. (5.6), we have $I - BA_{\lambda l} \in \Phi_+^b(X)$, and by using Theorem 2.2.5, we have $(I - BA_{\lambda l})(\lambda - A) \in \Phi_+^b(X)$ and

$$\begin{aligned} i[(I - BA_{\lambda l})(\lambda - A)] &= i(I - BA_{\lambda l}) + i(\lambda - A) \\ &= i(\lambda - A) \leq 0. \end{aligned}$$

By using both Eq. (5.9) and Lemma 2.3.1 (ii), we get

$$\lambda - A - B \in \Phi_+(X)$$

and

$$i(\lambda - A - B) = i(\lambda - A) \leq 0.$$

Hence, $\lambda \notin \sigma_{e7}(A + B)$. Conversely, let $\lambda \notin \sigma_{e7}(A + B)$, then by Proposition 3.1.3 (i), we have

$$\lambda - A - B \in \Phi_+^b(X)$$

and

$$i(\lambda - A - B) \leq 0.$$

Since $\|BA_{\lambda l}\| < 1$, and by Eq. (5.9), the operator $\lambda - A$ can be written in the form

$$\lambda - A = (I - BA_{\lambda l})^{-1}(\lambda - A - B + BF). \tag{5.10}$$

Using Lemma 2.3.1 (ii), we get

$$\lambda - A - B + BF \in \Phi_+^b(X)$$

and

$$i(\lambda - A - B + BF) = i(\lambda - A - B) \leq 0.$$

Since $I - BA_{\lambda l}$ is boundedly invertible, then by using Eq. (5.10), we have

$$\lambda - A \in \Phi_+^b(X)$$

and

$$i(\lambda - A) \leq 0.$$

This proves that $\lambda \notin \sigma_{e7}(A)$ and, we conclude that

$$\sigma_{e7}(A + B) = \sigma_{e7}(A).$$

(ii) Let $\lambda \notin \sigma_{e8}(A)$, then according to Proposition 3.1.3 (ii), we get

$$\lambda - A \in \Phi_-^b(X)$$

and

$$i(\lambda - A) \geq 0.$$

Since $A_{\lambda r}$ is a right Fredholm inverse of $\lambda - A$, then there exists $F \in \mathscr{F}^b(X)$ such that

$$(\lambda - A)A_{\lambda r} = I - F \text{ on } X. \tag{5.11}$$

By using Eq. (5.11), the operator $\lambda - A - B$ can be written in the form

$$\lambda - A - B = \lambda - A - ((\lambda - A)A_{\lambda r} + F)B = (\lambda - A)(I - A_{\lambda r}B) - FB. \tag{5.12}$$

A similar proof as (i), it suffices to replace $\Phi_+^b(\cdot)$, $\sigma_{e7}(\cdot)$, Eqs. (5.6), (5.9) and Lemma 2.3.1 (ii) by $\Phi_-^b(\cdot)$, $\sigma_{e8}(\cdot)$, Eqs. (5.7), (5.12) and Lemma 2.3.1 (iii), respectively. Hence, we show that

$$\sigma_{e8}(A + B) \subset \sigma_{e8}(A).$$

Conversely, let $\lambda \notin \sigma_{e8}(A + B)$, then by Proposition 3.1.3 (ii), we have

$$\lambda - A - B \in \Phi_-(X)$$

and

$$i(\lambda - A - B) \geq 0.$$

Since $\|A_{\lambda_r}B\| < 1$, and in view of Eq. (5.12), the operator $\lambda - A$ can be written in the form

$$\lambda - A = (\lambda - A - B + FB)(I - A_{\lambda_r}B)^{-1}. \qquad (5.13)$$

According to Lemma 2.3.1 (iii), we get

$$\lambda - A - B + FB \in \Phi_-^b(X)$$

and

$$i(\lambda - A - B + FB) = i(\lambda - A - B) \geq 0.$$

So, $I - A_{\lambda_r}B$ is boundedly invertible. Hence, by Eq. (5.13), we have $\lambda - A \in \Phi_-^b(X)$ and $i(\lambda - A) \geq 0$. This proves that $\lambda \notin \sigma_{e8}(A)$. Thus,

$$\sigma_{e8}(A + B) = \sigma_{e8}(A),$$

which completes the proof. Q.E.D.

5.2.2 Invariance of Essential Spectra

The purpose of this section is to show the following useful stability result for the essential approximate point spectrum and the essential defect spectrum of a closed, densely defined linear operator on a Banach space X. We begin with the following useful result.

Theorem 5.2.2 *Let X be a Banach space and let A, $B \in \Phi(X)$. Assume that there are A_0, $B_0 \in \mathscr{L}(X)$ and F_1, $F_2 \in \mathscr{PF}(X)$ such that*

$$AA_0 = I - F_1, \qquad (5.14)$$

and

$$BB_0 = I - F_2. \qquad (5.15)$$

(i) *If $0 \in \Phi_A \bigcap \Phi_B$, $A_0 - B_0 \in \mathscr{F}_+^b(X)$ and $i(A) = i(B)$, then*

$$\sigma_{e7}(A) = \sigma_{e7}(B).$$

(ii) *If $0 \in \Phi_A \bigcap \Phi_B$, $A_0 - B_0 \in \mathscr{F}_-^b(X)$ and $i(A) = i(B)$, then*

$$\sigma_{e7}(A) = \sigma_{e7}(B). \qquad \diamondsuit$$

Proof. Let $\lambda \in \mathbb{C}$, Eqs. (5.14) and (5.15) imply

$$(\lambda - A)A_0 - (\lambda - B)B_0 = F_1 - F_2 + \lambda(A_0 - B_0). \tag{5.16}$$

(i) Let $\lambda \notin \sigma_{e7}(B)$, then by Proposition 3.1.3 (i), we have

$$\lambda - B \in \Phi_+(X)$$

and

$$i(\lambda - B) \leq 0.$$

It is clearly that $B \in \mathscr{L}(X_B, X)$, where $X_B = (\mathscr{D}(B), \|\cdot\|_B)$ is a Banach space for the graph norm $\|\cdot\|_B$. We can regard B as operator from X_B into X. This will be denoted by $\widehat{B}$. Then,

$$\lambda - \widehat{B} \in \Phi_+^b(X_B, X)$$

and

$$i(\lambda - \widehat{B}) \leq 0.$$

Moreover, as $F_2 \in \mathscr{PF}(X)$, Eq. (5.15), and both Theorems 2.5.5 and 2.2.22 imply that $B_0 \in \Phi^b(X, X_B)$, and consequently,

$$(\lambda - \widehat{B})B_0 \in \Phi_+^b(X_B, X).$$

Using both Eq. (5.16) and Lemma 2.3.1, the operator $A_0 - B_0 \in \mathscr{F}_+^b(X)$ imply that

$$(\lambda - \widehat{A})A_0 \in \Phi_+^b(X) \tag{5.17}$$

and

$$i((\lambda - \widehat{A})A_0) = i((\lambda - \widehat{B})B_0). \tag{5.18}$$

A similar reasoning as before by combining Eq. (5.14), Theorem 2.5.5 and Theorem 2.2.22, we show that $A_0 \in \Phi^b(X, X_A)$, where $X_A = (\mathscr{D}(A), \|\cdot\|_A)$. Now, according to Theorem 2.2.3, we can write

$$A_0 T = I - F \text{ on } X_A, \tag{5.19}$$

where

$$T \in \mathscr{L}(X_A, X) \text{ and } F \in \mathscr{F}(X_A).$$

By Eq. (5.19), we have

$$(\lambda - \widehat{A})A_0T = \lambda - \widehat{A} - (\lambda - \widehat{A})F. \quad (5.20)$$

By using Eq. (5.19) and Theorem 2.2.22, we have $T \in \Phi^b(X_A, X)$. According to both Eq. (5.17) and Theorem 2.2.5, we have

$$(\lambda - \widehat{A})A_0T \in \Phi_+^b(X_A, X).$$

Using both Eq. (5.20) and Lemma 2.3.1 (ii), we get

$$\lambda - \widehat{A} \in \Phi_+^b(X_A, X).$$

Hence, Eq. (2.6) gives

$$\lambda - A \in \Phi_+(X). \quad (5.21)$$

As, F_1, $F_2 \in \mathscr{PF}(X)$, Eqs. (5.14), (5.15) and both Theorems 2.5.5 and 2.2.5 give

$$i(A) + i(A_0) = i(I - F_1) = 0$$

and

$$i(B) + i(B_0) = i(I - F_1) = 0.$$

Since $i(A) = i(B)$, then $i(A_0) = i(B_0)$. Using Eq. (5.18), we can write

$$i(\lambda - A) + i(A_0) = i(\lambda - B) + i(B_0).$$

Therefore,

$$i(\lambda - A) \leq 0. \quad (5.22)$$

Using both Eqs. (5.21) and (5.22), we get $\lambda \notin \sigma_{e7}(A)$. Hence,

$$\sigma_{e7}(A) \subset \sigma_{e7}(B).$$

The opposite inclusion follows by symmetry and, we obtain

$$\sigma_{e7}(A) = \sigma_{e7}(B).$$

(ii) The proof of (ii) may be checked in a similar way to that in (i). It suffices to replace $\sigma_{e7}(\cdot)$, $\Phi_+(\cdot)$, and $i(\cdot) \leq 0$ by $\sigma_{e8}(\cdot)$, $\Phi_-(\cdot)$, and $i(\cdot) \geq 0$, respectively. Q.E.D.

5.3 CONVERGENCE

In this section we gather some notation and results of each of convergence compactly and the convergence in generalized sense, that we need to prove our results later.

5.3.1 Convergence Compactly

In this section we investigate the essential spectra, $\sigma_{ei}(\cdot)$, $i = 1,\cdots,5$, of the sequence of linear operators in a Banach space X.

Theorem 5.3.1 *Let $(T_n)_n$ be a bounded linear operators mapping on X, and let T and B be two operators in $\mathscr{L}(X)$, $\lambda_0 \in \mathbb{C}$, and $\mathscr{U} \subseteq \mathbb{C}$ is open.*

(i) If $((\lambda_0 - T_n - B) - (\lambda_0 - T - B))_n$ converges to zero compactly, and $0 \in \mathscr{U}$, then there exists $n_0 \in \mathbb{N}$ such that, for all $n \geq n_0$,

$$\sigma_{ei}(\lambda_0 - T_n - B) \subseteq \sigma_{ei}(\lambda_0 - T - B) + \mathscr{U}, \quad i = 1,\cdots,5$$

and,

$$\delta\left(\sigma_{ei}(\lambda_0 - T_n - B), \sigma_{ei}(\lambda_0 - T - B)\right) = 0, \quad i = 1,\cdots,5,$$

where $\delta(\cdot)$ is the gap between two sets.

(ii) If $(\lambda_0 - T_n - B)_n$ converges to zero compactly, then there exists $n_0 \in \mathbb{N}$ such that for all $n \geq n_0$

$$\sigma_{ei}((\lambda_0 - T - B) + (\lambda_0 - T_n - B)) \subseteq \sigma_{ei}(\lambda_0 - T - B), \quad i = 1,\cdots,5$$

and,

$$\delta\left(\sigma_{ei}((\lambda_0 - T - B) + (\lambda_0 - T_n - B)), \sigma_{ei}(\lambda_0 - T - B)\right) = 0, \quad i = 1,\cdots,5.$$

Proof. (i) For $i = 1$. Assume that the assertion fails. Then, by passing to a subsequence, it may be deduced that, for each n, there exists $\lambda_n \in \sigma_{e1}(\lambda_0 - T_n - B)$ such that

$$\lambda_n \notin \sigma_{e1}(\lambda_0 - T - B) + \mathscr{U}.$$

It is clear (if necessary pass to a subsequence) that

$$\lim_{n \to +\infty} \lambda_n = \lambda$$

since $(\lambda_n)_n$ is bounded. This implies that

$$\lambda \notin \sigma_{e1}(\lambda_0 - T - B) + \mathscr{U}.$$

By using the fact that $0 \in \mathscr{U}$, we infer that $\lambda \notin \sigma_{e1}(\lambda_0 - T - B)$. Therefore,

$$\lambda - (\lambda_0 - T - B) \in \Phi_+^b(X).$$

Let $A_n = \lambda_n - \lambda + (\lambda_0 - T - B) - (\lambda_0 - T_n - B)$. Since A_n converges to zero compactly, writing

$$\lambda_n - (\lambda_0 - T_n - B) = \lambda - (\lambda_0 - T - B) + A_n$$

and according to Theorem 2.3.9, we infer that, there exists $n_0 \in \mathbb{N}$ such that for all $n \geq n_0$, we have $\lambda_n - (\lambda_0 - T_n - B) \in \Phi_+^b(X)$ and

$$\begin{aligned} i(\lambda_n - (\lambda_0 - T_n - B)) &= i(\lambda - (\lambda_0 - T - B) + A_n) \\ &= i(\lambda - (\lambda_0 - T - B)). \end{aligned}$$

So, $\lambda_n \notin \sigma_{e1}(\lambda_0 - T_n - B)$, which is a contradiction. Then, for all $n \geq n_0$, we have

$$\sigma_{e1}(\lambda_0 - T_n - B) \subseteq \sigma_{e1}(\lambda_0 - T - B) + \mathscr{U}.$$

Since $0 \in \mathscr{U}$, we obtain

$$\sigma_{e1}(\lambda_0 - T_n - B) \subseteq \sigma_{e1}(\lambda_0 - T - B).$$

Hence, in view of Remark 2.8.1 (i) (b), we get

$$\delta\big(\sigma_{e1}(\lambda_0 - T_n - B), \sigma_{e1}(\lambda_0 - T - B)\big) = 0, \ \text{ for all } n \geq n_0.$$

For $i = 2, 3, 4$, by using a similar reasoning as the proof of the case $i = 1$, by replacing $\sigma_{e1}(\cdot)$, and $\Phi_+^b(X)$ by $\sigma_{e2}(\cdot)$, or $\sigma_{e3}(\cdot)$, or $\sigma_{e4}(\cdot)$, and $\Phi_-^b(X)$, or $\Phi_-^b(X) \bigcup \Phi_+^b(X)$, or $\Phi^b(X)$, respectively, we get if

$$((\lambda_0 - T_n - B) - (\lambda_0 - T - B))_n$$

converges to zero compactly, and $0 \in \mathscr{U}$, then there exists $n_0 \in \mathbb{N}$ such that, for all $n \geq n_0$.

$$\sigma_{ei}(\lambda_0 - T_n - B) \subseteq \sigma_{ei}(\lambda_0 - T - B) + \mathscr{U}, \ \ i = 2, \cdots, 4.$$

Furthermore,

$$\delta\left(\sigma_{ei}(\lambda_0 - T_n - B), \sigma_{ei}(\lambda_0 - T - B)\right) = 0, \ \ i = 2, \cdots, 4.$$

For $i = 5$. Assume that the assertion fails. Then, by passing to a subsequence, it may be deduced that, for each n, there exists $\lambda_n \in \sigma_{e5}(\lambda_0 - T_n - B)$ such that

$$\lambda_n \notin \sigma_{e5}(\lambda_0 - T - B) + \mathscr{U}.$$

It is clear (if necessary pass to a subsequence) that

$$\lim_{n \to +\infty} \lambda_n = \lambda$$

since $(\lambda_n)_n$ is bounded, this implies that

$$\lambda \notin \sigma_{e5}(\lambda_0 - T - B) + \mathscr{U}.$$

Using the fact that $0 \in \mathscr{U}$, we have $\lambda \notin \sigma_{e5}(\lambda_0 - T - B)$ and therefore, $\lambda - (\lambda_0 - T - B) \in \Phi^b(X)$ and

$$i(\lambda - (\lambda_0 - T - B)) = 0.$$

Let $A_n = \lambda_n - \lambda + (\lambda_0 - T - B) - (\lambda_0 - T_n - B)$. Since $(A_n)_n$ converges to zero compactly, writing

$$\lambda_n - (\lambda_0 - T_n - B) = \lambda - (\lambda_0 - T - B) + A_n$$

and according to Theorem 2.3.9, we infer that, there exists $n_0 \in \mathbb{N}$ such that for all $n \geq n_0$, we have $\lambda_n - (\lambda_0 - T_n - B) \in \Phi^b(X)$ and

$$\begin{aligned} i(\lambda_n - (\lambda_0 - T_n - B)) &= i(\lambda - (\lambda_0 - T - B) + A_n) \\ &= i(\lambda - (\lambda_0 - T - B)) \\ &= 0. \end{aligned}$$

So, $\lambda_n \notin \sigma_{e5}(\lambda_0 - T_n - B)$, which is a contradiction. Then, for all $n \geq n_0$, we have

$$\sigma_{e5}(\lambda_0 - T_n - B) \subseteq \sigma_{e5}(\lambda_0 - T - B) + \mathscr{U}.$$

Since $0 \in \mathscr{U}$, we have

$$\sigma_{e5}(\lambda_0 - T_n - B) \subseteq \sigma_{e5}(\lambda_0 - T - B).$$

Hence, by Remark 2.8.1 (i) (b), we have for all $n \geq n_0$,

$$\delta\big(\sigma_{e5}(\lambda_0 - T_n - B), \sigma_{e5}(\lambda_0 - T - B)\big) = 0.$$

(ii) For $i = 1$. Let $\lambda \notin \sigma_{e1}(\lambda_0 - T - B)$. Then, $\lambda - (\lambda_0 - T - B) \in \Phi_+^b(X)$. Since $(\lambda_0 - T_n - B)_n$ converges to zero compactly and applying Theorem 2.3.9 to the operators $\lambda_0 - T - B$ and $\lambda_0 - T_n - B$, we prove that, there exists $n_0 \in \mathbb{N}$ such that

$$\lambda - (\lambda_0 - T - B) + (\lambda_0 - T_n - B) \in \Phi_+^b(X)$$

for all $n \geq n_0$. Hence,

$$\lambda \notin \sigma_{e1}((\lambda_0 - T - B) + (\lambda_0 - T_n - B)).$$

We conclude that

$$\sigma_{e1}((\lambda_0 - T - B) + (\lambda_0 - T_n - B)) \subseteq \sigma_{e1}(\lambda_0 - T - B).$$

Now applying Remark 2.8.1 (i) (b), we obtain for all $n \geq n_0$,

$$\delta\big(\sigma_{e1}((\lambda_0 - T - B) + (\lambda_0 - T_n - B)), \sigma_{e1}(\lambda_0 - T - B)\big) = 0.$$

For $i = 2, 3, 4$, by using a similar reasoning as the proof of the case $i = 1$, by replacing $\sigma_{e1}(\cdot)$, and $\Phi_+^b(X)$ by $\sigma_{e2}(\cdot)$, or $\sigma_{e3}(\cdot)$, or $\sigma_{e4}(\cdot)$, and $\Phi_-^b(X)$, or $\Phi_-^b(X) \bigcup \Phi_+^b(X)$, or $\Phi^b(X)$, respectively, we get if $(\lambda_0 - T_n - B)_n$ converges to zero compactly, then there exists $n_0 \in \mathbb{N}$ such that for all $n \geq n_0$,

$$\sigma_{ei}((\lambda_0 - T + B) + (\lambda_0 - T_n - B)) \subseteq \sigma_{ei}(\lambda_0 - T - B).$$

Furthermore, for all $n \geq n_0$, we have

$$\delta\big(\sigma_{ei}((\lambda_0 - T - B) + (\lambda_0 - T_n - B)), \sigma_{ei}(\lambda_0 - T - B)\big) = 0.$$

For $i = 5$. Let $\lambda \notin \sigma_{e5}(\lambda_0 - T - B)$. Then, $\lambda - (\lambda_0 - T - B) \in \Phi^b(X)$ and

$$i(\lambda - (\lambda_0 - T - B)) = 0.$$

Since $(\lambda_0 - T_n - B)_n$ converges to zero compactly and by applying the Theorem 2.3.9 to the both operators $\lambda_0 - T - B$ and $\lambda_0 - T_n - B$, we prove that, there exists $n_0 \in \mathbb{N}$ such that for all $n \geq n_0$, we have

$$\lambda - (\lambda_0 - T - B) + (\lambda_0 - T_n - B) \in \Phi^b(X).$$

Hence,

$$\lambda \notin \sigma_{e5}((\lambda_0 - T - B) + (\lambda_0 - T_n - B)).$$

We conclude that

$$\sigma_{e5}((\lambda_0 - T - B) + (\lambda_0 - T_n - B)) \subseteq \sigma_{e5}(\lambda_0 - T - B).$$

Now applying Remark 2.8.1 (i) (b), we have for all $n \geq n_0$

$$\delta\big(\sigma_{e5}((\lambda_0 - T - B) + (\lambda_0 - T_n - B)), \sigma_{e5}(\lambda_0 - T - B)\big) = 0. \quad \text{Q.E.D.}$$

The following results may be found in [28].

Theorem 5.3.2 *Let $(T_n)_n$ be a sequence in $\mathscr{L}(X)$ and let T be a bounded linear operator on X.*

(i) If $(T_n)_n$ converges to T compactly, $\mathscr{U} \subseteq \mathbb{C}$ is open and $0 \in \mathscr{U}$, then there exists $n_0 \in \mathbb{N}$ such that, for every $n \geq n_0$

$$\sigma_{e7}(T_n) \subseteq \sigma_{e7}(T) + \mathscr{U}, \tag{5.23}$$

and

$$\sigma_{e8}(T_n) \subseteq \sigma_{e8}(T) + \mathscr{U}. \tag{5.24}$$

(ii) If $(T_n)_n$ converges to zero compactly, then there exists $n_0 \in \mathbb{N}$ such that, for every $n \geq n_0$

$$\sigma_{e7}(T + T_n) \subseteq \sigma_{e7}(T), \tag{5.25}$$

and

$$\sigma_{e8}(T + T_n) \subseteq \sigma_{e8}(T). \tag{5.26}$$

$\diamondsuit$

Proof. (i) The proof by contradiction, assume that the inclusion is fails. Then, by passing to a subsequence (if necessary) it may be assumed that, for each n, there exists $\lambda_n \in \sigma_{e7}(T_n)$ such that

$$\lambda_n \notin \sigma_{e7}(T) + \mathscr{U},$$

since $(\lambda_n)_n$ is bounded, we suppose (if necessary pass to a subsequence) that

$$\lim_{n \to +\infty} \lambda_n = \lambda,$$

which implies that $\lambda \notin \sigma_{e7}(T) + \mathscr{U}$. Using the fact that $0 \in \mathscr{U}$, we have $\lambda \notin \sigma_{e7}(T)$. Therefore, $\lambda - T \in \Phi_+^b(X)$ and

$$i(\lambda - T) \leq 0.$$

In other hand $(\lambda_n - T_n) - (\lambda - T) \xrightarrow{c} 0$, which implies, by Theorem 2.3.9 (i) and (iv), that $\lambda_n - T_n \in \Phi_+^b(X)$, and

$$i(\lambda_n - T_n) = i(\lambda - T) \leq 0.$$

Hence, $\lambda_n \notin \sigma_{e7}(T_n)$, which is a contradiction. So, the inclusion (5.23) holds. The statement (5.24) for the essential defect spectrum can be proved similarly.

(ii) Let $\lambda \notin \sigma_{e7}(T)$. Then, $\lambda - T \in \Phi_+^b(X)$, and

$$i(\lambda - T) \leq 0.$$

Since $(T_n)_n$ converge to 0 compactly, then if we apply Theorem 2.3.9 (i) and (iv), we obtain that there exists $n_0 \in \mathbb{N}$ such that for all $n > n_0$,

$$(\lambda - T) - T_n = \lambda - (T + T_n) \in \Phi_+(X),$$

and

$$i(\lambda - T) = i\big(\lambda - (T + T_n)\big) \leq 0,$$

which implies that $\lambda \notin \sigma_{e7}(T + T_n)$. Hence, the inclusion (5.25) is valid. For the inclusion (5.26), the proof is similarly. Q.E.D.

From the above result, we have the following.

Corollary 5.3.1 *Let T be a closed linear operator and let $(T_n)_n$ be a sequence of closed linear operators on X such that $\rho(T_n) \bigcap \rho(T) \neq \emptyset$, and let $\eta \in \rho(T_n) \bigcap \rho(T)$. If $(T_n - \eta)^{-1} - (T - \eta)^{-1}$ converges to zero compactly, then there exists $n_0 \in \mathbb{N}$ such that, for all $n \geq n_0$*

$$\sigma_{e7}(T_n - \eta) \subseteq \sigma_{e7}(T - \eta),$$

and

$$\sigma_{e8}(T_n - \eta) \subseteq \sigma_{e8}(T - \eta). \qquad \diamond$$

Proof. If we put $K_n = (T_n - \eta)^{-1} - (T - \eta)^{-1}$, then

$$(T_n - \eta)^{-1} = (T - \eta)^{-1} + K_n,$$

and K_n converges to zero compactly. From the inclusions (5.25) and (5.26), respectively, we have

$$\sigma_{e7}\big((T_n - \eta)^{-1}\big) = \sigma_{e7}\big((T - \eta)^{-1} + K_n\big) \subseteq \sigma_{e7}\big((T - \eta)^{-1}\big),$$

and

$$\sigma_{e8}\big((T_n - \eta)^{-1}\big) = \sigma_{e8}\big((T - \eta)^{-1} + K_n\big) \subseteq \sigma_{e8}\big((T - \eta)^{-1}\big).$$

Then, by using Lemma 3.1.1, we have

$$\sigma_{e7}(T_n - \eta) \subseteq \sigma_{e7}(T - \eta),$$

and

$$\sigma_{e8}(T_n - \eta) \subseteq \sigma_{e8}(T - \eta),$$

which completes the proof. Q.E.D.

5.3.2 Convergence in the Generalized Sense

The first main result is embodied in the following theorem.

Theorem 5.3.3 *Let $(T_n)_n$ be a sequence of closed linear operators mapping on Banach spaces X and let $T \in \mathscr{C}(X)$, and let B and L be two operators in $\mathscr{L}(X)$, $\lambda_0 \in \mathbb{C}$ such that $(T_n)_n$ converges in the generalized sense to T, $\lambda_0 \in \rho(T+B)$, and $\mathscr{U} \subseteq \mathbb{C}$ is open.*

(i) If $0 \in \mathscr{U}$, then there exists $n_0 \in \mathbb{N}$ such that, for every $n \geq n_0$, we have

$$\sigma_{ei}(\lambda_0 - T_n - B) \subseteq \sigma_{ei}(\lambda_0 - T - B) + \mathscr{U}, \ \ i = 1, \cdots, 5. \tag{5.27}$$

Furthermore,

$$\delta\Big(\sigma_{ei}(\lambda_0 - T_n - B), \sigma_{ei}(\lambda_0 - T - B)\Big) = 0, \ \ i = 1, \cdots, 5.$$

(ii) Let $i = 1, \cdots, 5$. Then, there exist $\varepsilon > 0$ and $n \in \mathbb{N}$ such that, for all $\|L\| < \varepsilon$, we have

$$\sigma_{ei}(\lambda_0 - T_n - B + L) \subseteq \sigma_{ei}(\lambda_0 - T - B) + \mathscr{U}, \text{ for all } n \geq n_0. \tag{5.28}$$

Furthermore, for $i = 1, \cdots, 5$, we have

$$\delta\Big(\sigma_{ei}(\lambda_0 - T_n - B + L), \sigma_{ei}(\lambda_0 - T - B)\Big) = \delta\Big(\sigma_{ei}(\lambda_0 - T - B + L), \sigma_{ei}(\lambda_0 - T - B)\Big) = 0. \quad \diamondsuit$$

Proof. (i) For $i = 1$. In view of both $B - \lambda_0$ is a bounded operator and $\lambda_0 \in \rho(T+B)$ and according to Theorem 2.8.2 (i) and (iii), the sequence $(\lambda_0 - T_n - B)_n$ converges in the generalized sense to $\lambda_0 - T - B$, and $\lambda_0 \in \rho(T_n + B)$ for a sufficiently large n and $(\lambda_0 - T_n - B)^{-1}$ converges to $(\lambda_0 - T - B)^{-1}$. Now, to prove Eq. (5.27), it suffices to prove the existence of $n_0 \in \mathbb{N}$, such that for all $n \geq n_0$, we have

$$\sigma_{e1}\left((\lambda_0 - T_n - B)^{-1}\right) \subseteq \sigma_{e1}\left((\lambda_0 - T - B)^{-1}\right) + \mathscr{U}. \tag{5.29}$$

In first step, by an indirect proof, we suppose that the (5.29) does not hold, and for each $n \in \mathbb{N}$ there exists $\lambda_n \in \sigma_{e1}\left((\lambda_0 - T_n - B)^{-1}\right)$ such that

$$\lambda_n \notin \sigma_{e1}\left((\lambda_0 - T - B)^{-1}\right) + \mathscr{U}.$$

It is clear (if necessary pass to a subsequence) that

$$\lim_{n \to +\infty} \lambda_n = \lambda$$

since $(\lambda_n)_n$ is bounded. This implies that

$$\lambda \notin \sigma_{e1}\left((\lambda_0 - T - B)^{-1}\right) + \mathscr{U}.$$

In view of $0 \in \mathscr{U}$, we have $\lambda \notin \sigma_{e1}\left((\lambda_0 - T - B)^{-1}\right)$. Therefore,

$$\lambda - (\lambda_0 - T - B)^{-1} \in \Phi_+^b(X).$$

Applying Theorem 2.8.2 (ii), we conclude that

$$\widehat{\delta}(\lambda_n - (\lambda_0 - T_n - B)^{-1}, \lambda - (\lambda_0 - T - B)^{-1}) \to 0, \text{ as } n \to \infty.$$

Let $\delta = \widetilde{\gamma}(\lambda - (\lambda_0 - T - B)^{-1}) > 0$. Then, there exists $N \in \mathbb{N}$ such that, for all $n \geq N$, we have

$$\widehat{\delta}(\lambda_n - (\lambda_0 - T_n - B)^{-1}, \lambda - (\lambda_0 - T - B)^{-1}) \leq \frac{\delta}{\sqrt{1 + \delta^2}}.$$

According to Theorem 2.8.1 (iv), we infer that

$$\lambda_n - (\lambda_0 - T_n - B)^{-1} \in \Phi_+^b(X).$$

Then, we obtain

$$\lambda_n \notin \sigma_{e1}((\lambda_0 - T_n - B)^{-1}),$$

which contradicts our assumption. Hence, (5.29) holds.

Now, let $\lambda \in \sigma_{e1}(\lambda_0 - T_n - B)$, then

$$\frac{1}{\lambda} \in \sigma_{e1}((\lambda_0 - T_n - B)^{-1}).$$

In view of (5.29), we conclude that

$$\frac{1}{\lambda} \in \sigma_{e1}((\lambda_0 - T - B)^{-1}) + \mathscr{U}. \tag{5.30}$$

Since $0 \in \mathscr{U}$, then (5.30) implies that

$$\frac{1}{\lambda} \in \sigma_{e1}((\lambda_0 - T - B)^{-1}).$$

We have to prove

$$\lambda \in \sigma_{e1}(\lambda_0 - T - B) + \mathscr{U}. \tag{5.31}$$

We will proceed by contradiction, we suppose that

$$\lambda \notin \sigma_{e1}(\lambda_0 - T - B) + \mathscr{U}.$$

The fact that

$$0 \in \mathscr{U}$$

implies that $\lambda \notin \sigma_{e1}(\lambda_0 - T - B)$ and so,

$$\frac{1}{\lambda} \notin \sigma_{e1}((\lambda_0 - T - B)^{-1})$$

which contradicts our assumption. This proves Eq. (5.31). Therefore, (5.27) holds. Since $\mathscr{U}$ is an arbitrary neighborhood of 0 and by using the relation (5.27), we have

$$\sigma_{e1}(\lambda_0 - T_n - B) \subseteq \sigma_{e1}(T + B - \lambda_0),$$

for all $n \geq n_0$. Hence, by Remark 2.8.1 (i) (b), we have for all $n \geq n_0$,

$$\delta\Big(\sigma_{e1}(\lambda_0 - T_n - B), \sigma_{e1}(\lambda_0 - T - B)\Big) = \delta\Big(\overline{\sigma_{e1}(\lambda_0 - T_n - B)}, \overline{\sigma_{e1}(\lambda_0 - T + B)}\Big) = 0.$$

This ends the proof for $i = 1$.

For $i = 2, 3, 4$, by using a similar reasoning as the proof of ((i) for $i = 1$), by replacing $\sigma_{e1}(\cdot)$, and $\Phi_+(X)$ by $\sigma_{e2}(\cdot)$, or $\sigma_{e3}(\cdot)$, or $\sigma_{e4}(\cdot)$, and $\Phi_-(X)$, or $\Phi_-(X) \bigcup \Phi_+(X)$, or $\Phi(X)$, respectively, we get

$$\sigma_{ei}(\lambda_0 - T_n - B) \subseteq \sigma_{ei}(\lambda_0 - T - B) + \mathscr{U}.$$

Furthermore, for all $n \geq n_0$, we have

$$\delta\Big(\sigma_{ei}(\lambda_0 - T_n - B), \sigma_{ei}(\lambda_0 - T - B)\Big) = 0.$$

For $i = 5$. In view of both $\lambda_0 - B$ is a bounded operator and $\lambda_0 \in \rho(T + B)$, and according to Theorem 2.8.2 (i) and (iii), the sequence

$$(\lambda_0 - T_n - B)_n$$

converges in the generalized sense to $\lambda_0 - T - B$, and $\lambda_0 \in \rho(T_n + B)$ for a sufficiently large n and $(\lambda_0 - T_n - B)^{-1}$ converges to $(\lambda_0 - T - B)^{-1}$. Now, to prove Eq. (5.27), it suffices to prove there exist $n_0 \in \mathbb{N}$, such that for all $n \geq n_0$, we have

$$\sigma_{e5}\left((\lambda_0 - T_n - B)^{-1}\right) \subseteq \sigma_{e5}\left((\lambda_0 - T - B)^{-1}\right) + \mathscr{U}. \tag{5.32}$$

In first step, by an indirect proof, we suppose that the inclusion (5.32) does not hold, and for each $n \in \mathbb{N}$, there exists $\lambda_n \in \sigma_{e5}\left((\lambda_0 - T_n - B)^{-1}\right)$ such that

$$\lambda_n \notin \sigma_{e5}\left((\lambda_0 - T - B)^{-1}\right) + \mathscr{U}.$$

It is clear (if necessary pass to a subsequence) that

$$\lim_{n \to +\infty} \lambda_n = \lambda$$

since $(\lambda_n)_n$ is bounded. This implies that

$$\lambda \notin \sigma_{e5}\left((\lambda_0 - T - B)^{-1}\right) + \mathscr{U}.$$

In view of $0 \in \mathscr{U}$, we have

$$\lambda \notin \sigma_{e5}\left((\lambda_0 - T - B)^{-1}\right).$$

Therefore, $\lambda - (\lambda_0 - T - B)^{-1} \in \Phi^b(X)$ and

$$i(\lambda - (\lambda_0 - T - B)^{-1}) = 0.$$

Applying Theorem 2.8.2 (ii), we conclude that

$$\widehat{\delta}(\lambda_n - (\lambda_0 - T_n - B)^{-1}, \lambda - (\lambda_0 - T - B)^{-1}) \to 0 \; as \; n \to \infty.$$

Let $\delta = \widetilde{\gamma}(\lambda - (\lambda_0 - T - B)^{-1}) > 0$. Then, there exists $N \in \mathbb{N}$ such that, for all $n \geq N$, we have

$$\widehat{\delta}(\lambda_n - (\lambda_0 - T_n - B)^{-1}, \lambda - (\lambda_0 - T - B)^{-1}) \leq \frac{\delta}{\sqrt{1+\delta^2}}.$$

According to Theorem 2.8.1 (iv), we infer

$$\lambda_n - (\lambda_0 - T_n - B)^{-1} \in \Phi^b(X)$$

and

$$i(\lambda_n - (\lambda_0 - T_n - B)^{-1}) = i(\lambda - (\lambda_0 - T - B)^{-1}) = 0.$$

Then, we obtain $\lambda_n \notin \sigma_{e5}((\lambda_0 - T_n - B)^{-1})$, which is a contradicts. Hence, (5.32) holds. Now, if $\lambda \in \sigma_{e5}(\lambda_0 - T_n - B)$, then

$$\frac{1}{\lambda} \in \sigma_{e5}((\lambda_0 - T_n - B)^{-1}).$$

In view of Eq. (5.32), we conclude that

$$\frac{1}{\lambda} \in \sigma_{e5}((\lambda_0 - T - B)^{-1}) + \mathscr{U}. \tag{5.33}$$

Since $0 \in \mathscr{U}$, then (5.33) implies that

$$\frac{1}{\lambda} \in \sigma_{e5}\left((\lambda_0 - T - B)^{-1}\right).$$

We have to prove

$$\lambda \in \sigma_{e5}(\lambda_0 - T - B) + \mathscr{U}. \tag{5.34}$$

We will proceed by contradiction, we suppose that

$$\lambda \notin \sigma_{e5}(\lambda_0 - T - B) + \mathscr{U}.$$

The fact that $0 \in \mathscr{U}$ implies that

$$\lambda \notin \sigma_{e5}(\lambda_0 - T - B)$$

and so, $\frac{1}{\lambda} \notin \sigma_{e5}((\lambda_0 - T - B)^{-1})$ which contradicts our assumption. This proves Eq. (5.34). Therefore (5.27) holds. Since $\mathscr{U}$ is an arbitrary neighborhood of 0 and by using (5.27), we have for all $n \geq n_0$

$$\sigma_{e5}(\lambda_0 - T_n - B) \subseteq \sigma_{e5}(\lambda_0 - T - B).$$

Hence, by Remark 2.8.1 (i) (b), we have for all $n \geq n_0$,

$$\delta\Big(\sigma_{e5}(\lambda_0 - T_n - B), \sigma_{e5}(\lambda_0 - T - B)\Big) = \delta\Big(\overline{\sigma_{e5}(\lambda_0 - T_n - B)}, \overline{\sigma_{e5}(\lambda_0 - T - B)}\Big) = 0.$$

This ends the proof of (i).

(ii) For $i = 1$. Since $\lambda_0 \in \rho(T+B)$, then $(T+B-\lambda_0)^{-1}$ exists and bounded. Let $\varepsilon_1 = \frac{1}{\|(\lambda_0 - T - B)^{-1}\|}$ and let $L \in \mathscr{L}(X)$ such that $\|L\| < \varepsilon_1$. Then,

$$\|L(\lambda_0 - T - B)^{-1}\| < 1.$$

According to Theorem 2.8.2 (i), the sequence $(\lambda_0 - T_n - B + L)_n$ converges in the generalized sense to $\lambda_0 - T - B + L$, and the Neumann series

$$\sum_{k=0}^{\infty}(-L(\lambda_0 - T - B)^{-1})^k$$

converges to $(I + L(\lambda_0 - T - B)^{-1})^{-1}$ and

$$\|(I + L(\lambda_0 - T - B)^{-1})^{-1}\| < \frac{1}{1 - \|L\|\|(\lambda_0 - T - B)^{-1}\|}.$$

Since

$$\lambda_0 - T - B + L = (I + L(\lambda_0 - T - B)^{-1})(\lambda_0 - T - B),$$

then $\lambda_0 \in \rho(T + B + L)$. Now, applying (i) for $i = 1$, we deduce that there exists $n_0 \in \mathbb{N}$ such that for all $n \geq n_0$, we have

$$\sigma_{e1}(\lambda_0 - T_n - B + L) \subseteq \sigma_{e1}(\lambda_0 - T - B + L) + \mathscr{U}.$$

Let $\lambda \notin \sigma_{e1}(\lambda_0 - T - B)$. Then,

$$\lambda - (\lambda_0 - T - B) \in \Phi_+(X).$$

By applying Theorem 2.2.15, there exists $\varepsilon_2 > 0$ such that for $\|L\| < \varepsilon_2$, one has

$$\lambda - (\lambda_0 - T - B) - L \in \Phi_+(X)$$

and, this implies that $\lambda \notin \sigma_{e1}(\lambda_0 - T - B + L)$. From above and if we take $\varepsilon = \min(\varepsilon_1, \varepsilon_2)$, then for all $\|L\| < \varepsilon$, there exists $n_0 \in \mathbb{N}$ such that for all $n \geq n_0$, we have

$$\sigma_{e1}(\lambda_0 - T_n - B + L) \subseteq \sigma_{e1}(\lambda_0 - T - B) + \mathscr{U}.$$

Therefore, Eq. (5.28) holds for $i = 1$. Since $0 \in \mathscr{U}$, then we have

$$\delta\big(\sigma_{e1}(\lambda_0 - T_n - B + L), \sigma_{e1}(\lambda_0 - T - B)\big) = 0$$

and

$$\delta\big(\sigma_{e1}(\lambda_0 - T - B + L), \sigma_{e1}(\lambda_0 - T - B)\big) = 0.$$

For $i = 2, 3, 4$. By using a similar reasoning as the proof of (ii) for $i = 1$, by replacing $\sigma_{e1}(\cdot)$, and $\Phi_+(X)$ by $\sigma_{e2}(\cdot)$, or $\sigma_{e3}(\cdot)$, or $\sigma_{e4}(\cdot)$, and $\Phi_-(X)$, or $\Phi_-(X) \bigcup \Phi_+(X)$, or $\Phi(X)$, respectively, we get there exist $\varepsilon > 0$ and $n \in \mathbb{N}$ such that, for all $\|L\| < \varepsilon$, we have for all $n \geq n_0$,

$$\sigma_{ei}(\lambda_0 - T_n - B + L) \subseteq \sigma_{ei}(\lambda_0 - T - B) + \mathscr{U}.$$

Furthermore,

$$\delta\Big(\sigma_{ei}(\lambda_0 - T_n - B + L), \sigma_{ei}(\lambda_0 - T - B)\Big) = \delta\Big(\sigma_{ei}(\lambda_0 - T - B + L), \sigma_{ei}(\lambda_0 - T - B)\Big) = 0.$$

For $i=5$. Since $\lambda_0 \in \rho(T+B)$, then $(\lambda_0-T-B)^{-1}$ exists and bounded. We put $\varepsilon_1 = \frac{1}{\|(\lambda_0-T-B)^{-1}\|}$. Let $L \in \mathscr{L}(X)$ such that $\|L\| < \varepsilon_1$. This implies

$$\|L(\lambda_0 - T - B)^{-1}\| < 1.$$

By according to Theorem 2.8.2 (i), we have $(\lambda_0 - T_n - B + L)_n$ converges in the generalized sense to $\lambda_0 - T - B + L$, and the Neumann series

$$\sum_{k=0}^{\infty} (-L(\lambda_0 - T - B)^{-1})^k$$

converges to $(I + L(\lambda_0 - T - B)^{-1})^{-1}$ and

$$\|(I + L(\lambda_0 - T - B)^{-1})^{-1}\| < \frac{1}{1 - \|L\| \|(\lambda_0 - T - B)^{-1}\|}.$$

Since

$$\lambda_0 - T - B + L = (I + L(\lambda_0 - T - B)^{-1})(\lambda_0 - T - B),$$

then $\lambda_0 \in \rho(T + B + L)$. Now, applying (i) for $i = 5$, we deduce that there exists $n_0 \in \mathbb{N}$ such that for all $n \geq n_0$, we have

$$\sigma_{e5}(\lambda_0 - T_n - B + L) \subseteq \sigma_{e5}(\lambda_0 - T - B + L) + \mathscr{U}.$$

Let $\lambda \notin \sigma_{e5}(\lambda_0 - T - B)$. Then, $\lambda - (\lambda_0 - T - B) \in \Phi(X)$ and

$$i(\lambda - (\lambda_0 - T - B)) = 0.$$

By applying Theorem 2.2.15, there exists $\varepsilon_2 > 0$ such that for $\|L\| < \varepsilon_2$, one has

$$\lambda - (\lambda_0 - T - B) - L \in \Phi(X)$$

and

$$i(\lambda - (\lambda_0 - T - B - L)) = i(\lambda - (\lambda_0 - T - B) = 0.$$

This implies that $\lambda \notin \sigma_{e5}(\lambda_0 - T - B + L)$. From above and if we take $\varepsilon = \min(\varepsilon_1, \varepsilon_2)$, then for all $\|L\| < \varepsilon$, there exists $n_0 \in \mathbb{N}$ such that

$$\sigma_{e5}(\lambda_0 - T_n - B + L) \subseteq \sigma_{e5}(\lambda_0 - T - B) + \mathscr{U},$$

for all $n \geq n_0$. Since $0 \in \mathscr{U}$, then we have

$$\delta\big(\sigma_{e5}(\lambda_0 - T_n - B + L), \sigma_{e5}(\lambda_0 - T - B)\big) =$$
$$\delta\big(\sigma_{e5}(\lambda_0 - T - B + L), \sigma_{e5}(\lambda_0 - T - B)\big) = 0.$$

Therefore, (ii) holds for $i = 5$. Q.E.D.

The next main results in the rest of this section is embodied in the following theorem, when we discuss and study the essential approximate point spectrum, and the essential defect spectrum of a sequence of closed linear operators perturbed by a bounded operator, and converges in the generalized sense to a closed linear operator in a Banach space.

Theorem 5.3.4 *Let X be Banach space, let $(T_n)_n$ be a sequence of closed linear operators which converges in the generalized sense, in $\mathscr{C}(X)$, to a closed linear operator T, and let B be a bounded linear operator on X such that $\rho(T+B) \neq \emptyset$. If $\lambda_0 \in \rho(T+B)$, then*

(i) there exists $n_0 \in \mathbb{N}$ such that, for every $n \geq n_0$, we have

$$\sigma_{e7}(T_n + B - \lambda_0) \subseteq \sigma_{e7}(T + B - \lambda_0) + \mathscr{U} \tag{5.35}$$

and

$$\sigma_{e8}(T_n + B - \lambda_0) \subseteq \sigma_{e8}(T + B - \lambda_0) + \mathscr{U}, \tag{5.36}$$

where $\mathscr{U} \subset \mathbb{C}$ is an open containing 0. In particular, for all $n \geq n_0$

$$\delta\Big(\sigma_{e7}(T_n + B - \lambda_0), \sigma_{e7}(T + B - \lambda_0)\Big) =$$
$$\delta\Big(\sigma_{e8}(T_n + B - \lambda_0), \sigma_{e8}(T + B - \lambda_0)\Big) = 0,$$

(ii) there exist $\varepsilon > 0$, and $n_0 \in \mathbb{N}$, such that, for all $S \in \mathscr{L}(X)$, and $\|S\| < \varepsilon$, we have for all $n \geq n_0$,

$$\sigma_{e7}(T_n + B + S - \lambda_0) \subseteq \sigma_{e7}(T + B - \lambda_0) + \mathscr{U},$$

and

$$\sigma_{e8}(T_n + B + S - \lambda_0) \subseteq \sigma_{e8}(T + B - \lambda_0) + \mathscr{U},$$

where $\mathscr{U} \subset \mathbb{C}$ is an open containing 0. In particular, for all $n \geq n_0$

$$\delta\Big(\sigma_{e7}(T_n + B + S - \lambda_0), \sigma_{e7}(T + B - \lambda_0)\Big) =$$
$$\delta\Big(\sigma_{e8}(T_n + B + S - \lambda_0), \sigma_{e8}(T + B - \lambda_0)\Big) = 0. \quad \diamondsuit$$

Proof. For (i), before proof, we make some preliminary observations. Since $T_n \xrightarrow{g} T$, then by Theorem 2.8.2 (i),

$$T_n + B - \lambda_0 \xrightarrow{g} T + B - \lambda_0.$$

Furthermore, we have $(T + B - \lambda_0)^{-1} \in \mathcal{L}(X)$, which implies, according to Theorem 2.8.2 (iii), that $\lambda_0 \in \rho(T_n + B)$ for a sufficiently large n and $(T_n + B - \lambda_0)^{-1}$ converges to $(T + B - \lambda_0)^{-1}$. We recall that the essential approximate point spectrum of a bounded operator is compact. But this property is not valid for the case of unbounded operators, for this reason, using the compactness of $\sigma_{e7}((T + B - \lambda_0)^{-1})$ because $(T + B - \lambda_0)^{-1}$ is bounded, as a first step, we will prove the existence of $n_0 \in \mathbb{N}$, such that for all $n \geq n_0$, we have

$$\sigma_{e7}((T_n + B - \lambda_0)^{-1}) \subseteq \sigma_{e7}((T + B - \lambda_0)^{-1}) + \mathcal{U}. \tag{5.37}$$

The proof by contradiction. Suppose that (5.37) does not hold. Then, by studying a subsequence (if necessary), we may assume that, for each n, there exists

$$\lambda_n \in \sigma_{e7}((T_n + B - \lambda_0)^{-1})$$

such that

$$\lambda_n \notin \sigma_{e7}((T + B - \lambda_0)^{-1}) + \mathcal{U}.$$

Since $(\lambda_n)_n$ is bounded, we may assume (if necessary pass to a subsequence) that

$$\lim_{n \to +\infty} \lambda_n = \lambda,$$

which implies that

$$\lambda \notin \sigma_{e7}(T + B - \lambda_0)^{-1} + \mathcal{U}.$$

Since $0 \in \mathcal{U}$, then

$$\lambda \notin \sigma_{e7}((T + B - \lambda_0)^{-1}).$$

Therefore,

$$\lambda - (T + B - \lambda_0)^{-1} \in \Phi_+^b(X)$$

and

$$i\big(\lambda - (T+B-\lambda_0)^{-1}\big) \leq 0.$$

In other hand, $\lambda_n - (T_n+B-\lambda_0)^{-1}$ converges to $\lambda - (T+B-\lambda_0)^{-1}$, we deduce that

$$\widehat{\delta}\big(\lambda_n - (T_n+B-\lambda_0)^{-1}, \lambda - (T+B-\lambda_0)^{-1}\big) \to 0$$

as $n \to \infty$. Let $\delta = \widetilde{\gamma}\big(\lambda - (T+B-\lambda_0)^{-1}\big) > 0$. Then, there exists $n_0 \in \mathbb{N}$ such that, for all $n \geq n_0$, we have

$$\widehat{\delta}\big(\lambda_n - (T_n+B-\lambda_0)^{-1}, \lambda - (T+B-\lambda_0)^{-1}\big) \leq \frac{\delta}{\sqrt{1+\delta^2}}.$$

By using Theorem 2.8.1 (iv), we infer that

$$\lambda_n - (T_n+B-\lambda_0)^{-1} \in \Phi_+^b(X)$$

and

$$i\big(\lambda_n - (T_n+B-\lambda_0)^{-1}\big) \leq 0.$$

Furthermore, there exists $b > 0$ such that

$$\widehat{\delta}\big(\lambda_n - (T_n+B-\lambda_0)^{-1}, \lambda - (T+B-\lambda_0)^{-1}\big) < b,$$

which implies

$$i\big(\lambda_n - (T_n+B-\lambda_0)^{-1}\big) = i\big(\lambda - (T+B-\lambda_0)^{-1}\big) \leq 0.$$

Then, we obtain

$$\lambda_n \notin \sigma_{e7}\big((T_n+B-\lambda_0)^{-1}\big),$$

which is a contradiction. Hence, (5.37) holds. Now, we assume that $\lambda \in \sigma_{e7}(T_n+B-\lambda_0)$, then

$$\frac{1}{\lambda} \in \sigma_{e7}\big((T_n+B-\lambda_0)^{-1}\big).$$

By using the inclusion (5.37), we have

$$\frac{1}{\lambda} \in \sigma_{e7}\big((T+B-\lambda_0)^{-1}\big) + \mathscr{U},$$

which implies that

$$\frac{1}{\lambda} \in \sigma_{e7}\big((T+B-\lambda_0)^{-1}\big)$$

because $0 \in \mathscr{U}$, Now, we claim that $\lambda \in \sigma_{e7}(T+B-\lambda_0)+\mathscr{U}$. In fact, let us assume that $\lambda \notin \sigma_{e7}(T+B-\lambda_0)+\mathscr{U}$. The fact that $0 \in \mathscr{U}$ implies that $\lambda \notin \sigma_{e7}(T+B-\lambda_0)$ and so,

$$\frac{1}{\lambda} \notin \sigma_{e7}\big((T_n+B-\lambda_0)^{-1}\big)$$

which is a contradiction. This proves the claim and so,

$$\lambda \in \sigma_{e7}(T+B-\lambda_0)+\mathscr{U}.$$

This implies that (5.35) holds. Since $\mathscr{U}$ is an arbitrary neighborhood of 0 and by using (5.35), we get

$$\sigma_{e7}(T_n+B-\lambda_0) \subseteq \sigma_{e7}(T+B-\lambda_0),$$

for all $n \geq n_0$. Hence, by Remark 2.8.1 (ii),

$$\delta\Big(\sigma_{e7}(T_n+B-\lambda_0), \sigma_{e7}(T+B-\lambda_0)\Big) = \delta\Big(\overline{\sigma_{e7}(T_n+B-\lambda_0)}, \overline{\sigma_{e7}(T+B-\lambda_0)}\Big) = 0,$$

for all $n \geq n_0$. With the same procession as we do for (5.35), and using Proposition 3.1.3 (ii), the inclusion (5.36) yields. Therefore, (i) holds.

(ii) Since S is bounded, it is clear by using Theorem 2.8.2 (i), that the sequence

$$A_n = T_n+B+S-\lambda_0$$

converges in the generalized sense to the operator $A = T+B+S-\lambda_0$, then we need, for applying (i), to prove that

$$\rho(T+B) \subset \rho(T+B+S).$$

Let $\lambda_0 \in \rho(T+B)$, for $S \in \mathscr{L}(X)$ such that

$$\|S\| < \frac{1}{\|(T+B-\lambda_0)^{-1}\|} = \varepsilon_1,$$

we have

$$\|S(T+B-\lambda_0)^{-1}\| < 1,$$

which gives that $\left(I+S(T+B-\lambda_0)^{-1}\right)^{-1}$ exists and bounded, when the existence is given by the convergence of the Neumann serie

$$\sum_{k=0}^{\infty}\left(-S(T+B-\lambda_0)^{-1}\right)^k,$$

and the boundedness is immediately from the inequality

$$\|\left(I+S(T+B-\lambda_0)^{-1}\right)^{-1}\| < \frac{1}{1-\|S\|\|(T+B-\lambda_0)^{-1}\|},$$

which implies that the operator

$$\left((T+B-\lambda_0)+S\right)^{-1} = (T+B-\lambda_0)^{-1}\left(I+S(T+B-\lambda_0)^{-1}\right)^{-1}$$

exists and bounded, then $0 \in \rho(T+B+S-\lambda_0)$. Now, applying (i) to A_n and A, we deduce that there exists $n_0 \in \mathbb{N}$, such that

$$\sigma_{e7}(T_n+B+S-\lambda_0) \subseteq \sigma_{e7}(T+B+S-\lambda_0)+\mathscr{U},$$

for all $n \geq n_0$, where $\mathscr{U} \subset \mathbb{C}$ is an open set containing 0. Now, we will prove, by contradiction, the following

$$\sigma_{e7}(T+B+S-\lambda_0) \subseteq \sigma_{e7}(T+B-\lambda_0).$$

Let $\lambda \notin \sigma_{e7}(T+B-\lambda_0)$, then $\lambda-(T+B-\lambda_0) \in \Phi_+(X)$ and

$$i\big(\lambda-(T+B-\lambda_0)\big) \leq 0.$$

From Theorem 2.2.16, we deduce that there exists $\varepsilon_2 > 0$ such that for $\|S\| < \varepsilon_2$, one has $\lambda-(T+B+S-\lambda_0) \in \Phi_+(X)$ and

$$i\big(\lambda-(T+B+S-\lambda_0)\big) = i\big(\lambda-(T+B-\lambda_0)\big) \leq 0.$$

This implies that $\lambda \notin \sigma_{e7}(T+B+S-\lambda_0)$. Then, by transitivity

$$\sigma_{e7}(T_n+B+S-\lambda_0) \subseteq \sigma_{e7}(T+B-\lambda_0).$$

From what has been mentioned and if we take $\varepsilon = \min(\varepsilon_1, \varepsilon_2)$, then for all $\|S\| < \varepsilon$, there exists $n_0 \in \mathbb{N}$ such that

$$\sigma_{e7}(T_n + B + S - \lambda_0) \subseteq \sigma_{e7}(T + B - \lambda_0) + \mathscr{U},$$

for all $n \geq n_0$. With the same procedure of the previous prove, the inclusion concerned in $\sigma_{e8}(\cdot)$ yields. Since $\mathscr{U}$ is an arbitrary neighborhood of the origin, then we have

$$\delta\Big(\sigma_{e7}(T_n + S + B - \lambda_0), \sigma_{e7}(T + B + S - \lambda_0)\Big) = \delta\Big(\sigma_{e7}(T + S + B - \lambda_0), \sigma_{e7}(T + B - \lambda_0)\Big) = 0.$$

Therefore, (ii) holds. Q.E.D.

A particular case is obtained from Theorem 5.3.4, if we replace B by 0, and λ_0 by 0, which requires that $0 \in \rho(T)$, then we have the following corollary.

Corollary 5.3.2 *Let $(T_n)_n$ be a sequence of closed linear operators and T be a closed operator such that $(T_n)_n \xrightarrow{g} T$, we suppose that $0 \in \rho(T)$. If $\mathscr{U} \subset \mathbb{C}$ is open and $0 \in \mathscr{U}$, then there exists $n_0 \in \mathbb{N}$ such that, for every $n \geq n_0$, we have*

$$\sigma_{e7}(T_n) \subseteq \sigma_{e7}(T) + \mathscr{U},$$

and

$$\sigma_{e8}(T_n) \subseteq \sigma_{e8}(T) + \mathscr{U}.$$

In particular, for all $n \geq n_0$

$$\delta\Big(\sigma_{e7}(T_n), \sigma_{e7}(T)\Big) = \delta\Big(\sigma_{e8}(T_n), \sigma_{e8}(T)\Big) = 0. \qquad \diamondsuit$$

Chapter 6

S-Essential Spectra of Closed Linear Operator on a Banach Space

In this chapter, we give a characterization of S-essential spectra of the closed, densely define linear operator A on a Banach space X.

6.1 S-ESSENTIAL SPECTRA

6.1.1 Characterization of S-Essential Spectra

In the next, we will suppose that S is not invertible.

Lemma 6.1.1 *Let $A \in \mathscr{C}(X)$ and $S \in \mathscr{L}(X)$. Then,*

(i) $\sigma_{e5,S}(A) = \bigcap_{K \in \mathscr{F}_0(X)} \sigma_S(A+K) = \bigcap_{K \in \mathscr{F}(X)} \sigma_S(A+K)$,

(ii) $\sigma_{eap,S}(A) = \bigcap_{K \in \mathscr{F}_0(X)} \sigma_{ap,S}(A+K) = \bigcap_{K \in \mathscr{F}_+(X)} \sigma_{ap,S}(A+K)$, *and*

(iii) $\sigma_{e\delta,S}(A) = \bigcap_{K \in \mathscr{F}_0(X)} \sigma_{\delta,S}(A+K) = \bigcap_{K \in \mathscr{F}_-(X)} \sigma_{\delta,S}(A+K)$,

where $\mathscr{F}_0(X)$ stands for the ideal of finite rank operators and $\sigma_{\delta,S}(A) := \{\lambda \in \mathbb{C}$ such that $\lambda S - A$ is not surjective$\}$. ◇

Proof. (i) Let $\lambda \notin \mathscr{O} = \bigcap_{K \in \mathscr{F}_0(X)} \sigma_S(A+K)$. Then, there exists $K \in \mathscr{F}_0(X)$ such that $\lambda \in \rho_S(A+K)$. Hence,

$$A+K-\lambda S \in \Phi(X)$$

and

$$i(A+K-\lambda S) = 0.$$

Now, the operator $A-\lambda S$ can be written in the form

$$A-\lambda S = A+K-\lambda S-K.$$

By Theorem 2.2.14, we have $A-\lambda S \in \Phi(X)$ and

$$i(A-\lambda S) = 0.$$

Then,

$$\lambda \notin \sigma_{e5,S}(A).$$

Conversely, we suppose that $\lambda \notin \sigma_{e5,S}(A)$, then

$$A-\lambda S \in \Phi(X)$$

and

$$i(A-\lambda S) = 0.$$

Let $n = \alpha(A-\lambda S) = \beta(A-\lambda S)$, $\{x_1, \cdots, x_n\}$ being the basis for $N((A-\lambda S)^*)$ and $\{y'_1, \cdots, y'_n\}$ being the basis for annihilator $R(A-\lambda S)^{\perp}$. By Lemma 2.1.1, there are functionals $x'_1, \cdots, x'_n$ in X^* (the adjoint space of X) and elements $y_1, \cdots, y_n$ such that

$$x'_j(x_k) = \delta_{jk} \text{ and } y'_j(y_k) = \delta_{jk}, \ 1 \leq j, \ k \leq n,$$

where $\delta_{jk} = 0$ if $j \neq k$ and $\delta_{jk} = 1$ if $j = k$. The operator K is defined by

$$Kx = \sum_{k=1}^{n} x'_k(x) y_k, \ x \in X.$$

Clearly, K is a linear operator defined everywhere on X. It is bounded, since

$$\|Kx\| \leq \|x\| \Big(\sum_{k=1}^{n} \|x'_k\| \|y_k\| \Big).$$

Moreover, the range of K is contained in a finite dimensional subspace of X. Then, K is a finite rank operator in X. In the next, we will prove that

$$N(A - \lambda S) \bigcap N(K) = \{0\} \text{ and } R(A - \lambda S) \bigcap R(K) = \{0\}. \tag{6.1}$$

Let $x \in N(A - \lambda S)$, then

$$x = \sum_{k=1}^{n} \alpha_k x_k,$$

therefore

$$x'_j(x) = \alpha_j, \ 1 \leq j \leq n.$$

On the other hand, if $x \in N(K)$, then

$$x'_j(x) = 0, \ 1 \leq j \leq n.$$

This proves the first relation in Eq. (6.1). The second inclusion is similar. In fact, if $y \in R(K)$, then

$$y = \sum_{k=1}^{n} \alpha_k y_k,$$

and hence,

$$y_j(y) = \alpha_j, \ 1 \leq j \leq n.$$

But, if $y \in R(A - \lambda S)$, then

$$y'_j(y) = 0, \ 1 \leq j \leq n.$$

This gives the second relation in Eq. (6.1). Using the fact that K is a compact operator, we deduce from Theorem 2.2.14 that $\lambda \in \Phi_{A+K,S}$. If

$x \in N(A - \lambda S + K)$, then $(A - \lambda S)x$ is in $R(A - \lambda S) \bigcap R(K)$ which implies that $x \in N(A - \lambda S) \bigcap N(K)$ and hence, $x = 0$. Thus,

$$\alpha(A - \lambda S + K) = 0.$$

In the same way, one proves that

$$R(A - \lambda S + K) = X.$$

We conclude that $\lambda \notin \mathcal{O}$. Also,

$$\sigma_{e5,S}(A) = \bigcap_{K \in \mathscr{F}_0(X)} \sigma_S(A+K).$$

Now, let $\mathcal{O}_1 := \bigcap_{F \in \mathscr{F}(X)} \sigma_S(A+F)$. Since $\mathscr{F}_0(X) \subset \mathscr{F}(X)$, we infer that

$$\mathcal{O} \subset \sigma_{e5,S}(A).$$

Conversely, let $\lambda \notin \mathcal{O}_1$, then there exist $F \in \mathscr{F}(X)$ such that

$$\lambda \notin \sigma_S(A+F).$$

Then, $\lambda \in \rho_S(A+F)$. So,

$$A + F - \lambda S \in \Phi(X)$$

and

$$i(A + F - \lambda S) = 0.$$

The use of Lemma 2.3.1 makes us conclude that

$$A - \lambda S \in \Phi(X)$$

and

$$i(A - \lambda S) = 0.$$

Then,

$$\lambda \notin \sigma_{e5,S}(A).$$

So,

$$\sigma_{e5,S}(A) = \bigcap_{K \in \mathscr{F}_0(X)} \sigma_S(A+K) = \bigcap_{K \in \mathscr{F}(X)} \sigma_S(A+K).$$

Using the following relations $\mathscr{F}_0(X) \subset \mathscr{K}(X) \subset \mathscr{F}(X)$, we have

$$\sigma_{e5,S}(A) = \bigcap_{K\in\mathscr{F}(X)} \sigma_S(A+K) \subset \bigcap_{K\in\mathscr{K}(X)} \sigma_S(A+K) \subset \bigcap_{K\in\mathscr{F}_0(X)} \sigma_S(A+K) = \sigma_{e5,S}(A).$$

Statements (ii) and (iii) can be checked in a similar way as the assertion (i). Q.E.D.

Lemma 6.1.2 *Let $A \in \mathscr{C}(X)$ and $S \in \mathscr{L}(X)$. If $\Phi_{A,S}$ is connected and $\rho_S(A) \neq \emptyset$, then*

(i) $\sigma_{e1,S}(A) = \sigma_{eap,S}(A)$.

(ii) $\sigma_{e2,S}(A) = \sigma_{e\delta,S}(A)$. ◊

Proof. (i) It is easy to check that

$$\sigma_{e1,S}(A) \subset \sigma_{eap,S}(A).$$

For the second inclusion we take $\lambda \in \mathbb{C}\backslash\sigma_{e1,S}(A)$, then

$$\lambda \in \Phi_{A,S}\bigcup(\Phi_{+A,S}\backslash\Phi_{A,S}).$$

Hence, we will discuss the following two cases:

1st *case* : If $\lambda \in \Phi_{A,S}$, then

$$i(A-\lambda S) = 0.$$

Indeed, let $\lambda_0 \in \rho_S(A)$, then $\lambda_0 \in \Phi_{A,S}$ and

$$i(A-\lambda_0 S) = 0.$$

It follows from Proposition 2.2.1 that $i(A-\lambda S)$ is constant on any component of $\Phi_{A,S}$. Therefore,

$$\rho_S(A) \subseteq \Phi_{A,S}.$$

Hence,

$$i(A-\lambda S) = 0$$

for all $\lambda \in \Phi_{A,S}$. This shows that $\lambda \in \rho_{eap,S}(A)$.

$\underline{2^{nd}\ case}$: If $\mu \in \Phi_{+A,S} \backslash \Phi_{A,S}$, then

$$\alpha(A - \lambda S) < \infty$$

and

$$\beta(A - \mu S) = +\infty.$$

So,

$$i(A - \lambda S) = -\infty < 0.$$

Thus, we obtain from the above

$$\sigma_{eap,S}(A) \subset \sigma_{e1,S}(A).$$

Statement (ii) can be checked similarly from the assertion (ii). Q.E.D.

6.1.2 Stability of S-Essential Spectra of Closed Linear Operator

Theorem 6.1.1 *Let $A \in \mathscr{C}(X)$, $S \in \mathscr{L}(X)$, $\lambda \in \mathbb{C}$, and let B be a bounded operator on X. Then,*

(i) If $A - \lambda S \in \Phi(X)$ and $B \in A\mathscr{F}(X)$, then $A + B - \lambda S \in \Phi(X)$ and

$$i(A + B - \lambda S) = i(A - \lambda S).$$

(ii) If $A - \lambda S \in \Phi_+(X)$ and $B \in A\mathscr{F}_+(X)$, then $A + B - \lambda S \in \Phi_+(X)$.
(iii) If $A - \lambda S \in \Phi_-(X)$ and $B \in A\mathscr{F}_-(X)$, then $A + B - \lambda S \in \Phi_-(X)$.
(iv) $A - \lambda S \in \Phi_\pm(X)$ and $B \in A\mathscr{F}_+(X) \bigcap A\mathscr{F}_-(X)$, then

$$A + B - \lambda S \in \Phi_\pm(X). \qquad \diamond$$

Proof. Assume that $A - \lambda S \in \Phi(X)$. Then, using (2.14), we infer that

$$\widehat{A} - \lambda \widehat{S} \in \Phi^b(X_A, X).$$

Hence, it follows from Theorem 2.2.3 that there exist $A_0 \in \mathscr{L}(X, X_A)$, $K_1 \in \mathscr{K}(X)$ and $K_2 \in \mathscr{K}(X_A)$ such that

$$(\widehat{A} - \lambda \widehat{S})A_0 = I - K_1, \quad \text{on } X, \tag{6.2}$$

$$A_0(\widehat{A}-\lambda\widehat{S})=I-K_2, \quad \text{on } X_A. \tag{6.3}$$

Thus,

$$A_0(\widehat{A}+\widehat{B}-\lambda\widehat{S})=I-K_2+A_0\widehat{B}, \quad \text{on } X. \tag{6.4}$$

Next, using both Eq. (6.2) and Theorem 2.2.13, we get

$$(\widehat{A}-\lambda\widehat{S})A_0 \in \Phi^b(X)$$

and

$$i\big[(\widehat{A}-\lambda\widehat{S})A_0\big]=0.$$

So, by using Eqs. (6.2), (6.3), Lemma 2.2.3 and both Theorems 2.2.13 and 2.2.5, we deduce that $A_0 \in \Phi^b(X_A,X)$ and

$$i(\widehat{A}-\lambda\widehat{S})=-i(A_0).$$

Since $B \in A\mathscr{F}(X)$ and $A_0 \in \mathscr{L}(X)$, and by applying Lemma 2.3.2, we have $A_0\widehat{B} \in \mathscr{F}^b(X)$. So,

$$K_2-A_0\widehat{B} \in \mathscr{F}^b(X).$$

Using Eq. (6.4), we get

$$A_0(\widehat{A}+\widehat{B}-\lambda\widehat{S}) \in \Phi^b(X)$$

and

$$i(A_0(\widehat{A}+\widehat{B}-\lambda\widehat{S}))=0.$$

As, $A_0 \in \Phi^b(X,X_A)$, and according to both Theorems 2.2.7 and 2.2.5, we have

$$\widehat{A}+\widehat{B}-\lambda\widehat{S} \in \Phi^b(X_{A+B},X)$$

and

$$i(\widehat{A}+\widehat{B}-\lambda\widehat{S})=-i(A_0). \tag{6.5}$$

Now, by using Eqs. (2.14), (6.2) and (6.5), we affirm that

$$i(A+B-\lambda S)=i(A-\lambda S),$$

which completes the proof of (i).

The assertion (ii), the first part of (iii) and (iv) are immediate. To prove the second part of (iii) we proceed as follows. Let $A-\lambda S \in \Phi_-(X)$. By Theorem 2.2.2, we infer that

$$(A-\lambda S)^* = A^* - \overline{\lambda} S^* \in \Phi_+(X^*),$$

the fact that $B^* \in A\mathscr{F}_+(X^*)$ implies that

$$(A+B-\lambda S)^* = A^* + B^* - \overline{\lambda} S^* \in \Phi_+(X^*).$$

According to Theorem 2.2.2, we get $A+B-\lambda S \in \Phi_-(X)$. Q.E.D.

Corollary 6.1.1 *Let $A \in \mathscr{C}(X)$, $S \in \mathscr{L}(X)$, $\lambda \in \mathbb{C}$, and let B be a bounded operator on X. Then,*

(i) If $A-\lambda S \in \Phi_+(X)$ and $B \in A\mathscr{S}(X)$, then $A+B-\lambda S \in \Phi_+(X)$.

(ii) If $A-\lambda S \in \Phi_-(X)$ and $B \in A\mathscr{S}\mathscr{C}(X)$, then $A+B-\lambda S \in \Phi_-(X)$.

◊

Theorem 6.1.2 *Let $A \in \mathscr{C}(X)$ and let B be an operator on X. The following statements are satisfied.*

(i) If $B \in A\mathscr{F}_+(X)$, then

$$\sigma_{e1,S}(A+B) = \sigma_{e1,S}(A).$$

If, in addition, we suppose that the sets $\Phi_{A,S}$ and $\Phi_{A+B,S}$ are connected and the sets $\rho_S(A)$ and $\rho_S(A+B)$ are not empty, then

$$\sigma_{eap,S}(A+B) = \sigma_{eap,S}(A).$$

(ii) If $B \in A\mathscr{F}_-(X)$, then

$$\sigma_{e2,S}(A+B) = \sigma_{e2,S}(A).$$

If, in addition, we suppose that the sets $\Phi_{A,S}$ and $\Phi_{A+B,S}$ are connected and both the sets $\rho_S(A)$ and $\rho_S(A+B)$ are not empty, then

$$\sigma_{e\delta,S}(A+B) = \sigma_{e\delta,S}(A).$$

(iii) If $B \in A\mathscr{F}_+(X) \bigcap A\mathscr{F}_-(X)$*, then*

$$\sigma_{e3,S}(A+B) = \sigma_{e3,S}(A).$$

(iv) If $B \in A\mathscr{F}(X)$*, then*

$$\sigma_{ei,S}(A+B) = \sigma_{ei,S}(A), \quad i = 4,\ 5.$$

Moreover, if $\mathbb{C} \backslash \sigma_{e5,S}(A)$ *is connected. If neither* $\rho_S(A)$ *nor* $\rho_S(A+B)$ *is empty, then*

$$\sigma_{e6,S}(A+B) = \sigma_{e6,S}(A). \qquad \diamondsuit$$

Proof. (i) Let $\lambda \notin \sigma_{e1,S}(A)$, then

$$\lambda \in \Phi_{+A,S}.$$

Since $B \in A\mathscr{F}_+(X)$, applying Theorem 6.1.1 (ii), we infer that

$$\lambda S - A - B \in \Phi_+(X).$$

Thus,

$$\lambda \notin \sigma_{e1,S}(A+B).$$

Conversely, let $\lambda \notin \sigma_{e1,S}(A+B)$, then

$$\lambda S - A - B \in \Phi_+(X).$$

Using both Theorem 6.1.1 (ii) and the fact that $-B \in A\mathscr{F}_+(X)$, we get $\lambda \in \Phi_{+A,S}$. So,

$$\lambda \notin \sigma_{e1,S}(A).$$

We deduce that

$$\sigma_{e1,S}(A+B) = \sigma_{e1,S}(A).$$

Now, if $\Phi_{A,S}$ and $\Phi_{A+B,S}$ are connected and the sets $\rho_S(A)$ and $\rho_S(A+B)$ are not empty, then by Lemma 6.1.2, we have

$$\sigma_{eap,S}(A) = \sigma_{e1,S}(A)$$

and

$$\sigma_{eap,S}(A+B) = \sigma_{e1,S}(A+B).$$

Hence,

$$\sigma_{eap,S}(A+B) = \sigma_{eap,S}(A).$$

The proof of (ii) is analogues to the previous one.

(iii) This statement is an immediate consequence of the item (i).

(iv) For $i = 5$. Let $\lambda \notin \sigma_{e5,S}(A)$, then $\lambda \in \Phi_{A,S}$ and

$$i(\lambda S - A) = 0.$$

Since $B \in A\mathscr{F}(X)$, applying Theorem 6.1.1 (i), we infer that $\lambda \in \Phi_{A+B,S}$ and

$$i(\lambda S - A - B) = 0,$$

and therefore

$$\lambda \notin \sigma_{e5,S}(A+B).$$

Thus,

$$\sigma_{e5,S}(A+B) \subseteq \sigma_{e5,S}(A).$$

Similarly, If $\lambda \notin \sigma_{e5,S}(A+B)$, then using Theorem 6.1.1 (i) and arguing as above we derive the opposite inclusion

$$\sigma_{e5,S}(A) \subseteq \sigma_{e5,S}(A+B).$$

Now, we get

$$\mathbb{C} \backslash \sigma_{e5,S}(A+B) = \mathbb{C} \backslash \sigma_{e5,S}(A),$$

which is connected by hypothesis. Thus, by Lemma 3.3.2, we have

$$\sigma_{e5,S}(A) = \sigma_{e6,S}(A) \text{ and } \sigma_{e5,S}(A+B) = \sigma_{e6,S}(A+B).$$

We deduce that

$$\sigma_{e6,S}(A+B) = \sigma_{e6,S}(A). \qquad \text{Q.E.D.}$$

Theorem 6.1.3 *Let $A \in \mathscr{C}(X)$ and let $\mathscr{I}_i(X)$, $i \in \{1, 2, 3\}$ be any subset of operators satisfying*

(i) $\mathscr{K}(X) \subseteq \mathscr{I}_1(X) \subseteq A\mathscr{F}(X)$. Then,

$$\sigma_{e5,S}(A) = \bigcap_{B \in \mathscr{I}_1(X)} \sigma_S(A+B).$$

(ii) $\mathscr{K}(X) \subseteq \mathscr{I}_2(X) \subseteq A\mathscr{F}_+(X)$. Then,

$$\sigma_{eap,S}(A) = \bigcap_{B \in \mathscr{I}_2(X)} \sigma_{ap,S}(A+B).$$

(iii) $\mathscr{K}(X) \subseteq \mathscr{I}_3(X) \subseteq A\mathscr{F}_-(X)$. Then,

$$\sigma_{e\delta,S}(A) = \bigcap_{B \in \mathscr{I}_3(X)} \sigma_{\delta,S}(A+B). \qquad \diamondsuit$$

Proof. (i) Let $\mathscr{O} = \bigcap_{B \in \mathscr{I}_1(X)} \sigma_S(A+B)$. According to Remark 2.3.2, we have

$$\mathscr{K}(X) \subseteq A\mathscr{K}(X) \subseteq A\mathscr{F}(X).$$

Hence,

$$\mathscr{O} \subseteq \sigma_{e5,S}(A).$$

So, we have only to prove that

$$\sigma_{e5,S}(A) \subseteq \mathscr{O}.$$

Let $\lambda_0 \notin \mathscr{O}$, then there exists $B \in \mathscr{I}_1(X)$ such that

$$\lambda_0 \in \rho_S(A+B).$$

Let $x \in X$ and put

$$y = (\lambda_0 S - A - B)^{-1}x.$$

It follows from the estimate

$$\begin{aligned}
\|y\|_{A+B} &= \|y\| + \|(\widehat{A}+\widehat{B})y\| \\
&= \|y\| + \|x - \lambda_0 \widehat{S}y\| \\
&= \|(\lambda_0\widehat{S} - \widehat{A} - \widehat{B})^{-1}x\| + \|x - \lambda_0\widehat{S}(\lambda_0\widehat{S} - \widehat{A} - \widehat{B})^{-1}x\| \\
&\leq (1 + (1 + |\lambda_0|\|\widehat{S}\|)\|(\lambda_0 S - \widehat{A} - \widehat{B})^{-1}\|)\|x\|.
\end{aligned}$$

Thus,

$$(\lambda_0\widehat{S}-\widehat{A}-\widehat{B})^{-1}\in\mathscr{L}(X,X_{A+B}).$$

Since $B\in\mathscr{I}_1(X)\subseteq A\mathscr{F}(X)$, and the use of Lemma 2.3.2 allows us to conclude that

$$(\lambda_0\widehat{S}-\widehat{A}-\widehat{B})^{-1}\widehat{B}\in\mathscr{F}^b(X_A,X_{A+B}).$$

Let $\mathfrak{J}$ denote the imbedding operator which maps every $x\in X_A$ onto the same element $x\in X_{A+B}$. Clearly, we have $N(\mathfrak{J})=0$ and $R(\mathfrak{J})=X_{A+B}$. So,

$$\begin{array}{rcl}\|\mathfrak{J}(x)\| &=& \|x\|_{A+B}\\ &\leq& \|x\|+\|Ax\|+\|Bx\|\\ &\leq& \left(1+\|B\|_{\mathscr{L}(X,X_{A+B})}\right)\|x\|_{X_A},\quad \forall x\in X_A.\end{array}$$

Thus, $\mathfrak{J}\in\Phi^b(X_A,X_{A+B})$ and $i(\mathfrak{J})=0$. Next, since $(\lambda_0\widehat{S}-\widehat{A}-\widehat{B})^{-1}\widehat{B}\in\mathscr{F}^b(X_A,X_{A+B})$, we get by Theorem 6.1.1 (i),

$$\mathfrak{J}+(\lambda_0\widehat{S}-\widehat{A}-\widehat{B})^{-1}\widehat{B}\in\Phi^b(X_A,X_{A+B})\text{ and }i(\mathfrak{J}+(\lambda_0\widehat{S}-\widehat{A}-\widehat{B})^{-1}\widehat{B})=0. \tag{6.6}$$

On the other hand, since $\lambda_0\in\rho_S(A+B)$, it follows from Eq. (2.14) that

$$\lambda_0\widehat{S}-\widehat{A}-\widehat{B}\in\Phi^b(X_A,X_{A+B})\text{ and }i(\lambda_0\widehat{S}-\widehat{A}-\widehat{B})=0. \tag{6.7}$$

Writing $\lambda_0\widehat{S}-\widehat{A}$ in the from

$$\lambda_0\widehat{S}-\widehat{A}=(\lambda_0\widehat{S}-\widehat{A}-\widehat{B})(\mathfrak{J}+(\lambda_0\widehat{S}-\widehat{A}-\widehat{B})^{-1}\widehat{B}). \tag{6.8}$$

Using Eqs. (6.6), (6.7), and (6.8) together with Theorem 2.2.5, we get

$$\lambda_0\widehat{S}-\widehat{A}\in\Phi^b(X_A,X)\text{ and }i(\lambda_0\widehat{S}-\widehat{A})=0.$$

Now, using (2.14) we infer that

$$\lambda_0 S-A\in\Phi(X)\text{ and }i(\lambda_0 S-A)=0.$$

Hence,

$$\sigma_{e5,S}(A)\subseteq\mathscr{O}.$$

The proof of (ii) is analogues to the previous one.

(iii) This statement is an immediate consequence of the item (i).

Q.E.D.

6.2 S-LEFT AND S-RIGHT ESSENTIAL SPECTRA

6.2.1 Stability of S-Left and S-Right Fredholm Spectra

This section concerns the stability of S-left (resp. S-right) Fredholm spectrum of a bounded linear operator on a Banach space X.

Lemma 6.2.1 *Let $A \in \mathscr{L}(X)$ and S be an invertible operator on X such that $0 \in \rho(A)$. Then, for $\lambda \neq 0$, we have*

$$\lambda \in \sigma_{ei,S}(A) \text{ if, and only if, } \frac{1}{\lambda} \in \sigma_{ei,S^{-1}}(A^{-1}),\ i = l,\ r,\ wl,\ wr. \qquad \diamondsuit$$

Proof. For $\lambda \neq 0$, assume that $\frac{1}{\lambda} \in \Phi_{A^{-1},S^{-1}}$, then

$$\lambda^{-1}S^{-1} - A^{-1} \in \Phi^b(X).$$

The operator $\lambda S - A$ can be written in the form

$$\lambda S - A = -\lambda S(\lambda^{-1}S^{-1} - A^{-1})A. \tag{6.9}$$

Since A and S are one-to-one and onto, then by Eq. (6.9), we have $N(\lambda S - A)$ and $N(\lambda^{-1}S^{-1} - A^{-1})$ are isomorphic and, $R(\lambda S - A)$ and $R(\lambda^{-1}S^{-1} - A^{-1})$ are isomorphic. This shows that $\lambda \in \Phi_{+A,S}$ (resp. $\Phi_{-A,S}$) and $R(\lambda S - A)$ (resp. $N(\lambda S - A)$) is complemented if, and only if, $\lambda^{-1} \in \Phi_{+A^{-1},S^{-1}}$ (resp. $\Phi_{-A^{-1},S^{-1}}$) and $R(\lambda^{-1}S^{-1} - A^{-1})$ (resp. $N(\lambda^{-1}S^{-1} - A^{-1})$) is complemented. Therefore,

$$\lambda \in \Phi^l_{A,S} \ (\text{resp. } \Phi^r_{A,S}) \text{ if, and only if, } \frac{1}{\lambda} \in \Phi^l_{A^{-1},S^{-1}} \ (\text{resp. } \Phi^r_{A^{-1},S^{-1}}).$$

Since $0 \in \rho(A)$ and S is invertible, then

$$i(A) = i(S) = 0.$$

Using both Theorem 2.2.5 and Eq. (6.9), we conclude that

$$\begin{aligned} i(\lambda S - A) &= i(A) + i(S) + i(\lambda^{-1}S^{-1} - A^{-1}) \\ &= i(\lambda^{-1}S^{-1} - A^{-1}). \end{aligned}$$

So, $\lambda \in \Phi^l_{A,S}$ (resp. $\Phi^r_{A,S}$) and $i(\lambda S - A) \leq 0$ (resp. $i(\lambda S - A) \geq 0$) if, and only if, $\lambda^{-1} \in \Phi^l_{A^{-1},S^{-1}}$ (resp. $\Phi^r_{A^{-1},S^{-1}}$) and $i(\lambda^{-1}S^{-1} - A^{-1}) \leq 0$

$\left(\text{resp. } i(\lambda^{-1}S^{-1} - A^{-1}) \geq 0\right)$. Hence, $\lambda \in \sigma_{ei,S}(A)$, if, and only if, $\frac{1}{\lambda} \in \sigma_{ei,S^{-1}}(A^{-1})$, $i = l,\ r,\ wl,\ wr$, which completes the proof. Q.E.D.

Theorem 6.2.1 *Let* $A,\ B \in \mathscr{L}(X)$ *and* S *be an invertible operator on* X. *If, for some,* $\lambda \in \rho_S(A) \bigcap \rho_S(B)$,

$$(\lambda S - A)^{-1} - (\lambda S - B)^{-1} \in \mathscr{F}^b(X),$$

then

$$\sigma_{ei,S}(A) = \sigma_{ei,S}(B), \ \ i = l,\ r,\ wl,\ wr. \qquad \diamondsuit$$

Proof. Without loss of generality, we suppose that $\lambda = 0$. Since

$$A^{-1} - B^{-1} \in \mathscr{F}^b(X),$$

then it follows from Corollary 2.3.1 (i) that

$$\Phi^l_{A^{-1},S^{-1}} = \Phi^l_{B^{-1},S^{-1}}$$

$$\left(\text{resp. } \mathscr{W}^l_{A^{-1},S^{-1}} = \mathscr{W}^l_{B^{-1},S^{-1}}\right).$$

Thus, the use of Lemma 6.2.1 makes us conclude that

$$\Phi^l_{A,S} = \Phi^l_{B,S}$$

$$\left(\text{resp. } \mathscr{W}^l_{A,S} = \mathscr{W}^l_{B,S}\right).$$

Hence,

$$\sigma_{ei,S}(A) = \sigma_{ei,S}(B), \ \ i = l,\ wl.$$

Similarly, we can show that

$$\sigma_{ei,S}(A) = \sigma_{ei,S}(B), \ \ i = r,\ wr,$$

which completes the proof. Q.E.D.

The Fredholm perturbation operators in the previous theorem can be replaced by finite rank operators or compact operators but, in general, it is not true for Riesz operators.

Theorem 6.2.2 *Let A, $B \in \mathscr{L}(X)$ and S be an invertible operator on X such that $S \neq A$ and $S \neq B$. Suppose $\lambda_0 \in \rho_S(A) \bigcap \rho_S(B)$ and*

$$(\lambda_0 S - A)^{-1} - (\lambda_0 S - B)^{-1} \in \mathscr{R}(X).$$

If

$$(\lambda_0 S - A)^{-1}(\lambda_0 S - B)^{-1} - (\lambda_0 S - B)^{-1}(\lambda_0 S - A)^{-1} \in \mathscr{F}^b(X),$$

then

$$\sigma_{ei,S}(A) = \sigma_{ei,S}(B), \;\; i = l,\, r,\, wl,\, wr. \qquad \diamondsuit$$

Proof. Without loss of generality, we suppose that $\lambda_0 = 0$. Then, $0 \in \rho_S(A) \bigcap \rho_S(B)$ such that

$$R = A^{-1} - B^{-1} \in \mathscr{R}(X),$$

therefore

$$RB^{-1} - B^{-1}R = A^{-1}B^{-1} - B^{-1}A^{-1}.$$

Since

$$A^{-1}B^{-1} - B^{-1}A^{-1} \in \mathscr{F}^b(X),$$

we deduce that

$$RB^{-1} - B^{-1}R \in \mathscr{F}^b(X).$$

From Theorem 2.3.4 (i), we infer that

$$\Phi^l_{A^{-1},S^{-1}} = \Phi^l_{B^{-1},S^{-1}}$$

and, applying Lemma 6.2.1, we conclude that

$$\sigma_{el,S}(A) = \sigma_{el,S}(B).$$

Similarly, we can show that

$$\sigma_{ei,S}(A) = \sigma_{ei,S}(B), \;\; i = r,\, wl,\, wr,$$

which completes the proof. Q.E.D.

Theorem 6.2.3 *Let A, $S \in \mathscr{L}(X)$ such that $S \neq A$ and $E \in \mathscr{R}(X)$ such that $AE - EA \in \mathscr{F}^b(X)$ and $SE - ES \in \mathscr{F}^b(X)$. Then,*

(i) $\sigma_{el,S}(A) = \sigma_{el,S}(A+E)$,
(ii) $\sigma_{er,S}(A) = \sigma_{er,S}(A+E)$,
(iii) $\sigma_{ewl,S}(A) = \sigma_{ewl,S}(A+E)$, *and*
(iv) $\sigma_{ewr,S}(A) = \sigma_{ewr,S}(A+E)$. ◇

Proof. (i) Let $\lambda \notin \sigma_{el,S}(A)$, then

$$\lambda S - A \in \Phi_l(X).$$

The operator $(\lambda S - A)E - E(\lambda S - A)$ can be written in the form

$$\begin{aligned}(\lambda S - A)E - E(\lambda S - A) &= EA + \lambda SE - \lambda ES - AE \\ &= EA - AE + \lambda(SE - ES). \quad (6.10)\end{aligned}$$

Since $AE - EA \in \mathscr{F}^b(X)$ and $SE - ES \in \mathscr{F}^b(X)$, then from Theorem 2.3.4 (i), it follows that

$$\lambda S - (E + A) \in \Phi_l(X).$$

Hence,

$$\lambda \notin \sigma_{el,S}(A+E).$$

Therefore,

$$\sigma_{el,S}(A+E) \subseteq \sigma_{el,S}(A). \quad (6.11)$$

To prove the reverse inclusion of Eq. (6.11), it suffices to replace A and E by $A+E$ and $-E$, respectively.

The proof of (ii) is analogues to the previous one.

(iii) Assume that $\lambda \notin \sigma_{ewl,S}(A)$, then

$$\lambda S - A \in \mathscr{W}_l(X).$$

Now, we can written the operator

$$(\lambda S - A)E - E(\lambda S - A)$$

in the form (6.10). Since

$$AE - EA \in \mathscr{F}^b(X)$$

and

$$SE - ES \in \mathscr{F}^b(X),$$

then from Theorem 2.3.4 (iii), it follows that

$$\lambda S - (A+E) \in \mathscr{W}_l(X)$$

and hence

$$\lambda \notin \sigma_{ewl,S}(A+E).$$

Therefore,

$$\sigma_{ewl,S}(A+E) \subseteq \sigma_{ewl,S}(A). \tag{6.12}$$

The opposite inclusion of Eq. (6.12) follows by symmetry, it suffices to replace A and E by $A+E$ and $-E$, respectively.

The proof of (iv) may be checked in a similar way to that in (iii). Q.E.D.

Corollary 6.2.1 *Let A, $S \in \mathscr{L}(X)$ such that $S \neq A$ and $E \in \mathscr{F}^b(X)$. Then,*

(i) $\sigma_{el,S}(A) = \sigma_{el,S}(A+E)$,

(ii) $\sigma_{er,S}(A) = \sigma_{er,S}(A+E)$,

(iii) $\sigma_{ewl,S}(A) = \sigma_{ewl,S}(A+E)$, *and*

(iv) $\sigma_{ewr,S}(A) = \sigma_{ewr,S}(A+E)$. ◇

Remark 6.2.1 *The perturbation assumption in the Corollary 6.2.1 which cannot be relaxed, even nilpotent operators. For example, let $\mathscr{A}$, $\mathscr{T}$, and $\mathscr{K}$ in $\mathscr{L}(l_2 \times l_2)$ are defined by*

$$\mathscr{A} = \begin{pmatrix} V & 0 \\ 0 & U \end{pmatrix}, \ \mathscr{T} = \begin{pmatrix} U & 0 \\ 0 & V \end{pmatrix}, \text{ and } \mathscr{K} = \begin{pmatrix} 0 & I \\ 0 & 0 \end{pmatrix},$$

where U and V are the forward and the backward unilateral shifts. U and V are linear bounded operators on l_2 defined by

$$U(x_1, x_2, x_3, \cdots) = (x_2, x_3, \cdots)$$

and

$$V(x_1,x_2,x_3,\cdots) = (0,x_1,x_2,\cdots).$$

Then, $\mathscr{A}$ and $\mathscr{T}$ are both Weyl, while $\mathscr{T}+\mathscr{K}$ is not right Weyl and $\mathscr{A}-\mathscr{K}$ is not left Weyl. Indeed,

$$(\mathscr{A}-\mathscr{K})(\mathscr{T}+\mathscr{K}) = \begin{pmatrix} I & 0 \\ 0 & I \end{pmatrix} - \begin{pmatrix} 0 & 0 \\ 0 & I-UV \end{pmatrix}$$

and

$$(\mathscr{T}+\mathscr{K})(\mathscr{A}-\mathscr{K}) = \begin{pmatrix} I & 0 \\ 0 & I \end{pmatrix} - \begin{pmatrix} I-UV & 0 \\ 0 & 0 \end{pmatrix}.$$

Since $(\mathscr{A}-\mathscr{K})(\mathscr{T}+\mathscr{K})$ and $(\mathscr{T}+\mathscr{K})(\mathscr{A}-\mathscr{K})$ are Fredholm operators of index zero. Hence, the products are Weyl. We can check that $\mathscr{T}+\mathscr{K}$ is one-to-one and $\mathscr{A}-\mathscr{K}$ is onto, with

$$i(\mathscr{A}-\mathscr{K}) = 1 = -i(\mathscr{T}+\mathscr{K}).$$ ◇

The concept of the following theorem gives a refinement of the definition of some S-essential spectra on a Banach space.

Theorem 6.2.4 *Let $A \in \mathscr{L}(X)$ and $S \in \mathscr{L}(X)$ such that $S \neq 0$. Then,*

(i)

$$\sigma_{ewl,S}(A) = \bigcap_{E \in \widetilde{\mathscr{R}}_{A,S}(X)} \sigma_{l,S}(A+E), \tag{6.13}$$

(ii) $\sigma_{ewr,S}(A) = \bigcap_{E \in \widetilde{\mathscr{R}}_{A,S}(X)} \sigma_{R,S}(A+E)$, *and*

(iii) $\sigma_{e5,S}(A) = \bigcap_{E \in \widetilde{\mathscr{R}}_{A,S}(X)} \sigma_S(A+E)$,

where $\widetilde{\mathscr{R}}_{A,S}(X) = \{E \in \mathscr{R}(X)$ such that $AE-EA \in \mathscr{F}^b(X)$ and $SE-ES \in \mathscr{F}^b(X)\}$. ◇

Proof. (i) To prove the inclusion " $\subseteq$ " in Eq. (6.13). Suppose that

$$\lambda \notin \bigcap_{E \in \widetilde{\mathscr{R}}_{A,S}(X)} \sigma_{l,S}(A+E),$$

then there exists $E \in \widetilde{\mathscr{R}}_{A,S}(X)$ such that

$$\lambda \notin \sigma_{l,S}(A+E),$$

so

$$\lambda S - (E+A) \in \mathscr{G}_l(X)$$

and hence

$$\lambda S - (E+A) \in \mathscr{W}_l(X).$$

It follows from Theorem 6.2.3 (iii) that

$$\lambda S - A \in \mathscr{W}_l(X),$$

i.e.,

$$\lambda \notin \sigma_{ewl,S}(A).$$

To prove the inverse inclusion in Eq. (6.13), assume that $\lambda \notin \sigma_{ewl,S}(A)$, then $\lambda S - A$ is left Weyl. Using Lemma 2.2.7, there exists a compact operator E such that $\lambda S - (E+A)$ is injective, and by Theorem 2.2.12, it follows that $\lambda S - (E+A)$ is left invertible and hence,

$$\lambda \notin \sigma_{l,S}(A+E).$$

The proof of (ii) is analogues to the previous one.

(iii) This statement is an immediate consequence of both the items (i) and (ii). Q.E.D.

6.2.2 Stability of S-Left and S-Right Browder Spectra

The purpose of this section is to discuss the S-left and S-right Browder spectra of a bounded linear operator on a Banach space X. We first prove the following theorem.

Theorem 6.2.5 *Let $A,\ S \in \mathscr{L}(X)$ such that $S \neq A$ and E be a Riesz operator commuting with A and S, then*

(i) $\sigma_{bl,S}(A) = \sigma_{bl,S}(A+E)$, and

(ii) $\sigma_{br,S}(A) = \sigma_{br,S}(A+E)$. ◊

Proof. (i) Let $\lambda \notin \sigma_{bl,S}(A)$, then

$$\lambda S - A \in \mathscr{B}^l(X).$$

Since $AE = EA$ and $SE = ES$, then

$$E(\lambda S - A) = (\lambda S - A)E.$$

To use of Theorem 2.5.1 (i) leads to

$$\lambda S - (E + A) \in \mathscr{B}^l(X).$$

Thus,

$$\lambda \notin \sigma_{bl,S}(A+E).$$

Hence,

$$\sigma_{bl,S}(A+E) \subseteq \sigma_{bl,S}(A).$$

The opposite inclusion follows by symmetry, it suffices to replace A and E by $A+E$ and $-E$, respectively, to find

$$\sigma_{bl,S}(A) \subseteq \sigma_{bl,S}(A+E).$$

(ii) Using the same reasoning, we can show that

$$\sigma_{br,S}(A) = \sigma_{br,S}(A+E).$$ Q.E.D.

Remark 6.2.2 *The commutativity assumption in Theorem 6.2.5 cannot generally be relaxed even bounded operators. For example, let*

$$A = I - \frac{1}{2}U,$$

$$E = I + U,$$

and

$$S = U,$$

where U is the forward unilateral shift on l_2. Then,

$$AE = EA,$$

$$SE = ES$$

and E is not Riesz operator. On the other hand, since

$$\frac{1}{2}S - (A+E) = -2I,$$

then

$$\frac{1}{2} \notin \sigma_{bl,S}(A+E).$$

So,

$$\frac{1}{2} \in \sigma_{bl,S}(A),$$

since

$$\frac{1}{2}S - A = U - I \notin \mathscr{B}^l(X),$$

which implies that

$$\sigma_{bl,S}(A) \neq \sigma_{bl,S}(A+E). \qquad \diamondsuit$$

In the following theorem, we extend to S-left and S-right Browder spectra.

Theorem 6.2.6 *Let A, $S \in \mathscr{L}(X)$. Then,*

(*i*) $\sigma_{bl,S}(A) \subseteq \bigcap\{\sigma_{l,S}(A+E)$ *such that* $E \in \mathscr{R}(X)$, $AE = EA$ and $SE = ES\}$, (*ii*) $\sigma_{br,S}(A) \subseteq \bigcap\{\sigma_{R,S}(A+E)$ *such that* $E \in \mathscr{R}(X)$, $AE = EA$ and $SE = ES\}$, *and*

(*iii*) $\sigma_{e6,S}(A) \subseteq \bigcap\{\sigma_S(A+E)$ *such that* $E \in \mathscr{R}(X)$, $AE = EA$ and $SE = ES\}$. ♢

Proof. (*i*) Assume that $\lambda \in \sigma_{bl,S}(A)$, so by Theorem 6.2.5 (*i*), $\lambda \in \sigma_{bl,S}(A+E)$ for all $E \in \mathscr{R}(X)$ such that $AE = EA$ and $SE = ES$. On the other hand, we have

$$\sigma_{bl,S}(A+E) \subset \sigma_{l,S}(A+E),$$

then

$$\lambda \in \sigma_{l,S}(A+E)$$

for all $E \in \mathscr{R}(X)$ such that $AE = EA$ and $SE = ES$. Therefore,

$$\sigma_{bl,S}(A) \subseteq \bigcap\{\sigma_{l,S}(A+E) \text{ such that } E \in \mathscr{R}(X),\ AE = EA \text{ and } SE = ES\}.$$

(*ii*) Can be checked in the same way as (*i*).

The assertion (*iii*) is an immediate consequence of (*i*) and (*ii*). Q.E.D.

Chapter 7

S-Essential Spectrum and Measure of Non-Strict-Singularity

One of the central questions in the study of the S-essential spectra of closed densely defined linear operators consists in showing when different notions of essential spectrum coincide with the studying of the invariance of a certain class of perturbations. The purpose of this chapter is to characterize the S-essential spectrum by mean of measure of non-strict-singularity and give some application to matrix operator.

7.1 A CHARACTERIZATION OF THE S-ESSENTIAL SPECTRUM

In this section we will give a fine description of the S-essential spectrum of a closed densely defined linear operator by mean of the measure of non-strict-singularity.

Let $A \in \mathscr{C}(X)$ and S, $K \in \mathscr{L}(X)$ and let $f(\cdot)$ be a measure of non-strict-singularity, given in (2.22). Consider the following sets

$$\mathcal{S}^1_{A,S}(X) = \Big\{K \in \mathcal{L}(X) \text{ such that } f\Big([(\lambda S - A - K)^{-1}K]^n\Big) < 1, \\ n \in \mathbb{N},\ \forall \lambda \in \rho_S(A+K)\Big\}$$

and

$$\mathcal{S}^2_{A,S}(X) = \Big\{K \in \mathcal{L}(X) \text{ such that } f\Big([K(\lambda S - A - K)^{-1}]^n\Big) < 1, \\ n \in \mathbb{N},\ \forall \lambda \in \rho_S(A+K)\Big\}.$$

(i) If $(\lambda S - A)^{-1}K \in \mathcal{S}(X)$ (resp. $K(\lambda S - A)^{-1} \in \mathcal{S}(X)$) for some $\lambda \in \rho_S(A)$, then $(\lambda S - A)^{-1}K \in \mathcal{S}(X)$ (resp. $K(\lambda S - A)^{-1} \in \mathcal{S}(X)$) for all $\lambda \in \rho_S(A)$. Indeed, for all $\lambda,\ \mu \in \rho_S(A)$, we have

$$(\lambda S - A)^{-1}K - (\mu S - A)^{-1}K = (\mu - \lambda)(\mu S - A)^{-1}S(\lambda S - A)^{-1}K,$$

$$\Big(\text{resp. } K(\lambda S - A)^{-1} - K(\mu S - A)^{-1} = (\mu - \lambda)K(\mu S - A)^{-1}S(\lambda S - A)^{-1}\Big).$$

(ii) Now, if we consider the following sets

$$H_{A,S}(X) = \Big\{K \in \mathcal{L}(X) \text{ such that } (\lambda S - A)^{-1}K \in \mathcal{S}(X) \\ \text{for some (hence for all) } \lambda \in \rho_S(A)\Big\}$$

and

$$\mathcal{F}_{A,S}(X) = \Big\{K \in \mathcal{L}(X) \text{ such that } K(\lambda S - A)^{-1} \in \mathcal{S}(X) \\ \text{for some (hence for all) } \lambda \in \rho_S(A)\Big\}.$$

Then,

$$H_{A,S}(X) \subset \mathcal{S}^1_{A,S}(X)$$

and

$$\mathcal{F}_{A,S}(X) \subset \mathcal{S}^2_{A,S}(X).$$

Indeed, let $K \in H_{A,S}(X)$, then there exists $\lambda \in \rho_S(A)$ such that $(\lambda S - A)^{-1}K$ is strictly singular. For $\mu \in \rho_S(A+K)$, we have

$$(\mu S - A - K)^{-1}K = \Big[I + (\mu S - A - K)^{-1}((\lambda - \mu)S + K)\Big]\Big[(\lambda S - A)^{-1}K\Big].$$

By the ideal's propriety of $\mathcal{S}(X)$, we deduce that $(\mu S - A - K)^{-1}K$ is strictly singular. Then, by using Proposition 2.6.3, we have

$$f((\mu S - A - K)^{-1}K) = 0,$$

where $f(\cdot)$ is a measure of non-strict-singularity, given in (2.22). Therefore,

$$K \in \mathscr{S}^1_{A,S}(X).$$

So,

$$H_{A,S}(X) \subset \mathscr{S}^1_{A,S}(X).$$

A similar reasoning allows us to deduce that $\mathscr{F}_{A,S}(X) \subset \mathscr{S}^2_{A,S}(X)$.

We begin with the following theorem which gives a refined definition of S-Schechter essential spectrum.

Theorem 7.1.1 *Let $A \in \mathscr{C}(X)$, and $S \in \mathscr{L}(X)$. Then,*

$$\sigma_{e5,S}(A) = \bigcap_{K \in \mathscr{S}^1_{A,S}(X)} \sigma_S(A+K). \qquad \diamondsuit$$

Proof. We first claim that

$$\sigma_{e5,S}(A) \subset \bigcap_{K \in \mathscr{S}^1_{A,S}(X)} \sigma_S(A+K).$$

Indeed, if

$$\lambda \notin \bigcap_{K \in \mathscr{S}^1_{A,S}(X)} \sigma_S(A+K),$$

then there exists $K \in \mathscr{S}^1_{A,S}(X)$ such that

$$\lambda \notin \sigma_S(A+K).$$

So,

$$f([(\lambda S - A - K)^{-1}K]^n) < 1$$

for some $n \in \mathbb{N}$, where $f(\cdot)$ is a measure of non-strict-singularity, given in (2.22). Hence, by Proposition 2.6.4, we get

$$I + (\lambda S - A - K)^{-1}K \in \Phi^b(X)$$

and

$$i\Big(I + (\lambda S - A - K)^{-1}K\Big) = 0.$$

Writing

$$\lambda S-A=\left(\lambda S-A-K\right)\left[I+(\lambda S-A-K)^{-1}K\right],$$

we can deduce that

$$\lambda S-A\in\Phi(X)$$

and

$$i(\lambda S-A)=0.$$

This shows that

$$\lambda\notin\sigma_{e5,S}(A).$$

Conversely, since $\mathscr{K}(X)\subset\mathscr{S}^1_{A,S}(X)$, then

$$\bigcap_{K\in\mathscr{S}^1_{A,S}(X)}\sigma_S(A+K)\subset\sigma_{e5,S}(A).$$

Hence,

$$\sigma_{e5,S}(A)=\bigcap_{K\in\mathscr{S}^1_{A,S}(X)}\sigma_S(A+K),$$

which completes the proof of theorem. Q.E.D.

Theorem 7.1.2 *Let $A\in\mathscr{C}(X)$ and $S\in\mathscr{L}(X)$. Then,*

$$\sigma_{e5,S}(A)=\bigcap_{K\in\mathscr{S}^2_{A,S}(X)}\sigma_S(A+K).$$ ◊

Proof. Since $\mathscr{K}(X)\subset\mathscr{S}^2_{A,S}(X)$, then

$$\bigcap_{K\in\mathscr{S}^2_{A,S}(X)}\sigma_S(A+K)\subset\sigma_{e5,S}(A).$$

Now, we may prove that

$$\sigma_{e5,S}(A)\subset\bigcap_{K\in\mathscr{S}^2_{A,S}(X)}\sigma_S(A+K).$$

Indeed, if

$$\lambda\notin\bigcap_{K\in\mathscr{S}^2_{A,S}(X)}\sigma_S(A+K),$$

then there exists $K \in \mathscr{S}^2_{A,S}(X)$ such that

$$\lambda \notin \sigma_S(A+K).$$

So,

$$f([K(\lambda S - A - K)^{-1}]^n) < 1$$

for some $n \in \mathbb{N}$, where $f(\cdot)$ is a measure of non-strict-singularity, given in (2.22). Hence, by applying Proposition 2.6.4, we get

$$I + K(\lambda S - A - K)^{-1} \in \Phi^b(X)$$

and

$$i(I + K(\lambda S - A - K)^{-1}) = 0.$$

Writing

$$\lambda S - A = [I + K(\lambda S - A - K)^{-1}](\lambda S - A - K),$$

we can deduce that

$$\lambda S - A \in \Phi(X)$$

and

$$i(\lambda S - A) = 0.$$

This shows that $\lambda \notin \sigma_{e5,M}(A)$. Then,

$$\sigma_{e5,S}(A) \subset \bigcap_{K \in \mathscr{S}^2_{A,S}(X)} \sigma_S(A+K).$$

Hence,

$$\sigma_{e5,S}(A) = \bigcap_{K \in \mathscr{S}^2_{A,S}(X)} \sigma_S(A+K),$$

which completes the proof of theorem. Q.E.D.

Corollary 7.1.1 *Let $A \in \mathscr{C}(X)$, $S \in \mathscr{L}(X)$, and $\mathscr{M}(X)$ be any subset of $\mathscr{L}(X)$ satisfying*

$$\mathscr{K}(X) \subset \mathscr{M}(X) \subset \mathscr{S}^1_{A,S}(X)$$

or

$$\mathscr{K}(X) \subset \mathscr{M}(X) \subset \mathscr{S}^2_{A,S}(X).$$

Then,

$$\sigma_{e5,S}(A) = \bigcap_{K \in \mathcal{M}(X)} \sigma_S(A+K). \qquad \diamondsuit$$

7.2 THE S-ESSENTIAL SPECTRA OF 2×2 BLOCK OPERATOR MATRICES

During the last years, the following papers [39, 111, 120, 146] were devoted to the study of the $\mathcal{I}$-essential spectra of operators defined by a 2×2 block operator matrices that

$$\mathcal{L}_0 = \begin{pmatrix} A & B \\ C & D \end{pmatrix} \tag{7.1}$$

acts on the product $X \times Y$ of Banach spaces, where $\mathcal{I}$ is the identity operator defined on the product space $X \times Y$ by

$$\mathcal{I} = \begin{pmatrix} I & 0 \\ 0 & I \end{pmatrix}.$$

Let S be a bounded operator formally defined on the product space $X \times Y$ by

$$S = \begin{pmatrix} M_1 & M_2 \\ M_3 & M_4 \end{pmatrix},$$

where operator M_1 acts on X and everywhere defined and the intertwining operator M_2 (resp. M_3) acts on the Banach space Y (resp. on X) everywhere defined and are strictly singular. The operator M_4 acts on Y and everywhere defined and $\mathcal{L}_0$ is given by Eq. (7.1), where the operator A acts on X and has domain $\mathcal{D}(A)$, D is defined on $\mathcal{D}(D)$ and acts on the Banach space Y and the intertwining operator B (resp. C) is defined on the domain $\mathcal{D}(B)$ (resp. $\mathcal{D}(C)$) and acts on Y into X (resp. Y into Y). The purpose of this section is to discuss the S-essential spectrum of the 2×2 matrix operator $\mathcal{L}_0$.

In what follows, we will assume that the following conditions, introduced by M. Faierman, R. Mennicken and M. Muller in [69], are hold:

$(H1)$ A is closed, densely defined linear operator on X with non empty M_1-resolvent set $\rho_{M_1}(A)$.

$(H2)$ The operator B is densely defined linear operator on X and for some (hence for all) $\mu \in \rho_{M_1}(A)$, the operator $(A-\mu M_1)^{-1}B$ is closable.

$(H3)$ The operator C satisfies $\mathscr{D}(A) \subset \mathscr{D}(C)$, and for some (hence for all) $\mu \in \rho_{M_1}(A)$, the operator $C(A-\mu M_1)^{-1}$ is bounded (in particular, if C is closable, then $C(A-\mu M_1)^{-1}$ is bounded).

$(H4)$ The lineal $\mathscr{D}(B)\bigcap\mathscr{D}(D)$ is dense in Y, and for some (hence for all) $\mu \in \rho_{M_1}(A)$, the operator $D-C(A-\mu M_1)^{-1}B$ is closable, we will denote by $S(\mu)$ the closure of the operator $D-(C-\mu M_3)(A-\mu M_1)^{-1}(B-\mu M_2)$.

Remark 7.2.1 (i) *It follows, from the closed graph theorem that the operator*

$$G(\mu) := \overline{(A-\mu M_1)^{-1}(B-\mu M_2)}$$

is bounded on Y.

(ii) *We emphasize that neither the domain of $S(\mu)$ nor the property of being closable depend on μ. Indeed, consider λ, $\mu \in \rho_{M_1}(A)$, then we have*

$$S(\lambda)-S(\mu) = (\mu-\lambda)\Big[M_3G(\mu)+F(\lambda)M_2+F(\lambda)M_1G(\mu)\Big],$$

where

$$F(\lambda) = (C-\lambda M_3)(A-\lambda M_1)^{-1}.$$

Since the operators $F(\lambda)$ and $G(\mu)$ are bounded (see the condition $(H3)$ and (i), respectively), then the difference $S(\lambda)-S(\mu)$ is bounded. Therefore, neither the domain of $S(\mu)$ nor the property of being closable depend on μ. ◇

We recall the following result which describes the closure of the operator $\mathscr{L}_0$.

Theorem 7.2.1 *(M. Faierman, R. Mennicken, and M. Muller [69]) Let conditions* $(H1)$*-*$(H3)$ *be satisfied and the lineal* $\mathscr{D}(B)\bigcap\mathscr{D}(D)$ *be dense in* Y*. Then, the operator* $\mathscr{L}_0$ *is closable if, and only if, the operator* $D-C(A-\mu M_1)^{-1}B$*, is closable in* Y*, for some* $\mu\in\rho_{M_1}(A)$*. Moreover, the closure* $\mathscr{L}$ *of* $\mathscr{L}_0$ *is given by*

$$\mathscr{L}=\mu S+\begin{pmatrix} I & 0 \\ F(\mu) & I \end{pmatrix}\begin{pmatrix} A-\mu M_1 & 0 \\ 0 & S(\mu)-\mu M_4 \end{pmatrix}\begin{pmatrix} I & G(\mu) \\ 0 & I \end{pmatrix}. \tag{7.2}$$

◇

For $n\in\mathbb{N}$, let

$$\mathscr{I}_n(X)=\Big\{K\in\mathscr{L}(X)\text{ satisfying }f\Big((KB)^n\Big)<1\text{ for all }B\in\mathscr{L}(X)\Big\},$$

where $f(\cdot)$ is a measure of non-strict-singularity, given in (2.22). We have the following inclusion

$$\mathscr{S}(X)\subset\mathscr{I}_n(X).$$

Theorem 7.2.2 *(N. Moalla [122, Theorem 3.2 (i)]) Let* $A\in\Phi(X)$*, then for all* $K\in\mathscr{I}_n(X)$*, we have* $A+K\in\Phi(X)$ *and*

$$i(A+K)=i(A). \qquad \diamondsuit$$

Remark 7.2.2 *(i) If* $K\in\mathscr{I}_n(X)$ *and* $A\in\mathscr{L}(X)$*, then* $KA\in\mathscr{I}_n(X)$*.*

(ii) If $K\in\mathscr{I}_n(X)$ *and* $S\in\mathscr{S}(X)$*, then*

$$K+S\in\mathscr{I}_n(X). \qquad \diamondsuit$$

Let $f(\cdot)$ be a measure of non-strict-singularity, given in (2.22). In all that follows we will make the following assumption

$$(H5):\begin{cases} f\Big(M_1G(\mu)HM_1G(\mu)K\Big)<\frac{1}{4},\ f\Big(F(\mu)M_1HF(\mu)M_1K\Big)<\frac{1}{4}\\ f\Big(M_1G(\mu)HF(\mu)M_1K\Big)<\frac{1}{4},\ f\Big(F(\mu)M_1HM_1G(\mu)K\Big)<\frac{1}{4}\\ \text{for some }\mu\in\rho_{M_1}(A)\text{ and for all bounded operators }H\text{ and }K.\end{cases}$$

Remark 7.2.3 *(i) Note that if $G(\mu)$ and $F(\mu)$ are strictly singular operators, then hypothesis $(H5)$ is satisfied.*

(ii) If the hypothesis

$$f\Big(F(\mu)M_1HM_1G(\mu)K\Big) < \frac{1}{4} \tag{7.3}$$

for all bounded operators H and K, then $F(\mu)M_1G(\mu)$ is strictly singular. Indeed, since Eq. (7.3) is valid for all bounded operators K and H, we can consider

$$K = n^2 I_X$$

and

$$HM_1 = I_Y$$

(where $n \in \mathbb{N}^*$, I_X and I_Y denote the identity operator on X and Y, respectively). We obtain

$$f\Big(F(\mu)M_1G(\mu)\Big) < \frac{1}{4n^2}.$$

So,

$$f\Big(F(\mu)M_1G(\mu)\Big) = 0$$

and this implies that $F(\mu)M_1G(\mu)$ is strictly singular. ◇

Theorem 7.2.3 *Let the matrix operator $\mathscr{L}_0$ satisfy conditions $(H1)$-$(H4)$ and assume that hypothesis $(H5)$ is satisfied, then*

$$\sigma_{e5,S}(\mathscr{L}) \subseteq \sigma_{e5,M_1}(A)\bigcup \sigma_{e5,M_4}(S(\mu)).$$

Moreover, if Φ_{A,M_1} is connected, then

$$\sigma_{e5,S}(\mathscr{L}) = \sigma_{e5,M_1}(A)\bigcup \sigma_{e5,M_4}(S(\mu)).$$ ◇

Proof. Let $\mu \in \rho_{M_1}(A)$ be such that hypothesis $(H5)$ is satisfied and set λ be a complex number. It follows from Eq. (7.2) that

$$\lambda S - \mathcal{L} = UV(\lambda)W - (\lambda - \mu)\begin{pmatrix} 0 & M_1G(\mu) - M_2 \\ F(\mu)M_1 - M_3 & F(\mu)M_1G(\mu) \end{pmatrix}, \tag{7.4}$$

where

$$U = \begin{pmatrix} I & 0 \\ F(\mu) & I \end{pmatrix},$$

$$W = \begin{pmatrix} I & G(\mu) \\ 0 & I \end{pmatrix},$$

and

$$V(\lambda) = \begin{pmatrix} \lambda M_1 - A & 0 \\ 0 & \lambda M_4 - S(\mu) \end{pmatrix}.$$

Let

$$\mathcal{K} = \begin{pmatrix} K_1 & K_2 \\ K_3 & K_4 \end{pmatrix}$$

be a bounded operator, formally defined on the product space $X \times Y$. Then,

$$\left[\begin{pmatrix} 0 & M_1G(\mu) \\ F(\mu)M_1 & 0 \end{pmatrix}\mathcal{K}\right]^2 = \begin{pmatrix} \mathcal{J}_1 & \mathcal{J}_2 \\ \mathcal{J}_3 & \mathcal{J}_4 \end{pmatrix}$$

where

$$\mathcal{J}_1 = (M_1G(\mu)K_3)^2 + M_1G(\mu)K_4F(\mu)M_1K_1,$$

$$\mathcal{J}_2 = M_1G(\mu)K_3M_1G(\mu)K_4 + M_1G(\mu)K_4F(\mu)M_1K_2,$$

$$\mathcal{J}_3 = F(\mu)M_1K_1M_1G(\mu)K_3 + F(\mu)M_1K_2F(\mu)M_1K_1,$$

and

$$\mathcal{J}_4 = F(\mu)M_1K_1M_1G(\mu)K_4 + (F(\mu)M_1K_2)^2.$$

It follows from both hypothesis $(H5)$ and Lemma 2.6.2 that

$$g\left(\left[\begin{pmatrix} 0 & M_1G(\mu) \\ F(\mu)M_1 & 0 \end{pmatrix}\mathcal{K}\right]^2\right) < 1,$$

which implies that the operator

$$\begin{pmatrix} 0 & M_1G(\mu) \\ F(\mu)M_1 & 0 \end{pmatrix} \in \mathscr{I}_2(X\times Y),$$

where $g(\cdot)$ is a measure of non-strict-singularity on the space $\mathscr{L}(X\times Y)$, given in (2.23). Then, we can deduce from Remark 7.2.2 (ii) and the fact that $F(\mu)M_1G(\mu)$, M_2 and M_3 are strictly singular that

$$\begin{pmatrix} 0 & M_1G(\mu)-M_2 \\ F(\mu)M_1-M_3 & F(\mu)M_1G(\mu) \end{pmatrix} \in \mathscr{I}_2(X\times Y).$$

Now, if we use Eq. (7.4) and apply Theorem 7.2.2, we can conclude that the operator $\lambda S-\mathscr{L}$ is a Fredholm operator if, and only if, $UV(\lambda)W$ is a Fredholm operator. Also observe that the operator U and W are bounded and have bounded inverse. Hence, the operator $UV(\lambda)W$ is a Fredholm operator if, and only if, $V(\lambda)$ has this property if, and only if, λM_1-A and $\lambda M_4-S(\mu)$ are Fredholm operators. By Theorem 2.2.19, we have

$$\begin{aligned} i(\lambda S-\mathscr{L}) &= i(U)+i(V(\lambda))+i(W) \\ &= 0+i(V(\lambda))+0. \end{aligned}$$

So,

$$i(\lambda S-\mathscr{L}) = i(\lambda M_1-A)+i(\lambda M_4-S(\mu)). \tag{7.5}$$

Let $\lambda \notin \left(\sigma_{e5,M_1}(A)\bigcup\sigma_{e5,M_4}(S(\mu))\right)$. The use of Proposition 3.4.1, we get λM_1-A and $\lambda M_4-S(\mu)$ are Fredholm operators and

$$i(\lambda M_1-A) = i(\lambda M_4-S(\mu)) = 0.$$

Then, $\lambda S-\mathscr{L}$ is Fredholm and

$$i(\lambda S-\mathscr{L}) = 0.$$

So,

$$\lambda \notin \sigma_{e5,S}(\mathscr{L}).$$

This shows that

$$\sigma_{e5,S}(\mathscr{L}) \subseteq \sigma_{e5,M_1}(A)\bigcup\sigma_{e5,M_4}(S(\mu)).$$

Now, let $\lambda \notin \sigma_{e5,S}(\mathscr{L})$. The use of Proposition 3.4.1, we get

$$\lambda S - \mathscr{L}$$

is a Fredholm operator and

$$i(\lambda S - \mathscr{L}) = 0.$$

Then, $\lambda M_1 - A$ and $\lambda M_4 - S(\mu)$ are Fredholm operators. Since Φ_{A,M_1} is connected and $\rho_{M_1}(A) \neq \emptyset$ (see, hypothesis $(H1)$), then by using Corollary 3.4.1, we get

$$i(\lambda M_1 - A) = 0.$$

By, Eq. (7.5), we have

$$i(\lambda M_4 - S(\mu)) = 0.$$

This shows that

$$\sigma_{e5,S}(\mathscr{L}) = \sigma_{e5,M_1}(A) \bigcup \sigma_{e5,M_4}(S(\mu)). \qquad \text{Q.E.D.}$$

Chapter 8

S-Pseudospectra and Structured S-Pseudospectra

This chapter deals with the structured S-pseudospectra of a closed densely defined linear operator on a Banach spaces. Precisely, we define the structured S-pseudospectra and we give some characterizations of this set. In particular, we establish a relationship between structured S-essential pseudospectra and S-essential spectra.

8.1 STUDY OF THE *S*-PSEUDOSPECTRA

Theorem 8.1.1 *Let* $A \in \mathscr{C}(X)$, $S \in \mathscr{L}(X)$ *and* $\varepsilon > 0$, *then*

$$\sigma_{S,\,\varepsilon}(A) = \bigcup_{\|D\|<\varepsilon} \sigma_S(A+D). \qquad \diamondsuit$$

Proof. • Let $\lambda \in \sigma_{S,\,\varepsilon}(A)$, we try to find $D \in \mathscr{L}(X)$ such that $\|D\| < \varepsilon$ and

$$\lambda \in \sigma_S(A+D).$$

If $\lambda \in \sigma_S(A)$, in this case we take $D = 0$. If $\lambda \in \sigma_{S,\varepsilon}(A)\backslash\sigma_S(A)$, then $\lambda \in \rho_S(A)$ and

$$\|R_S(\lambda,A)\| > \frac{1}{\varepsilon}.$$

Since

$$\|R_S(\lambda,A)\| = \sup_{w\neq 0} \frac{\|R_S(\lambda,A)w\|}{\|w\|}$$

and according to the definition of the upper bound, we obtain that there exists $w \neq 0$ such that

$$\|R_S(\lambda,A)w\| > \frac{\|w\|}{\varepsilon}.$$

Therefore, we denote by $v = R_S(\lambda,A)w$, then $v \in \mathscr{D}(A)$ and

$$(\lambda S - A)v = w.$$

So,

$$\|v\| > \frac{\|(\lambda S - A)v\|}{\varepsilon}.$$

Since $v \neq 0$, then $\varepsilon > \|(\lambda S - A)u\|$ and

$$u = \frac{v}{\|v\|}.$$

We have $\|u\| = 1$ so $u \neq 0$ and by using the Hahn-Banach theorem (see Theorem 2.1.4), we obtain the existence of $\psi \in X^*$ such that

$$\psi(u) = \|u\|$$

and

$$\|\psi\| = 1.$$

In this case, we denote by

$$Dx = \psi(x)(\lambda S - A)u.$$

It is obviously clear that D is a linear operator. Furthermore, for $x \in X$, we have

$$\begin{aligned}
\|Dx\| &= \|\psi(x)(\lambda S - A)u\|, \\
&= |\psi(x)|\|(\lambda S - A)u\|, \\
&\leq \|\psi\|\|x\|\|(\lambda S - A)u\|, \\
&< \varepsilon\|x\|.
\end{aligned}$$

So, D is a bounded operator. We see immediately that

$$\|D\| < \varepsilon.$$

We have

$$Du = \psi(u)(\lambda S - A)u = (\lambda S - A)u.$$

Then, for $u \neq 0$, we have

$$[\lambda S - (A + D)]\, u = 0.$$

So, $\lambda S - (A + D)$ is not one-to-one. Consequently,

$$\lambda \in \sigma_S(A + D).$$

• Let $\lambda \in \bigcup_{\|D\|<\varepsilon} \sigma_S(A + D)$, then there exists $D \in \mathscr{L}(X)$ such that $\|D\| < \varepsilon$ and $\lambda \in \sigma_S(A + D)$. We prove that $\lambda \in \sigma_{S,\varepsilon}(A)$. We argue by contradiction. Suppose that $\lambda \in \rho_{S,\varepsilon}(A)$, then

$$\lambda S - A - D = (\lambda S - A)\left[I - R_S(\lambda, A)D\right].$$

Furthermore,

$$\|R_S(\lambda, A)D\| \leq \|R_S(\lambda, A)\| \|D\| < 1$$

this implies that $I - R_S(\lambda, A)D$ is invertible. Hence,

$$\lambda \in \rho_S(A + D)$$

which is a contradiction. Q.E.D.

Corollary 8.1.1 *Let $A \in \mathscr{C}(X)$, $S \in \mathscr{L}(X)$ and $\varepsilon > 0$ such that $\|S\| \leq 1$, then*

(i) $\sigma_S(A) + \mathbb{B}(0, \varepsilon) \subseteq \sigma_{S,\varepsilon}(A)$, where $\mathbb{B}(0, \varepsilon)$ is the open disk centered at $O(0,0)$ and with radius ε.

(ii) $\sigma_{S,\varepsilon}(A) + \mathbb{B}(0, \delta) \subseteq \sigma_{S,\varepsilon+\delta}(A)$. ◇

Proof. (*i*) Let $\lambda \in \sigma_S(A) + \mathbb{B}(0, \varepsilon)$, then there exists $\lambda_1 \in \sigma_S(A)$ and $\lambda_2 \in \mathbb{B}(0, \varepsilon)$ such that $\lambda = \lambda_1 + \lambda_2$. Since $\lambda_1 \in \sigma_S(A)$, then

$$\lambda_1 + \lambda_2 \in \sigma_S(A + \lambda_2 S).$$

Furthermore,

$$\|\lambda_2 S\| = |\lambda_2| \|S\| < \varepsilon.$$

In this case, we denote by $D := \lambda_2 S$. So, $D \in \mathscr{L}(X)$, $\|D\| < \varepsilon$ and

$$\lambda \in \sigma_S(A + D).$$

Consequently,

$$\lambda \in \sigma_{S,\varepsilon}(A).$$

(*ii*) Let $\lambda \in \sigma_{S,\varepsilon}(A) + \mathbb{B}(0, \delta)$, then there exists $\lambda_1 \in \sigma_{S,\varepsilon}(A)$ and $\lambda_2 \in \mathbb{B}(0, \delta)$ such that

$$\lambda = \lambda_1 + \lambda_2.$$

Since $\lambda_1 \in \sigma_{S,\varepsilon}(A)$, then there exists $B \in \mathscr{L}(X)$ such that $\|B\| < \varepsilon$ and $\lambda_1 \in \sigma_S(A + B)$. Therefore,

$$\lambda = \lambda_1 + \lambda_2 \in \sigma_S(A + B + \lambda_2 S).$$

In addition, we have $B + \lambda_2 S \in \mathscr{L}(X)$ with

$$\|B + \lambda_2 S\| \leq \|B\| + \|\lambda_2 S\| < \varepsilon + \delta.$$

Consequently,

$$\lambda \in \sigma_{S,\varepsilon+\delta}(A). \qquad \text{Q.E.D.}$$

Theorem 8.1.2 *Let $A \in \mathscr{C}(X)$, $S \in \mathscr{L}(X)$ and $\varepsilon > 0$, then*

$$\sigma_{S,\varepsilon}(A) = \sigma_S(A) \bigcup \Big\{\lambda \in \mathbb{C} \text{ such that } \exists\, x \in \mathscr{D}(A) \text{ and } \|(\lambda S - A)x\| < \varepsilon \|x\|\Big\}. \quad \diamond$$

Proof. • Let $\lambda \in \sigma_{S,\varepsilon}(A)$ this means that $\lambda \in \sigma_S(A)$ or

$$\|R_S(\lambda,A)\| > \frac{1}{\varepsilon}.$$

If $\lambda \in \sigma_{S,\varepsilon}(A)\backslash\sigma_S(A)$, then

$$\sup_{u\neq 0}\frac{\|R_S(\lambda,A)u\|}{\|u\|} > \frac{1}{\varepsilon}.$$

Using the definition of the upper bound, we obtain that there exists $u \neq 0$ such that

$$\|R_S(\lambda,A)u\| > \frac{\|u\|}{\varepsilon}.$$

Therefore, we denote by $v := R_S(\lambda,A)u$ this implies that

$$v \in \mathscr{D}(A) \text{ and } \|(\lambda S - A)v\| < \varepsilon\|v\|.$$

• Let $\lambda \in \mathbb{C}$ such that there is $x \in \mathscr{D}(A)$ and

$$\|(\lambda S - A)x\| < \varepsilon\|x\|$$

or $\lambda \in \sigma_S(A)$. If $\lambda \notin \sigma_S(A)$ and we set

$$v = (\lambda S - A)x$$

this implies that

$$x = R_S(\lambda,A)v.$$

In this case, we obtain that

$$\|v\| < \varepsilon\|R_S(\lambda,A)v\|.$$

Since $v \neq 0$ it gives that

$$\|R_S(\lambda,A)v\| > \frac{1}{\varepsilon}\|v\|.$$

We see that

$$\|R_S(\lambda,A)\| > \frac{1}{\varepsilon},$$

and so $\lambda \in \sigma_{S,\varepsilon}(A)$. Q.E.D.

Theorem 8.1.3 *Let $A \in \mathscr{C}(X)$, $S \in \mathscr{L}(X)$ and $\varepsilon > 0$, then*

$$\sigma_{S,\varepsilon}(A) = \sigma_S(A) \cup \Big\{\lambda \in \mathbb{C} : \exists\, x_n \in \mathscr{D}(A),\ \|x_n\| = 1 \text{ and } \lim_{n\to+\infty} \|(\lambda S - A)x_n\| < \varepsilon\Big\}. \quad \diamondsuit$$

Proof. • Let $\lambda \in \sigma_{S,\varepsilon}(A)\backslash\sigma_S(A)$, then $\lambda \in \rho_S(A)$ and

$$\|R_S(\lambda,A)\| > \frac{1}{\varepsilon}.$$

On the one hand,

$$\|R_S(\lambda,A)\| = \sup_{\|y\|=1} \|R_S(\lambda,A)y\|$$

means exactly that for every $n \in \mathbb{N}^*$, there exists $(y_n)_n$ such that $\|y_n\| = 1$ and

$$\|R_S(\lambda,A)\| - \frac{1}{n} < \|R_S(\lambda,A)y_n\| \leq \|R_S(\lambda,A)\|.$$

Consequently,

$$\lim_{n\to+\infty} \|R_S(\lambda,A)y_n\| = \|R_S(\lambda,A)\| > \frac{1}{\varepsilon}.$$

On the other hand, we denote by

$$x_n = \|R_S(\lambda,A)y_n\|^{-1} R_S(\lambda,A)y_n,$$

this implies that $\|x_n\| = 1$, $x_n \in \mathscr{D}(A)$ and

$$(\lambda S - A)x_n = \|R_S(\lambda,A)y_n\|^{-1} y_n.$$

Therefore,

$$\begin{aligned} \lim_{n\to+\infty} \|(\lambda S - A)x_n\| &= \lim_{n\to+\infty} \|R_S(\lambda,A)y_n\|^{-1}, \\ &= \left(\lim_{n\to+\infty} \|R_S(\lambda,A)y_n\|\right)^{-1}, \\ &< \varepsilon. \end{aligned}$$

• Let $\lambda \in \mathbb{C}$ such that there is $x_n \in \mathscr{D}(A)$, $\|x_n\| = 1$ and

$$\lim_{n\to+\infty} \|(\lambda S - A)x_n\| < \varepsilon$$

or $\lambda \in \sigma_S(A)$. If $\lambda \notin \sigma_S(A)$ and we denote by

$$y_n = \|(\lambda S - A)x_n\|^{-1}(\lambda S - A)x_n,$$

then $\|y_n\| = 1$,

$$R_S(\lambda, A)y_n = \|(\lambda S - A)x_n\|^{-1}x_n$$

and

$$\|R_S(\lambda, A)y_n\| = \|(\lambda S - A)x_n\|^{-1}.$$

We know that

$$\|R_S(\lambda, A)\| \geq \|R_S(\lambda, A)y_n\|$$

for $\|y_n\| = 1$, so

$$\lim_{n \to +\infty} \|R_S(\lambda, A)\| \geq \lim_{n \to +\infty} \|R_S(\lambda, A)y_n\|.$$

Therefore,

$$\|R_S(\lambda, A)\| \geq \lim_{n \to +\infty} \|(\lambda S - A)x_n\|^{-1}.$$

Thus,

$$\|R_S(\lambda, A)\| \geq \left(\lim_{n \to +\infty} \|(\lambda S - A)x_n\|\right)^{-1},$$

Consequently,

$$\|R_S(\lambda, A)\| > \tfrac{1}{\varepsilon}. \qquad \text{Q.E.D.}$$

8.2 CHARACTERIZATION OF THE STRUCTURED S-PSEUDOSPECTRA

Our first result is the following theorem.

Theorem 8.2.1 *Let $A \in \mathscr{C}(X)$, $S \in \mathscr{L}(X)$, $B \in \mathscr{L}(X, Y)$, $C \in \mathscr{L}(Z, X)$, and $\varepsilon > 0$, then*

$$\sigma_S(A, B, C, \varepsilon) = \sigma_S(A) \bigcup \left\{ z \in \mathbb{C} \text{ such that } \|BR_S(z, A)C\| > \frac{1}{\varepsilon} \right\}. \quad \diamond$$

Proof. We will discuss the following two cases:

1st case : If B or $C = 0$, then the result is immediately.

2nd case : If $B \neq 0$ and $C \neq 0$, then we can see immediately

$$\sigma_S(A) \subseteq \sigma_S(A,B,C,\varepsilon).$$

Let $z \notin \sigma_S(A)$. If $\|BR_S(z,A)C\| \leq \frac{1}{\varepsilon}$, then for every $\|D\| < \varepsilon$, we have

$$\|DBR_S(z,A)C\| < 1.$$

Therefore,

$$I - DBR_S(z,A)C$$

is invertible. In view of Theorem 2.1.6, we obtain that for any $\|D\| < \varepsilon$

$$1 \notin \sigma(DBR_S(z,A)C)$$

means exactly

$$1 \notin \sigma(CDBR_S(z,A)).$$

Since we can write

$$zS - A - CDB = (I - CDBR_S(z,A))\,(zS - A),$$

this implies

$$z \notin \sigma_S(A + CDB) \;\; \forall \; \|D\| < \varepsilon.$$

For the other inclusion, if $z \notin \sigma_S(A)$, then

$$\|BR_S(z,A)C\| > \frac{1}{\varepsilon}.$$

It follows

$$\|R_S(z,A)\| > \frac{1}{\varepsilon\|B\|\|C\|}.$$

According to the definition of the upper bound, we have the existence of $w \neq 0$ such that

$$\|R_S(z,A)w\| > \frac{\|w\|}{\varepsilon\|B\|\|C\|}.$$

Therefore, set $v = R_S(z,A)w$, then $v \in \mathscr{D}(A)$ and

$$(zS - A)v = w.$$

Which leads to

$$\|(zS - A)u\| < \varepsilon\|B\|\|C\| \text{ for } \|u\| = 1.$$

Using Hahn-Banach theorem (see Theorem 2.1.4), we obtain the existence of $\psi \in X^*$ such that

$$\psi(u) = \|u\|$$

and $\|\psi\| = 1$. Set

$$CDBx = \psi(x)(zS - A)u.$$

So,

$$\begin{aligned}\|CDBx\| &= |\psi(x)|\|(zS - A)u\| \\ &\leq \|\psi\|\|x\|\|(zS - A)u\| \\ &< \varepsilon\|x\|\|B\|\|C\|.\end{aligned}$$

This implies that $D \in \mathscr{L}(Y,Z)$. We see immediately that $\|D\| < \varepsilon$. In addition, for $u \neq 0$, we have

$$[zS - (A + CDB)]u = 0.$$

Consequently,

$$z \in \sigma_S(A,B,C,\varepsilon). \qquad \text{Q.E.D.}$$

Proposition 8.2.1 *Let $A \in \mathscr{C}(X)$, $S \in \mathscr{L}(X)$, $B \in \mathscr{L}(X,Y)$, $C \in \mathscr{L}(Z,X)$, and $\varepsilon > 0$ such that the operator B (resp. C) is not null. If B is one-to-one, then $\sigma_S(A,B,C,\varepsilon) \neq \emptyset$.* ◇

Proof. Suppose that $\sigma_S(A,B,C,\varepsilon)=\emptyset$, then

$$\rho_S(A)=\mathbb{C}$$

and

$$\left\{\lambda\in\mathbb{C} \text{ such that } \|BR_S(\lambda,A)C\|\leq\frac{1}{\varepsilon}\right\}=\mathbb{C}.$$

Consider the function

$$\begin{array}{rll}\varphi: & \mathbb{C} & \longrightarrow \mathscr{L}(Z,Y)\\ & \lambda & \longrightarrow BR_S(\lambda,A)C.\end{array}$$

Since φ is analytic on $\mathbb{C}$ and for every $\lambda\in\mathbb{C}$, we have

$$\|\varphi(\lambda)\|\leq\frac{1}{\varepsilon},$$

then φ is an entire bounded function. Therefore, using Liouville theorem (see Theorem 2.1.5), we obtain that φ is a constant function. It follows that $BR_S(\lambda,A)C$ is a null operator and so is C, which is impossible. Q.E.D.

Proposition 8.2.2 *Let $A\in\mathscr{C}(X)$, $S\in\mathscr{L}(X)$, $B\in\mathscr{L}(X,Y)$, $C\in\mathscr{L}(Z,X)$, and $\varepsilon>0$, then the structured S-pseudospectrum verifies the following properties:*

(i) $\sigma_S(A,B,C,\varepsilon)\neq\emptyset$,

(ii) the structured S-pseudospectra $\{\sigma_S(A,B,C,\varepsilon)\}_{\varepsilon>0}$ *are increasing sets in the sense of inclusion relative to the strictly positive parameter* ε, *and*

(iii) $\bigcap\limits_{\varepsilon>0}\sigma_S(A,B,C,\varepsilon)=\sigma_S(A).$ ◇

Proof. *(i)* Suppose that $\sigma_S(A,B,C,\varepsilon)=\emptyset$, then

$$\rho_S(A)=\mathbb{C}$$

and

$$\left\{\lambda\in\mathbb{C} \text{ such that } \|BR_S(\lambda,A)C\|\leq\frac{1}{\varepsilon}\right\}=\mathbb{C}.$$

Consider the function

$$\begin{array}{rll} \varphi: & \mathbb{C} & \longrightarrow \mathscr{L}(Z,Y) \\ & \lambda & \longrightarrow BR_S(\lambda,A)C. \end{array}$$

Since φ is analytic on $\mathbb{C}$ and for every $\lambda \in \mathbb{C}$, we have

$$\|\varphi(\lambda)\| \leq \frac{1}{\varepsilon},$$

then φ is an entire bounded function. Therefore, using Liouville theorem (see Theorem 2.1.5), we obtain that φ is a constant function. It follows that $R_S(\lambda,A)$ is a null operator, which is impossible.

(ii) Let ε_1, $\varepsilon_2 > 0$ such that $\varepsilon_1 < \varepsilon_2$. Let $\lambda \in \sigma_S(A,B,C,\varepsilon_1)$, then

$$\|BR_S(\lambda,A)C\| > \frac{1}{\varepsilon_1} > \frac{1}{\varepsilon_2}.$$

Therefore,

$$\lambda \in \sigma_S(A,B,C,\varepsilon_2).$$

(iii) We have

$$\begin{aligned} &\bigcap_{\varepsilon>0} \sigma_S(A,B,C,\varepsilon) \\ &\quad= \bigcap_{\varepsilon>0}\left(\sigma_S(A)\bigcup\left\{\lambda \in \mathbb{C} \text{ such that } \|BR_S(\lambda,A)C\| > \frac{1}{\varepsilon}\right\}\right) \\ &\quad= \sigma_S(A)\cup\left(\bigcap_{\varepsilon>0}\left\{\lambda \in \mathbb{C} \text{ such that } \|BR_S(\lambda,A)C\| > \frac{1}{\varepsilon}\right\}\right). \end{aligned}$$

It suffices to prove that

$$\bigcap_{\varepsilon>0}\left\{\lambda \in \mathbb{C} \text{ such that } \|BR_S(\lambda,A)C\| > \frac{1}{\varepsilon}\right\} = \{\lambda \in \mathbb{C} \text{ such that } \|BR_S(\lambda,A)C\| = +\infty\}.$$

For the inclusion in the direct sense: let $\lambda \in \mathbb{C}$ such that for all $\varepsilon > 0$

$$\|BR_S(\lambda,A)C\| > \frac{1}{\varepsilon}.$$

This implies that

$$\lim_{\varepsilon\to 0^+} \|BR_S(\lambda,A)C\| = +\infty.$$

Consequently,

$$\|BR_S(\lambda,A)C\| = +\infty.$$

For the other inclusion: let $\lambda \in \mathbb{C}$ such that

$$\|BR_S(\lambda,A)C\| = +\infty > \frac{1}{\varepsilon}$$

for all $\varepsilon > 0$. Hence,

$$\begin{aligned} \bigcap_{\varepsilon>0} \sigma_S(A,B,C,\varepsilon) &= \sigma_S(A) \bigcup \Big\{ \lambda \in \mathbb{C} \text{ such that } \|BR_S(\lambda,A)C\| = +\infty \Big\}, \\ &= \sigma_S(A), \end{aligned}$$

which completes the proof of proposition. Q.E.D.

Corollary 8.2.1 *Let $A \in \mathscr{C}(X)$, $S \in \mathscr{L}(Y,Z)$, $B \in \mathscr{L}(X,Y)$, $C \in \mathscr{L}(Z,X)$ and $\varepsilon > 0$ such that $\|S\| \leq 1$. If $S_1 = CSB$, then*

(i) $\sigma_{S_1}(A) + \mathbb{B}(0,\varepsilon) \subseteq \sigma_{S_1}(A,B,C,\varepsilon)$, where $\mathbb{B}(0,\varepsilon)$ is the open disk centered at $O(0,0)$ and with radius ε.

(ii) $\sigma_{S_1}(A,B,C,\varepsilon) + \mathbb{B}(0,\delta) \subseteq \sigma_{S_1}(A,B,C,\varepsilon+\delta)$. ◊

Proof. (i) Let $\lambda \in \sigma_{S_1}(A) + \mathbb{B}(0,\varepsilon)$, then there exists $\lambda_1 \in \sigma_{S_1}(A)$ and $\lambda_2 \in \mathbb{B}(0,\varepsilon)$ such that

$$\lambda = \lambda_1 + \lambda_2.$$

Since $\lambda_1 \in \sigma_{S_1}(A)$, we have

$$\lambda_1 + \lambda_2 \in \sigma_{S_1}(A + \lambda_2 S_1).$$

In this case, we take $D := \lambda_2 S$. Consequently,

$$\lambda \in \sigma_{S_1}(A,B,C,\varepsilon).$$

(ii) Let $\lambda \in \sigma_{S_1}(A,B,C,\varepsilon) + \mathbb{B}(0,\delta)$, then there exists $\lambda_1 \in \sigma_{S_1}(A,B,C,\varepsilon)$ and $\lambda_2 \in \mathbb{B}(0,\delta)$ such that

$$\lambda = \lambda_1 + \lambda_2.$$

Since $\lambda_1 \in \sigma_{S_1}(A,B,C,\varepsilon)$, then there exists $D \in \mathscr{L}(Y,Z)$ such that $\|D\| < \varepsilon$ and

$$\lambda_1 \in \sigma_{S_1}(A + CDB).$$

In addition, $\lambda = \lambda_1 + \lambda_2 \in \sigma_{S_1}(A + CD_2B)$ with $D_2 := D + \lambda_2 S$. Therefore,

$$\lambda \in \sigma_{S_1}(A,B,C,\varepsilon+\delta). \qquad \text{Q.E.D.}$$

Theorem 8.2.2 *Let $\varepsilon > 0$, $A \in \mathscr{C}(X)$, $S \in \mathscr{L}(X)$, $B \in \mathscr{L}(X,Y)$, $C \in \mathscr{L}(Z,X)$ such that $S \neq A$ and $S \neq 0$, then*

$$\sigma_S(A,B,C,\varepsilon) = \sigma_S(A) \bigcup \Big\{\lambda \in \mathbb{C} : \exists\, x \in \mathscr{D}(A) \text{ and } \|(\lambda S - A)x\| < \varepsilon\|x\|\|B\|\|C\|\Big\}. \quad \diamondsuit$$

Proof. We will discuss these two cases:

<u>1st *case*</u> : If B or $C = 0$, then $\sigma_S(A,B,C,\varepsilon) = \sigma_S(A,\varepsilon)$.

<u>2nd *case*</u> : If B and $C \neq 0$. Let $\lambda \in \sigma_S(A,B,C,\varepsilon)\backslash\sigma_S(A)$, then

$$\|BR_S(\lambda,A)C\| > \frac{1}{\varepsilon}.$$

Using the definition of the upper bound, we obtain that there exists $u \neq 0$ such that

$$\|R_S(\lambda,A)u\| > \frac{\|u\|}{\varepsilon\|B\|\|C\|}.$$

Therefore, we denote by $v := R_S(\lambda,A)u$ this implies that $v \in \mathscr{D}(A)$ and

$$\|(\lambda S - A)v\| < \varepsilon\|v\|\|B\|\|C\|.$$

• Let $\lambda \in \mathbb{C}$ such that there is $x \in \mathscr{D}(A)$ and

$$\|(\lambda S - A)x\| < \varepsilon\|x\|\|B\|\|C\|.$$

Therefore,

$$\|(\lambda S - A)u\| < \varepsilon\|B\|\|C\|$$

and $\|u\| = 1$. Using the Hahn-Banach theorem (see Theorem 2.1.4), we obtain the existence of $\psi \in X^*$ such that

$$\psi(u) = \|u\|$$

and $\|\psi\| = 1$. Set $CDBx = \psi(x)(\lambda S - A)u$. So,

$$\begin{aligned}
\|CDBx\| &= |\psi(x)|\|(\lambda S - A)u\|,\\
&\leq \|\psi\|\|x\|\|(\lambda S - A)u\|,\\
&< \varepsilon\|x\|\|B\|\|C\|.
\end{aligned}$$

This implies that $D \in \mathscr{L}(Y,Z)$. We see immediately that $\|D\| < \varepsilon$. Furthermore, for $u \neq 0$, we have

$$[\lambda S - (A + CDB)]u = 0.$$

Hence,

$$\lambda \in \sigma_S(A,B,C,\varepsilon). \qquad \text{Q.E.D.}$$

Theorem 8.2.3 *Let A, $S \in \mathscr{L}(X)$, $B \in \mathscr{L}(X,Y)$, $C \in \mathscr{L}(Z,X)$, and $\varepsilon > 0$ such that $S \neq A + CDB$ for all $D \in \mathscr{L}(Y,Z)$ with $\|D\| < \varepsilon$. Then,*

$$\sigma_{ap,S}(A,B,C,\varepsilon) = \sigma_{p,S}(A,B,C,\varepsilon) \cup \bigcup_{\|D\|<\varepsilon} \Big\{\lambda \in \mathbb{C} : \; R(\lambda S - A - CDB) \text{ is not closed}\Big\}. \quad \diamondsuit$$

Proof. • We shall first show

$$\bigcup_{\|D\|<\varepsilon} \{\lambda \in \mathbb{C} : \; R(\lambda S - A - CDB) \text{ is not closed}\Big\} \subset \sigma_{ap,S}(A,B,C,\varepsilon).$$

Let $\lambda \notin \sigma_{ap,S}(A,B,C,\varepsilon)$, then for every $\|D\| < \varepsilon$, there exists $\alpha > 0$ such that for any $x \in X$, we have

$$\|(\lambda S - A - CDB)x\| \geq \alpha \, \|x\|. \tag{8.1}$$

Let $y_n \in R(\lambda S - A - CDB)$ such that $y_n \to y$ as $n \to +\infty$. Hence, $(y_n)_n$ is a Cauchy sequence with

$$y_n = (\lambda S - A - CDB)x_n.$$

On the other hand, using Eq. (8.1), we obtain that $(x_n)_n$ is a Cauchy sequence in X and this implies that it is convergent to x. Since $\lambda S - A - CDB$ is a bounded operator, then for any $\|D\| < \varepsilon$, the sequence

$$((\lambda S - A - CDB)x_n)_n$$

converges to $(\lambda S - A - CDB)x$. By uniqueness of the limit, we obtain that for every $\|D\| < \varepsilon$, we have

$$y = (\lambda S - A - CDB)x.$$

• Let $\lambda \in \sigma_{ap,S}(A,B,C,\varepsilon)\backslash\sigma_{p,S}(A,B,C,\varepsilon)$. We assume that for any $\|D\| < \varepsilon$ the range $R(\lambda S - A - CDB)$ is closed. Therefore, for any $\|D\| < \varepsilon$, we have

$$\lambda S - A - CDB$$

is one-to-one and onto from X into $R(\lambda S - A - CDB)$ which is a Banach space. Then, using the Banach theorem, we get for every $\|D\| < \varepsilon$,

$$(\lambda S - A - CDB)^{-1}$$

is a bounded operator from $R(\lambda S - A - CDB)$ into X. This implies that for every $\|D\| < \varepsilon$, there exists $M > 0$ such that for any $y \in R(\lambda S - A - CDB)$, we have

$$\|(\lambda S - A - CDB)^{-1}y\| \leq M\|y\|.$$

Set $x := (\lambda S - A - CDB)^{-1}y$, then for every $\|D\| < \varepsilon$ and for any $x \in X$

$$\|x\| \leq M\|(\lambda S - A - CDB)x\|. \tag{8.2}$$

Since $\lambda \in \sigma_{ap,S}(A,B,C,\varepsilon)$, then there exists a sequence $(x_n)_n$ of unit vectors in X and

$$\lim_{n\to+\infty} \|(\lambda S - A - CDB)x_n\| = 0.$$

The use of Eq. (8.2) leads to $1 \leq 0$ which is absurd. Q.E.D.

8.3 CHARACTERIZATION OF THE STRUCTURED S-ESSENTIAL PSEUDOSPECTRA

In all the sequel, we shall suppose that for every $\|D\| < \varepsilon$, we have $S \neq CDB$ and $S \neq A + CDB$.

Theorem 8.3.1 *Let A, $S \in \mathcal{L}(X)$, $B \in \mathcal{L}(X,Y)$, $C \in \mathcal{L}(Z,X)$ and $\varepsilon > 0$. Then,*

(i) If for every $\|D\| < \varepsilon$, *we have* $ACDB \in \mathcal{F}_+^b(X)$ *and* $SA = AS$, *then*

$$\sigma_{e1,S}(A,B,C,\varepsilon)\backslash\{0\} \subset \Big[\sigma_{e1,S}(A) \bigcup \bigcup_{\|D\|<\varepsilon} \sigma_{e1,S}(CDB)\Big]\backslash\{0\}.$$

If further, $CDBA \in \mathcal{F}_+^b(X)$ *for all* $\|D\| < \varepsilon$ *and* $S \in \Phi_+^b(X)$, *then*

$$\sigma_{e1,S}(A,B,C,\varepsilon)\backslash\{0\} = \Big[\sigma_{e1,S}(A) \bigcup \bigcup_{\|D\|<\varepsilon} \sigma_{e1,S}(CDB)\Big]\backslash\{0\}.$$

(ii) If for every $\|D\| < \varepsilon$, *we have* $ACDB \in \mathcal{F}_-^b(X)$ *and* $SCDB = CDBS$, *then*

$$\sigma_{e2,S}(A,B,C,\varepsilon)\backslash\{0\} \subset \Big[\sigma_{e2,S}(A) \bigcup \bigcup_{\|D\|<\varepsilon} \sigma_{e2,S}(CDB)\Big]\backslash\{0\}.$$

If further, $CDBA \in \mathcal{F}_-^b(X)$ *for all* $\|D\| < \varepsilon$ *and* $S \in \Phi_-^b(X)$, *then*

$$\sigma_{e2,S}(A,B,C,\varepsilon)\backslash\{0\} = \Big[\sigma_{e2,S}(A) \cup \bigcup_{\|D\|<\varepsilon} \sigma_{e2,S}(CDB)\Big]\backslash\{0\}. \quad \diamondsuit$$

Proof. • Let $\lambda \notin \Big[\sigma_{e1,S}(A) \cup \bigcup_{\|D\|<\varepsilon} \sigma_{e1,S}(CDB)\Big]$ or $\lambda = 0$.

If $\lambda \neq 0$, then

$$\lambda S - A \in \Phi_+^b(X)$$

and for any $\|D\| < \varepsilon$, we have

$$\lambda S - CDB \in \Phi_+^b(X).$$

Using Theorem 2.2.5, we obtain

$$(\lambda S - A)(\lambda S - CDB) \in \Phi_+^b(X)$$

for every $\|D\| < \varepsilon$. Since $ACDB \in \mathcal{F}_+^b(X)$ and applying both Eq. (3.16) and Lemma 2.3.1, we get

$$S(\lambda S - A - CDB) \in \Phi_+^b(X)$$

for all $\|D\| < \varepsilon$. Furthermore, using Theorem 2.2.10, we obtain that

$$\lambda S - A - CDB \in \Phi_+^b(X)$$

for all $\|D\| < \varepsilon$. Consequently,

$$\lambda \notin \sigma_{e1,S}(A,B,C,\varepsilon)\backslash\{0\}.$$

• Let $\lambda \notin \sigma_{e1,S}(A,B,C,\varepsilon)$ or $\lambda = 0$. If $\lambda \neq 0$, then

$$\lambda S - A - CDB \in \Phi_+^b(X)$$

for all $\|D\| < \varepsilon$. Since $S \in \Phi_+(X)$, then for all $\|D\| < \varepsilon$, we have

$$S(\lambda S - A - CDB) \in \Phi_+^b(X)$$

and

$$(\lambda S - A - CDB)S \in \Phi_+^b(X).$$

Since $ACDB \in \mathscr{F}_+^b(X)$, $CDBA \in \mathscr{F}_+^b(X)$, and applying Eqs. (3.16), (3.17) and Lemma 2.3.1, we obtain $\forall\ \|D\| < \varepsilon$

$$(\lambda S - A)(\lambda S - CDB) \in \Phi_+^b(X)$$

and

$$(\lambda S - CDB)(\lambda S - A) \in \Phi_+^b(X).$$

By using Theorem 2.2.10, it is clear that

$$\lambda S - A \in \Phi_+^b(X)$$

and

$$\lambda S - CDB \in \Phi_+^b(X)\ \forall\ \|D\| < \varepsilon.$$

Hence,

$$\lambda \notin \left[\sigma_{e1,S}(A) \bigcup \bigcup_{\|D\|<\varepsilon} \sigma_{e1,S}(CDB)\right]\backslash\{0\}.$$

In the same way we prove the item (ii). Q.E.D.

Theorem 8.3.2 *Let A, $S \in \mathscr{L}(X)$, $B \in \mathscr{L}(X,Y)$, $C \in \mathscr{L}(Z,X)$ and $\varepsilon > 0$. If for every $\|D\| < \varepsilon$, we have $ACDB \in \mathscr{F}^b(X)$, $\alpha(S) < \infty$ and $SA = AS$, then*

$$\sigma_{e4,S}(A,B,C,\varepsilon)\backslash\{0\} \subset \left[\sigma_{e4,S}(A) \bigcup \bigcup_{\|D\|<\varepsilon} \sigma_{e4,S}(CDB)\right]\backslash\{0\}.$$

Further, if $CDBA \in \mathscr{F}^b(X)$ for all $\|D\| < \varepsilon$ and $\beta(S) < \infty$, then

$$\sigma_{e4,S}(A,B,C,\varepsilon)\backslash\{0\} = \left[\sigma_{e4,S}(A) \cup \bigcup_{\|D\|<\varepsilon} \sigma_{e4,S}(CDB)\right]\backslash\{0\}. \quad \diamondsuit$$

Proof. • Let $\lambda \notin \left[\sigma_{e4,S}(A) \cup \bigcup_{\|D\|<\varepsilon} \sigma_{e4,S}(CDB)\right]$ or $\lambda = 0$. If $\lambda \neq 0$, then

$$\lambda S - A \in \Phi^b(X)$$

and for any $\|D\| < \varepsilon$, we have

$$\lambda S - CDB \in \Phi^b(X).$$

Using Theorem 2.2.5, we obtain

$$(\lambda S - A)(\lambda S - CDB) \in \Phi^b(X) \ \forall \ \|D\| < \varepsilon.$$

Since for every $\|D\| < \varepsilon$, $ACDB \in \mathcal{F}^b(X)$, applying Eq. (3.16), we have

$$S(\lambda S - A - CDB) \in \Phi^b(X)$$

for all $\|D\| < \varepsilon$. In view of the nullity $\alpha(S) < \infty$, together with Theorem 2.2.7, we conclude that

$$\lambda S - A - CDB \in \Phi^b(X) \ \forall \ \|D\| < \varepsilon.$$

Consequently,

$$\lambda \notin \sigma_{e4,S}(A,B,C,\varepsilon)\backslash\{0\}.$$

• For the inverse inclusion: Let $\lambda \notin \sigma_{e4,S}(A + CDB)$ for every $\|D\| < \varepsilon$ or $\lambda = 0$. If $\lambda \neq 0$, then for every $\|D\| < \varepsilon$, we have

$$\lambda S - A - CDB \in \Phi^b(X).$$

Since $S \in \Phi^b(X)$, then

$$S(\lambda S - A - CDB) \in \Phi^b(X)$$

and

$$(\lambda S - A - CDB)S \in \Phi^b(X).$$

Furthermore, we have for every $\|D\| < \varepsilon$,

$$ACDB \in \mathscr{F}^b(X)$$

and

$$CDBA \in \mathscr{F}^b(X),$$

then by both Eqs. (3.16) and (3.17), we obtain for all $\|D\| < \varepsilon$

$$(\lambda S - A)(\lambda S - CDB) \in \Phi^b(X)$$

and

$$(\lambda S - CDB)(\lambda S - A) \in \Phi^b(X).$$

An immediate consequence of Lemma 2.2.3 that

$$\lambda S - A \in \Phi^b(X)$$

and

$$\lambda S - CDB \in \Phi^b(X) \ \forall \ \|D\| < \varepsilon.$$

Thus,

$$\lambda \notin \left[\sigma_{e4,S}(A) \cup \bigcup_{\|D\|<\varepsilon} \sigma_{e4,S}(CDB)\right] \backslash \{0\}.$$ Q.E.D.

Lemma 8.3.1 *Let* $A,\ S \in \mathscr{L}(X)$, $B \in \mathscr{L}(X,Y)$, $C \in \mathscr{L}(Z,X)$ *and* $\varepsilon > 0$. *Then,*

$$\sigma_{e5,S}(A,B,C,\varepsilon) = \sigma_{e4,S}(A,B,C,\varepsilon) \cup \bigcup_{\|D\|<\varepsilon} \left\{\lambda \in \mathbb{C} \text{ such that } i(\lambda S - A - CDB) \neq 0\right\}.$$ ◊

Proof. Using Theorem 3.3.4, we obtain that for every $\|D\| < \varepsilon$

$$\sigma_{e5,S}(A + CDB) = \sigma_{e4,S}(A + CDB) \bigcup \left\{\lambda \in \mathbb{C} : \ i(\lambda S - A - CDB) \neq 0\right\}.$$

Then, we can immediately deduce the result. Q.E.D.

Proposition 8.3.1 *Let $A,\ S \in \mathscr{L}(X)$, $B \in \mathscr{L}(X,Y)$, $C \in \mathscr{L}(Z,X)$ and $\varepsilon > 0$. If for every $\|D\| < \varepsilon$, we have $\mathbb{C}\backslash\sigma_{e4,S}(A+CDB)$ is connected and $\rho_S(A+CDB) \neq \emptyset$, then*

$$\sigma_{e5,S}(A,B,C,\varepsilon) = \sigma_{e4,S}(A,B,C,\varepsilon). \qquad \diamondsuit$$

Proof. By virtue of Proposition 3.3.5, it is easy to get the following estimation for every $\|D\| < \varepsilon$

$$\sigma_{e4,S}(A+CDB) = \sigma_{e5,S}(A+CDB).$$

This equality allows us to reach the desired result. Q.E.D.

Theorem 8.3.3 *Let $A,\ S \in \mathscr{L}(X)$, $B \in \mathscr{L}(X,Y)$, $C \in \mathscr{L}(Z,X)$ and $\varepsilon > 0$. If for every $\|D\| < \varepsilon$, $ACDB \in \mathscr{F}^b(X)$, $\alpha(S) = \beta(S) < \infty$ and $SA = AS$, then*

$$\sigma_{e5,S}(A,B,C,\varepsilon)\backslash\{0\} \subset \left[\sigma_{e5,S}(A) \bigcup \bigcup_{\|D\|<\varepsilon} \sigma_{e5,S}(CDB)\right]\backslash\{0\}.$$

If further, $\mathbb{C}\backslash\sigma_{e4,S}(A)$ is connected, $\rho_S(A) \neq \emptyset$ and for all $\|D\| < \varepsilon$, we have $CDBA \in \mathscr{F}^b(X)$, then

$$\sigma_{e5,S}(A,B,C,\varepsilon)\backslash\{0\} = \left[\sigma_{e5,S}(A) \cup \bigcup_{\|D\|<\varepsilon} \sigma_{e5,S}(CDB)\right]\backslash\{0\}. \qquad \diamondsuit$$

Proof. • Let $\lambda \notin \left[\sigma_{e5,S}(A) \cup \bigcup_{\|D\|<\varepsilon} \sigma_{e5,S}(CDB)\right]\backslash\{0\}$. Then,

$$\lambda S - A \in \Phi^b(X),$$

and

$$i(\lambda S - A) = 0$$

and for every $\|D\| < \varepsilon$, we have

$$\lambda S - CDB \in \Phi^b(X)$$

and

$$i(\lambda S - CDB) = 0.$$

Using Theorem 2.2.5, we infer that for all $\|D\| < \varepsilon$

$$(\lambda S - A)(\lambda S - CDB) \in \Phi^b(X)$$

and

$$i\big((\lambda S - A)(\lambda S - CDB)\big) = 0.$$

Moreover, since for any $\|D\| < \varepsilon, ACDB \in \mathscr{F}^b(X)$, we can apply Eq. (3.16) and Lemma 2.3.1 to obtain that for all $\|D\| < \varepsilon$

$$S(\lambda S - A - CDB) \in \Phi^b(X)$$

and

$$i\big(S(\lambda S - A - CDB)\big) = 0.$$

Then, for every $\|D\| < \varepsilon$

$$\lambda S - A - CDB \in \Phi^b(X)$$

and

$$i(\lambda S - A - CDB) = 0.$$

Hence,

$$\lambda \notin \sigma_{e5,S}(A,B,C,\varepsilon)\backslash\{0\}.$$

- Let $\lambda \notin \sigma_{e5,S}(A,B,C,\varepsilon)\backslash\{0\}$, then for every $\|D\| < \varepsilon$,

$$\lambda S - A - CDB \in \Phi^b(X)$$

and

$$i(\lambda S - A - CDB) = 0.$$

Since $ACDB \in \mathscr{F}^b(X)$,

$$CDBA \in \mathscr{F}^b(X)$$

and

$$\alpha(S) = \beta(S) < \infty,$$

it is easy to show that

$$\lambda S - A \in \Phi^b(X)$$

and for all $\|D\| < \varepsilon$, we have

$$\lambda S - CDB \in \Phi^b(X)$$

and

$$i(\lambda S - CDB) = -i(\lambda S - A).$$

Besides, $\mathbb{C}\backslash\sigma_{e4,S}(A)$ is connected and $\rho_S(A) \neq \emptyset$ so we can use Lemma 3.3.2 to obtain the result. Q.E.D.

Chapter 9

Structured Essential Pseudospectra

In this chapter, we study the structured essential pseudospectra of the closed, densely defined linear operator on Banach space.

9.1 ON A CHARACTERIZATION OF THE STRUCTURED WOLF, AMMAR-JERIBI, AND BROWDER ESSENTIAL PSEUDOSPECTRA

9.1.1 *Structured Ammar-Jeribi, and Browder Essential Pseudospectra*

Theorem 9.1.1 *Let* $A_1 \in \mathscr{C}(X)$, $A_2 \in \mathscr{L}(X)$, $B \in \mathscr{L}(X,Y)$, $C \in \mathscr{L}(Z,X)$ *and* $\varepsilon > 0$. *Suppose that there exist a positive integer* n *and* $F \in \mathscr{F}(X)$ *such that* $A_2 : \mathscr{D}(A_1^n) \longrightarrow \mathscr{D}(A_1)$ *and for all* $\|D\| < \varepsilon$, $x \in \mathscr{D}(A_1^n)$, *we have*

$$(A_1 + CDB)A_2 x = A_2(A_1 + CDB)x + Fx.$$

Thereby, we obtain

(*i*) $\sigma_{e4}(A_1 + A_2, B, C, \varepsilon) \subseteq \sigma_{e4}(A_1, B, C, \varepsilon) + \sigma_{e4}(A_2, B, C, \varepsilon)$. *Furthermore, if* $\sigma_{e4}(A_1, B, C, \varepsilon)$ *is empty set, then* $\sigma_{e4}(A_1, B, C, \varepsilon) + \sigma_{e4}(A_2, B, C, \varepsilon)$ *is also empty set.*

(ii) If, further, for every $\|D\| < \varepsilon$, we have $\mathbb{C}\backslash\sigma_{e4}(A_1 + CDB)$, $\mathbb{C}\backslash\sigma_{e4}(A_2 + CDB)$ and $\mathbb{C}\backslash\sigma_{e4}(A_1 + A_2 + CDB)$ are connected such that $\rho(A_1 + CDB) \neq \emptyset$ and $\rho(A_1 + A_2 + CDB) \neq \emptyset$, then

$$\sigma_{e5}(A_1 + A_2, B, C, \varepsilon) \subseteq \sigma_{e5}(A_1, B, C, \varepsilon) + \sigma_{e5}(A_2, B, C, \varepsilon).$$

(iii) Moreover, if for all $\|D\| < \varepsilon$, we assume that $\mathbb{C}\backslash\sigma_{e5}(A_1 + CDB)$, $\mathbb{C}\backslash\sigma_{e5}(A_2 + CDB)$ and $\mathbb{C}\backslash\sigma_{e5}(A_1 + A_2 + CDB)$ are connected with $\rho(A_1 + CDB) \neq \emptyset$ and $\rho(A_1 + A_2 + CDB) \neq \emptyset$, then

$$\sigma_{e6}(A_1 + A_2, B, C, \varepsilon) \subseteq \sigma_{e6}(A_1, B, C, \varepsilon) + \sigma_{e6}(A_2, B, C, \varepsilon). \qquad \diamondsuit$$

Proof. If $\sigma_{e4}(A_1, B, C, \varepsilon) + \sigma_{e4}(A_2, B, C, \varepsilon) = \mathbb{C}$, then the theorem is trivially true. We, therefore, assume that $\sigma_{e4}(A_1, B, C, \varepsilon) + \sigma_{e4}(A_2, B, C, \varepsilon)$ is not the entire plane. Let

$$\gamma \notin \sigma_{e4}(A_1, B, C, \varepsilon) + \sigma_{e4}(A_2, B, C, \varepsilon).$$

If $\lambda \in \sigma_{e4}(A_2, B, C, \varepsilon)$, then

$$\gamma - \lambda \notin \sigma_{e4}(A_1, B, C, \varepsilon).$$

Since $\overline{\sigma_{e4}(A_2, B, C, \varepsilon)}$ is compact, then there exists an open set

$$U \supset \overline{\sigma_{e4}(A_2, B, C, \varepsilon)}$$

such that ∂U, the boundary of U, is bounded and $\lambda \in U$. Set $A_3 := \gamma - A_1 - CDB$, we have for all $\|D\| < \varepsilon$, $\lambda \in \Phi_{A_3}$. Therefore,

$$\overline{\sigma_{e4}(A_2, B, C, \varepsilon)} \subset U \subset \Phi_{A_3}.$$

By Theorem 2.9.1, we obtain the existence of a bounded Cauchy domain D_1 such that

$$\sigma_{e4}(A_2, B, C, \varepsilon) \subset D_1$$

and

$$\overline{D_1} \subset U.$$

Since $\Phi^0(A_3)$ (resp. $\Phi^0(A_2)$) does not accumulate in Φ_{A_3} (resp. Φ_{A_2}), D_1 can be chosen such that $R'_\lambda(A_3)$ and $R'_\lambda(A_2)$ are analytic on ∂D_1. Define the operators N_1 and N_2 by

$$N_1 = -\frac{1}{2\pi i}\int_{+\partial D_1} R'_\lambda(A_3)R'_\lambda(A_2)\,d\lambda$$

and

$$N_2 = -\frac{1}{2\pi i}\int_{+\partial D_1} R'_\lambda(A_2)R'_\lambda(A_3)\,d\lambda.$$

$R'_\lambda(A_3)$ is of the form $TC(\lambda)$, where $C(\lambda)$ is a bounded operator valued analytic function of λ and T is a fixed bounded operator such that

$$T : X \longrightarrow \mathscr{D}(A_1).$$

Thereby, $N_1 : X \longrightarrow \mathscr{D}(A_1)$. We will now show that for all $\|D\| < \varepsilon$, there exist two Fredholm perturbations F'_1 and F'_2 such that

$$(\gamma - A_1 - A_2 - CDB)N_1 = I + F'_1$$

and

$$N_2(\gamma - A_1 - A_2 - CDB) = I + F'_2$$

on $\mathscr{D}(A_1)$. Hence, we can write

$$\gamma - A_1 - A_2 - CDB = -(\lambda - A_3) + (\lambda - A_2).$$

Thus,

$$\begin{aligned}(\gamma - A_1 - A_2 - CDB)N_1 &\\ = \frac{1}{2\pi i}\int_{+\partial D_1}&(\lambda - A_3)R'_\lambda(A_3)R'_\lambda(A_2)\,d\lambda - \\ &\frac{1}{2\pi i}\int_{+\partial D_1}(\lambda - A_2)R'_\lambda(A_3)R'_\lambda(A_2)\,d\lambda. \qquad (9.1)\end{aligned}$$

Since

$$(\lambda - A_3)R'_\lambda(A_3) = I + K(\lambda),$$

where $K(\lambda)$ is a bounded finite rank operator depending analytically on λ, then

$$\frac{1}{2\pi i}\int_{+\partial D_1}(\lambda - A_3)R'_\lambda(A_3)R'_\lambda(A_2)\,d\lambda = \frac{1}{2\pi i}\int_{+\partial D_1} R'_\lambda(A_2)d\lambda + \frac{1}{2\pi i}\int_{+\partial D_1} K(\lambda)R'_\lambda(A_2)d\lambda.$$

Using Theorem 2.9.5, we obtain

$$\frac{1}{2\pi i}\int_{+\partial D_1} R'_\lambda(A_2)d\lambda \text{ is of the form } I+K_1,$$

where $K_1 \in \mathscr{K}(X)$. So, the first integral of Eq. (9.1) is of the form $I+K_2$, $K_2 \in \mathscr{K}(X)$. By virtue of Lemma 2.9.1, we get

$$R'_\lambda(A_3)R'_\lambda(A_2) = R'_\lambda(A_2)R'_\lambda(A_3) + F(\lambda).$$

The second integral of Eq. (9.1) is equal to

$$\begin{aligned}
&-\frac{1}{2\pi i}\int_{+\partial D_1}(\lambda - A_2)R'_\lambda(A_3)R'_\lambda(A_2)\,d\lambda \\
&= -\frac{1}{2\pi i}\int_{+\partial D_1}(\lambda - A_2)R'_\lambda(A_2)R'_\lambda(A_3)\,d\lambda - \frac{1}{2\pi i}\int_{+\partial D_1}(\lambda - A_2)F(\lambda)d\lambda. \\
&= -\frac{1}{2\pi i}\int_{+\partial D_1}(I+G(\lambda))R'_\lambda(A_3)d\lambda - \frac{1}{2\pi i}\int_{+\partial D_1}(\lambda - A_2)F(\lambda)d\lambda.
\end{aligned}$$

Since

$$\frac{1}{2\pi i}\int_{+\partial D_1} R'_\lambda(A_3)d\lambda$$

is compact (see Lemma 2.9.2) and

$$\frac{1}{2\pi i}\int_{+\partial D_1}(\lambda - A_2)F(\lambda)d\lambda$$

is a Fredholm perturbation, then we obtain

$$-\frac{1}{2\pi i}\int_{+\partial D_1}(\lambda - A_2)R'_\lambda(A_3)R'_\lambda(A_2)\,d\lambda = F_1,$$

where $F_1 \in \mathscr{F}(X)$. Consequently, for all $\|D\| < \varepsilon$, we have

$$(\gamma - A_1 - A_2 - CDB)N_1 = I + F'_1,$$

where $F'_1 \in \mathscr{F}(X)$. By a similar argument, we get for all $\|D\| < \varepsilon$

$$N_2(\gamma - A_1 - A_2 - CDB) = I + F'_2,$$

where $F_2' \in \mathscr{F}(X)$. So, for all $\|D\| < \varepsilon$, we have

$$\gamma - A_1 - A_2 - CDB \in \Phi(X).$$

For the second result of (i), we argue by contradiction. We suppose that

$$\sigma_{e4}(A_1, B, C, \varepsilon) + \sigma_{e4}(A_2, B, C, \varepsilon)$$

is not empty, then there exists

$$\lambda \in \sigma_{e4}(A_1, B, C, \varepsilon) + \sigma_{e4}(A_2, B, C, \varepsilon).$$

This implies that

$$\lambda = \mu_1 + \mu_2,$$

where $\mu_1 \in \sigma_{e4}(A_1, B, C, \varepsilon)$ and $\mu_2 \in \sigma_{e4}(A_2, B, C, \varepsilon)$ which is absurd. Using Lemma 3.6.1, we obtain the items (ii) and (iii). Q.E.D.

Let $A_1 \in \mathscr{R}(X)$, $A_2 \in \mathscr{R}(X)$, $B \in \mathscr{L}(X,Y)$, $C \in \mathscr{L}(Z,X)$ and $\varepsilon > 0$ such that the operator B (resp. C) is not null. We assume that $\varepsilon = \frac{1}{\|B\|\|C\|\|R\|}$, where $R \in \mathscr{L}(X)$ satisfying

$$\|R\| > \max\{\|A_0\|, \|B_0\|\}$$

such that A_0 (resp. B_0) is a quasi-inverse operator of $\lambda - A_1$ (resp. $\lambda - A_2$) for all $\lambda \neq 0$. Now, we can see that for any $D \in \mathscr{L}(Y,Z)$ satisfying $\|D\| < \varepsilon$

$$A_1 + CDB \in \mathscr{R}(X)$$

and

$$A_2 + CDB \in \mathscr{R}(X).$$

If there exists $K \in \mathscr{K}(X)$ such that for all $\|D\| < \varepsilon$, $x \in X$, we have

$$(A_1 + CDB)A_2 x = A_2(A_1 + CDB)x + Kx,$$

then we can deduce, from Theorem 2.3.2, that

$$A_1 + A_2 + CDB \in \mathscr{R}(X).$$

This leads to

$$\sigma_{e4}(A_1,B,C,\varepsilon)=\{0\},$$

$$\sigma_{e4}(A_2,B,C,\varepsilon)=\{0\},$$

and

$$\sigma_{e4}(A_1+A_2,B,C,\varepsilon)=\{0\}.$$

Hence, the results of Theorem 9.1.1 are satisfied.

In all the sequels we give some properties of the structured Ammar-Jeribi essential pseudospectrum.

Theorem 9.1.2 *Let $A\in\mathscr{C}(X)$, $B\in\mathscr{L}(X,Y)$, $C\in\mathscr{L}(Z,X)$ and $\varepsilon>0$. Then,*

$$\sigma_{e5}(A,B,C,\varepsilon)=\bigcap_{K\in\mathscr{K}(X)}\sigma(A+K,B,C,\varepsilon). \qquad \diamondsuit$$

Proof. Let $\lambda\notin\sigma_{e5}(A,B,C,\varepsilon)$, then for all $\|D\|<\varepsilon$, we have $\lambda-A-CDB\in\Phi(X)$ and

$$i(\lambda-A-CDB)=0.$$

Thus, there exists $K\in\mathscr{K}(X)$ such that for any D satisfying $\|D\|<\varepsilon$, we have

$$\lambda-A-K-CDB\in\Phi(X)$$

and

$$i(\lambda-A-K-CDB)=0.$$

Therefore, we obtain

$$\sigma_{e5}(A,B,C,\varepsilon)=\bigcap_{K\in\mathscr{K}(X)}\sigma_{e5}(A+K,B,C,\varepsilon),$$

and so

$$\sigma_{e5}(A,B,C,\varepsilon)\subset\bigcap_{K\in\mathscr{K}(X)}\sigma(A+K,B,C,\varepsilon).$$

Let

$$\lambda\in\bigcap_{K\in\mathscr{K}(X)}\sigma(A+K,B,C,\varepsilon).$$

Without loss of generality, we may suppose that $\lambda = 0$. We argue by contradiction, we suppose that for all $\|D\| < \varepsilon$,

$$A + CDB \in \Phi(X)$$

and

$$i(A + CDB) = 0.$$

Let $d = \alpha(A + CDB) = \beta(A + CDB)$ and $\{u_1, \cdots, u_d\}$ (resp. $\{z'_1, \cdots, z'_d\}$) being the basis for $N(A + CDB)$ (resp. $N((A + CDB)^*)$). By Lemma 2.1.1, there are functionals $u'_1, \cdots, u'_d$ in X^* and elements $z_1, \cdots, z_d$ such that

$$u'_j(u_k) = \delta_{jk}$$

and

$$z'_j(z_k) = \delta_{jk},$$

$1 \leq j, k \leq d$, where

$$\delta_{jk} = \begin{cases} 0 & \text{if } j \neq k \\ 1 & \text{if } j = k. \end{cases}$$

Consider the operator K defined by

$$Ku = \sum_{i=1}^{d} u'_i(u)\, z_i, \; u \in X.$$

It is clear that K is a finite rank operator on X. Then, K is a compact operator on X. By virtue of Lemma 2.3.1 (i), we obtain that for all $\|D\| < \varepsilon$,

$$A + CDB + K \in \Phi(X)$$

and

$$\alpha(A + CDB + K) = \beta(A + CDB + K).$$

Obviously, if $u \in N(A + CDB + K)$, then

$$(A + CDB)u = K(-u).$$

Since

$$R(A + CDB) \bigcap R(K) = \{0\}$$

and

$$N(A+CDB)\bigcap N(K)=\{0\},$$

so we can deduce that $u=0$. Thus, we have

$$\alpha(A+CDB+K)=\beta(A+CDB+K)=0.$$

Consequently, for any D satisfying $\|D\|<\varepsilon$, we have

$$0\in\rho(A+CDB+K),$$

which completes the proof. Q.E.D.

As a consequence of the previous theorem and Proposition 3.3.1 for $S=I$, we may state:

Corollary 9.1.1 *Let $A\in\mathscr{C}(X)$, $B\in\mathscr{L}(X,Y)$, $C\in\mathscr{L}(Z,X)$ and $\varepsilon>0$. Then,*

(*i*) $\bigcap_{\varepsilon>0}\sigma_{e5}(A,B,C,\varepsilon)=\sigma_{e5}(A)$, *and*

(*ii*) *if $\varepsilon_1<\varepsilon_2$, then*

$$\sigma_{e5}(A,B,C,\varepsilon_1)\subset\sigma_{e5}(A,B,C,\varepsilon_2).\qquad\diamondsuit$$

Theorem 9.1.3 *Let $A\in\mathscr{C}(X)$, $B\in\mathscr{L}(X,Y)$, $C\in\mathscr{L}(Z,X)$ and $\varepsilon>0$. Then,*

$$\sigma_{e5}(A,B,C,\varepsilon)=\bigcap_{F\in\mathscr{F}(X)}\sigma(A+F,B,C,\varepsilon).\qquad\diamondsuit$$

Proof. Let $\lambda\notin\bigcap_{F\in\mathscr{F}(X)}\sigma(A+F,B,C,\varepsilon)$, then there exists $F\in\mathscr{F}(X)$ such that for any $D\in\mathscr{L}(X)$ satisfying $\|D\|<\varepsilon$, we have

$$\lambda\in\rho(A+F+CDB).$$

By Lemma 2.3.1 (*i*), we can deduce that for all $\|D\|<\varepsilon$

$$\lambda-A-CDB\in\Phi(X)$$

and

$$i(\lambda - A - CDB) = 0.$$

It is clear that

$$\bigcap_{F \in \mathscr{F}(X)} \sigma(A+F,B,C,\varepsilon) \subset \bigcap_{K \in \mathscr{K}(X)} \sigma(A+K,B,C,\varepsilon),$$

because $\mathscr{K}(X) \subset \mathscr{F}(X)$. Q.E.D.

Corollary 9.1.2 *Let $A \in \mathscr{C}(X)$, $B \in \mathscr{L}(X,Y)$, $C \in \mathscr{L}(Z,X)$ and $\varepsilon > 0$. If $E(X)$ be any subset of $\mathscr{L}(X)$ satisfying $\mathscr{K}(X) \subset E(X) \subset \mathscr{F}(X)$, then*

$$\sigma_{e5}(A,B,C,\varepsilon) = \bigcap_{F \in E(X)} \sigma(A+F,B,C,\varepsilon). \qquad \diamondsuit$$

Theorem 9.1.4 *Let A_1, A_2 be operators in $\mathscr{C}(X)$, $B \in \mathscr{L}(X,Y)$, $C \in \mathscr{L}(Z,X)$ and $\varepsilon > 0$. If A_2 is A_1-compact, then*

$$\sigma_{e5}(A_1,B,C,\varepsilon) = \sigma_{e5}(A_1+A_2,B,C,\varepsilon). \qquad \diamondsuit$$

Proof. Let us begin with the inclusion $\subset$. Assume that

$$\lambda \notin \sigma_{e5}(A_1+A_2,B,C,\varepsilon),$$

then for all $D \in \mathscr{L}(Y,Z)$ such that $\|D\| < \varepsilon$, we have

$$\lambda - A_1 - A_2 - CDB \in \Phi(X) \text{ and } i(\lambda - A_1 - A_2 - CDB) = 0.$$

Since A_2 is A_1-compact, then for any $D \in \mathscr{L}(Y,Z)$ satisfying $\|D\| < \varepsilon$, we have A_2 is (A_1+CDB)-compact. Thus, by Theorem 2.3.7, we get that A_2 is (A_1+A_2+CDB)-compact. So, we can see that A_2 is $(\lambda - A_1 - A_2 - CDB)$-compact. Therefore, from Theorem 2.3.6, we can deduce that for all $\|D\| < \varepsilon$

$$\lambda - A_1 - CDB \in \Phi(X)$$

and

$$i(\lambda - A_1 - CDB) = 0.$$

Let us prove now the converse inclusion. Let $\lambda \notin \sigma_{e5}(A_1,B,C,\varepsilon)$ this means that for any $D \in \mathscr{L}(Y,Z)$ such that $\|D\| < \varepsilon$, we have

$$\lambda - A_1 - CDB \in \Phi(X)$$

and

$$i(\lambda - A_1 - CDB) = 0.$$

Using the fact that A_2 is $(A_1 + CDB)$-compact, then for all $\|D\| < \varepsilon$, we obtain

$$\lambda - A_1 - A_2 - CDB \in \Phi(X)$$

and

$$i(\lambda - A_1 - A_2 - CDB) = 0,$$

which completes the proof. Q.E.D.

In addition, we have the following useful stability result for the structured Wolf and the structured Ammar-Jeribi essential pseudospectra.

Theorem 9.1.5 *Let A_1, $A_2 \in \mathscr{C}(X)$, $B \in \mathscr{L}(X,Y)$, $C \in \mathscr{L}(Z,X)$, $\mathscr{M}(X)$ be a non zero two-sided ideal of $\mathscr{L}(X)$ satisfying $\mathscr{K}(X) \subseteq \mathscr{M}(X) \subseteq \mathscr{F}(X)$ and $\varepsilon > 0$. Assume that there are B_1, $B_2 \in \mathscr{L}(X)$ and J_1, $J_2 \in \mathscr{M}(X)$ such that*

$$A_1B_1 = I - J_1, \tag{9.2}$$

$$A_2B_2 = I - J_2. \tag{9.3}$$

If $0 \in \Phi_{A_1} \bigcap \Phi_{A_2}$ and $B_1 - B_2 \in \mathscr{M}(X)$, then

$$\sigma_{e4}(A_1,B,C,\varepsilon) = \sigma_{e4}(A_2,B,C,\varepsilon).$$

If, further, $i(A_1) = i(A_2) = 0$, then

$$\sigma_{e5}(A_1,B,C,\varepsilon) = \sigma_{e5}(A_2,B,C,\varepsilon). \tag{9.4}$$

Proof. By Eqs. (9.2) and (9.3), we have for any scalar λ and $D \in \mathcal{L}(Y,Z)$

$$(\lambda - A_1 - CDB)B_1 - (\lambda - A_2 - CDB)B_2 = J_1 - J_2 + (\lambda - CDB)(B_1 - B_2). \tag{9.5}$$

If $\lambda \notin \sigma_{e4}(A_2,B,C,\varepsilon)$, then for any $D \in \mathcal{L}(Y,Z)$ satisfying $\|D\| < \varepsilon$, we have

$$\lambda - A_2 - CDB \in \Phi(X).$$

Since A_2 is closed, $\mathcal{D}(A_2)$ endowed with the graph norm is a Banach space denoted by X_{A_2}. Then, using Eq. (2.6), we infer that for all $\|D\| < \varepsilon$,

$$\lambda - \widehat{A_2} - \widehat{CDB} \in \Phi^b(X_{A_2}, X).$$

Furthermore, since $J_2 \in \mathcal{M}(X)$, it follows from Eq. (9.3) and Theorem 2.2.7 that $B_2 \in \Phi^b(X, X_{A_2})$. Thereby, for every $\|D\| < \varepsilon$, we have

$$(\lambda - \widehat{A_2} - \widehat{CDB})B_2 \in \Phi^b(X).$$

If $B_1 - B_2 \in \mathcal{M}(X)$, then Eq. (9.5) together with Lemma 2.3.1 (i) gives for all $\|D\| < \varepsilon$,

$$(\lambda - \widehat{A_1} - \widehat{CDB})B_1 \in \Phi^b(X)$$

and

$$i\Big[(\lambda - \widehat{A_1} - \widehat{CDB})B_1\Big] = i\Big[(\lambda - \widehat{A_2} - \widehat{CDB})B_2\Big]. \tag{9.6}$$

Since $A_1 \in \mathcal{C}(X)$, using Eq. (9.2) and arguing as above we conclude that $B_1 \in \Phi^b(X, X_{A_1})$. Therefore, for all $\|D\| < \varepsilon$,

$$\lambda - \widehat{A_1} - \widehat{CDB} \in \Phi^b(X_{A_1}, X).$$

This implies, for every $\|D\| < \varepsilon$, we have

$$\lambda - A_1 - CDB \in \Phi(X).$$

The opposite inclusion follows by symmetry. We now prove Eq. (9.4). If $\lambda \notin \sigma_{e5}(A_2,B,C,\varepsilon)$, then for all $\|D\| < \varepsilon$, we have

$$\lambda - A_2 - CDB \in \Phi(X) \text{ and } i(\lambda - A_2 - CDB) = 0.$$

Since J_1 and J_2 belong to $\mathcal{M}(X)$ and

$$i(A_1) = i(A_2) = 0,$$

applying Lemma 2.3.1 (i) to Eqs. (9.2) and (9.3) and using Atkinson theorem, we get

$$i(B_1) = i(B_2) = 0.$$

This result together with Eq. (9.6), the Atkinson theorem and Eq. (2.6) show that for all $\|D\| < \varepsilon$

$$i(\lambda - A_1 - CDB) = i(\lambda - A_2 - CDB) = 0.$$

Consequently,

$$\lambda \notin \sigma_{e5}(A_1, B, C, \varepsilon).$$

The opposite inclusion follows by symmetry. Q.E.D.

We end this section by the following result which provides accurate characterization of the structured Ammar-Jeribi essential pseudospectrum in particular space.

Theorem 9.1.6 *Let $A \in \mathcal{C}(X)$ such that $\rho(A) \neq \emptyset$, $B \in \mathcal{L}(X,Y)$, $C \in \mathcal{L}(Z,X)$ and $\varepsilon > 0$. Then,*

(i) If X has the Dunford-Pettis property, then

$$\sigma_{e5}(A,B,C,\varepsilon) = \bigcap_{S \in \mathcal{G}_A(X)} \sigma(A+S,B,C,\varepsilon),$$

where

$$\mathcal{G}_A(X) = \Big\{K \in \mathcal{L}(X) \text{ such that } (\lambda - A)^{-1}K \in \mathcal{W}(X) \text{ for some } \lambda \in \rho(A)\Big\}.$$

(ii) If X is isomorphic to one of the spaces $L_p(\Omega)$, $p > 1$, then

$$\sigma_{e5}(A,B,C,\varepsilon) = \bigcap_{S \in \mathcal{G}'_A(X)} \sigma(A+S,B,C,\varepsilon),$$

where

$$\mathcal{G}'_A(X) = \Big\{K \in \mathcal{L}(X) \text{ such that } (\lambda - A)^{-1}K \in \mathcal{S}(X) \text{ for some } \lambda \in \rho(A)\Big\}. \quad \diamondsuit$$

Proof. (i) Since $\mathscr{K}(X) \subset \mathscr{G}_A(X)$, then

$$\bigcap_{S\in\mathscr{G}_A(X)} \sigma(A+S,B,C,\varepsilon) \subset \sigma_{e5}(A,B,C,\varepsilon).$$

Let $\lambda \notin \bigcap_{S\in\mathscr{G}_A(X)} \sigma(A+S,B,C,\varepsilon)$, then there exists $S \in \mathscr{G}_A(X)$ such that for any $D \in \mathscr{L}(Y,Z)$ satisfying $\|D\| < \varepsilon$, we have

$$\lambda \in \rho(A+S+CDB).$$

Thus,

$$\lambda - A - S - CDB \in \Phi(X)$$

and

$$i(\lambda - A - S - CDB) = 0.$$

Let $\mu \in \rho(A)$, we have

$$(\lambda - A - S - CDB)^{-1}S = \left[I + (\lambda - A - S - CDB)^{-1}(\mu - \lambda + S + CDB)\right] \times (\mu - A)^{-1}S. \qquad (9.7)$$

Since $S \in \mathscr{G}_A(X)$ and in view of Eq. (9.7), we deduce that for any $\|D\| < \varepsilon$, $(\lambda - A - S - CDB)^{-1}S$ is weakly compact on X. Using the fact that the composed of two weakly compact operators is compact, so we can deduce from Lemma 2.2.5 that $I + (\lambda - A - S - CDB)^{-1}S$ is a Fredholm operator with

$$i(I + (\lambda - A - S - CDB)^{-1}S) = 0.$$

Since we can write

$$\lambda - A - CDB = (\lambda - A - S - CDB)\,[I + (\lambda - A - S - CDB)^{-1}S],$$

then for all $\|D\| < \varepsilon$, $\lambda - A - CDB \in \Phi(X)$ and

$$i(\lambda - A - CDB) = 0.$$

The prove of the item (ii) follows by the same reasoning as (i). Q.E.D.

Remark 9.1.1 *The Theorem 9.1.6 shows that for all $K \in \mathcal{G}_A(X)$ or $K \in \mathcal{G}'_A(X)$, we have*

$$\sigma_{e5}(A+K,B,C,\varepsilon) = \sigma_{e5}(A,B,C,\varepsilon). \qquad \diamondsuit$$

Corollary 9.1.3 *Let $A \in \mathcal{C}(X)$, $B \in \mathcal{L}(X,Y)$, $C \in \mathcal{L}(Z,X)$ and $\varepsilon > 0$. Then,*

$$\sigma_{e5}(A,B,C,\varepsilon) = \bigcap_{K \in \mathcal{G}''_A(X)} \sigma(A+K,B,C,\varepsilon),$$

where

$$\mathcal{G}''_A(X) = \Big\{K \in \mathcal{L}(X) \text{ such that } (\lambda - A)^{-1}K \in \mathcal{F}(X) \text{ for some } \lambda \in \rho(A)\Big\}. \qquad \diamondsuit$$

9.1.2 A Characterization of the Structured Browder Essential Pseudospectrum

Theorem 9.1.7 *Let $A \in \mathcal{C}(X)$, $B \in \mathcal{L}(X,Y)$, $C \in \mathcal{L}(Z,X)$ and $\varepsilon > 0$. Then,*

$$\sigma_{e6}(A,B,C,\varepsilon) = \bigcap_{R \in \mathcal{R}_\varepsilon(X)} \sigma(A+R,B,C,\varepsilon),$$

where

$$\mathcal{R}_\varepsilon(X) = \Big\{R \in \mathcal{R}(X) : \forall D \in \mathcal{L}(Y,Z) \text{ satisfying } \|D\| < \varepsilon, \text{ we have } R \text{ commutes with } A+CDB\Big\}. \qquad \diamondsuit$$

Proof. Suppose that $\lambda \notin \bigcap_{R \in \mathcal{R}_\varepsilon(X)} \sigma(A+R,B,C,\varepsilon)$, then there exists a Riesz operator R which commutes with $A+CDB$ for every $D \in \mathcal{L}(Y,Z)$ satisfying $\|D\| < \varepsilon$, and

$$\lambda \notin \sigma(A+R,B,C,\varepsilon).$$

It follows from Lemma 2.4.3 that for all $\|D\| < \varepsilon$, R commutes with $\lambda - A - R - CDB$. Since for every $\|D\| < \varepsilon$, we have $\lambda \in \rho(A+R+CDB)$, we can assert that

$$\alpha(\lambda - A - R - CDB) = \beta(\lambda - A - R - CDB) = 0$$

and

$$\operatorname{asc}(\lambda - A - R - CDB) = 0.$$

By virtue of Theorem 2.4.5, we obtain for all $\|D\| < \varepsilon$

$$\alpha(\lambda - A - CDB) = \beta(\lambda - A - CDB) < \infty$$

and

$$\operatorname{asc}(\lambda - A - CDB) = \operatorname{desc}(\lambda - A - CDB) < \infty.$$

Consequently, for all $\|D\| < \varepsilon$,

$$\lambda \notin \sigma_{e6}(A + CDB).$$

Conversely, let $\lambda \notin \sigma_{e6}(A, B, C, \varepsilon)$, then for any $\|D\| < \varepsilon$, we have

$$\lambda \notin \sigma_{e6}(A + CDB).$$

If $\lambda \in \sigma(A + CDB) \backslash \sigma_{e6}(A + CDB)$, then we may interchange T in Theorem 2.3.8 by $A + CDB$ to get that λ is a pole of the resolvent operator, and Theorem 2.4.2 implies that for any $\|D\| < \varepsilon$, we have

$$\operatorname{asc}(\lambda - A - CDB) < \infty.$$

Otherwise, for every $\|D\| < \varepsilon$

$$\alpha(\lambda - A - CDB) = \beta(\lambda - A - CDB) = 0$$

and

$$\operatorname{asc}(\lambda - A - CDB) = 0,$$

when

$$\lambda \in \bigcap_{\|D\| < \varepsilon} \rho(A + CDB).$$

Now, we see from Proposition 2.4.3 that there exists a finite dimensional range operator R which commutes with $\lambda - A - CDB$ for all $\|D\| < \varepsilon$, and

$$\lambda \notin \sigma(A + R + CDB).$$

Q.E.D.

Remark 9.1.2 *Theorem 9.1.7 remains true if "Riesz" is replaced by "compact". More precisely, if we set*

$$\mathcal{K}_{\varepsilon}(X) = \Big\{K \in \mathcal{K}(X) : \forall D \in \mathcal{L}(Y,Z) \ \text{satisfying} \ \|D\| < \varepsilon, \ K \ \text{commutes} \ \text{with} \ A + CDB\Big\}, \quad (9.8)$$

then

$$\sigma_{e6}(A,B,C,\varepsilon) = \bigcap_{K \in \mathcal{K}_{\varepsilon}(X)} \sigma(A+K,B,C,\varepsilon).$$

As a consequence of the previous identity that the structured Browder essential pseudospectrum, $\sigma_{e6}(A,B,C,\varepsilon)$, remains invariant under perturbations of A by operators in $\mathcal{K}_{\varepsilon}(X)$. ◇

Corollary 9.1.4 *Let $A \in \mathcal{C}(X)$, $B \in \mathcal{L}(X,Y)$, $C \in \mathcal{L}(Z,X)$ and $\varepsilon > 0$. Then,*

$$\sigma_{e6}(A,B,C,\varepsilon) = \bigcap_{F \in \mathcal{F}_{\varepsilon}(X)} \sigma(A+F,B,C,\varepsilon),$$

where

$$\mathcal{F}_{\varepsilon}(X) = \Big\{F \in \mathcal{F}(X) : \forall D \in \mathcal{L}(Y,Z) \ \text{satisfying} \ \|D\| < \varepsilon, \ K \ \text{commutes with} \ A + CDB\Big\}.$$ ◇

Remark 9.1.3 *We conclude by Corollary 9.1.4 that for every $F \in \mathcal{F}_{\varepsilon}(X)$, we have*

$$\sigma_{e6}(A+F,B,C,\varepsilon) = \sigma_{e6}(A,B,C,\varepsilon).$$ ◇

Theorem 9.1.8 *Let $A \in \mathcal{C}(X)$, $B \in \mathcal{L}(X,Y)$ and $C \in \mathcal{L}(Z,X)$. Then,*

$$\bigcap_{\varepsilon>0} \sigma_{e6}(A,B,C,\varepsilon) = \sigma_{e6}(A).$$ ◇

Proof. We begin by observing that, for all $\varepsilon > 0$,

$$\sigma_{e6}(A) \subset \bigcup_{\|D\|<\varepsilon} \sigma_{e6}(A+CDB).$$

Hence,

$$\sigma_{e6}(A) \subset \bigcap_{\varepsilon>0} \sigma_{e6}(A,B,C,\varepsilon).$$

On the other hand, let $\lambda \in \bigcap_{\varepsilon>0} \sigma_{e6}(A,B,C,\varepsilon)$, then for all $\varepsilon > 0$ and for every $R \in \mathscr{R}_\varepsilon(X)$, we have $\lambda \in \sigma(A+R,B,C,\varepsilon)$, where $\mathscr{R}_\varepsilon(X)$ is given in Theorem 9.1.7. Since for all $\|D\| < \varepsilon$, we have R commutes with $A+CDB$ this implies that $Rx \in \mathscr{D}(A+CDB) = \mathscr{D}(A)$ for any $x \in \mathscr{D}(A)$ and

$$R(A+CDB)x = (A+CDB)Rx \text{ for all } x \in \mathscr{D}((A+CDB)^2).$$

It follows that

$$\begin{aligned} \|RAx - ARx\| &\leq 2\,\|C\|\|D\|\|B\|\|R\|\|x\| \\ &< 2\,\varepsilon\|C\|\|B\|\|R\|\|x\|. \end{aligned}$$

Thus,

$$\lim_{\varepsilon\to 0} \|RAx - ARx\| = 0,$$

and so for every $x \in \mathscr{D}(A)$, we have $RAx = ARx$. Furthermore, using Proposition 8.2.2 for $S = I$, we get

$$\bigcap_{\varepsilon>0} \sigma(A+R,B,C,\varepsilon) = \sigma(A+R).$$

Whence

$$\lambda \in \bigcap_{R\in\mathscr{R}_c(X)} \sigma(A+R),$$

where

$$\mathscr{R}_c(X) = \Big\{ R \in \mathscr{R}(X) \text{ such that } R \text{ commutes with } A \Big\}.$$

Using Theorem 3.1.3, we deduce that $\lambda \in \sigma_{e6}(A)$. Q.E.D.

Before continuing with the properties of the structured Browder essential pseudospectrum, we also show that we have a similar result as the previous theorem for the structured Wolf essential pseudospectrum.

Theorem 9.1.9 *Let $A \in \mathscr{C}(X)$, $B \in \mathscr{L}(X,Y)$ and $C \in \mathscr{L}(Z,X)$. Then,*

$$\bigcap_{\varepsilon>0} \sigma_{e4}(A,B,C,\varepsilon) = \sigma_{e4}(A). \qquad \diamondsuit$$

Proof. 1st case : If $B = 0$ or $C = 0$, then the proof follows immediately from Definition 3.6.1.

2nd case : If $B \neq 0$ and $C \neq 0$, then for $\lambda \notin \sigma_{e4}(A)$, we have $\lambda - A \in \Phi(X)$. Set $\varepsilon = \frac{1}{\|B\|\|C\|\|A_0\|}$, where A_0 is a quasi-inverse of $\lambda - A$. Then, there exist $K_1, K_2 \in \mathscr{K}(X)$ such that

$$(\lambda - A)A_0 = I - K_1$$

whenever $R(A_0) \subset \mathscr{D}(A)$, and

$$A_0(\lambda - A) = I - K_2.$$

Thus, for every $\|D\| < \varepsilon$, we have

$$(\lambda - A - CDB)A_0 = \left[I - K_1(I - CDBA_0)^{-1}\right](I - CDBA_0),$$

whence

$$(\lambda - A - CDB)A_0(I - CDBA_0)^{-1} = I - K_1(I - CDBA_0)^{-1}.$$

In the same way, we can write that for any $\|D\| < \varepsilon$,

$$(I - A_0CDB)^{-1}A_0(\lambda - A - CDB) = I - (I - A_0CDB)^{-1}K_2.$$

Since $K_1(I - CDBA_0)^{-1} \in \mathscr{K}(X)$ and $(I - A_0CDB)^{-1}K_2 \in \mathscr{K}(X)$, then for any $\|D\| < \varepsilon$, we have $\lambda - A - CDB \in \Phi(X)$. Hence,

$$\lambda \notin \bigcap_{\varepsilon>0} \sigma_{e4}(A,B,C,\varepsilon).$$

Conversely, we have

$$\sigma_{e4}(A) \subset \bigcup_{\|D\|<\varepsilon} \sigma_{e4}(A + CDB),$$

and so,

$$\sigma_{e4}(A) \subset \bigcap_{\varepsilon>0} \sigma_{e4}(A,B,C,\varepsilon). \qquad \text{Q.E.D.}$$

Theorem 9.1.10 *Let $A \in \mathscr{C}(X)$, $B \in \mathscr{L}(X,Y)$, $C \in \mathscr{L}(Z,X)$ and $\varepsilon > 0$. Then,*

$$\sigma_{e6}(A,B,C,\varepsilon) = \bigcap_{K \in \mathscr{G}_\varepsilon(X)} \sigma(A+K,B,C,\varepsilon),$$

where

$$\mathscr{G}_\varepsilon(X) = \Big\{K \in \mathscr{L}(X) : \forall D \in \mathscr{L}(Y,Z) \text{ satisfying } \|D\| < \varepsilon, K \text{ commutes with } A+CDB \text{ and } \Delta_\psi\Big(K(\lambda - A - K - CDB)^{-1}\Big) < 1, \ \forall\, \lambda \in \rho(A+K+CDB)\Big\},$$

and $\Delta_\psi(\cdot)$ is the measure of non-strict-singularity given in (2.11). ◊

Proof. We first show that

$$\sigma_{e6}(A,B,C,\varepsilon) \subset \bigcap_{K \in \mathscr{G}_\varepsilon(X)} \sigma(A+K,B,C,\varepsilon).$$

Let $\lambda \notin \bigcap_{K \in \mathscr{G}_\varepsilon(X)} \sigma(A+K,B,C,\varepsilon)$. There exists $K \in \mathscr{G}_\varepsilon(X)$ such that for all $\|D\| < \varepsilon$, we have

$$\lambda \in \rho(A+K+CDB).$$

Therefore, for every $\|D\| < \varepsilon$, we have the following identity

$$\lambda - A - CDB = \left(I + K(\lambda - A - K - CDB)^{-1}\right)(\lambda - A - K - CDB). \quad (9.9)$$

Since

$$\Delta_\psi\Big(K(\lambda - A - K - CDB)^{-1}\Big) < 1 = \Gamma_\psi(I) \quad (9.10)$$

and by using Lemma 2.2.6, we can conclude that

$$I + K(\lambda - A - K - CDB)^{-1} \in \Phi_+^b(X)$$

and

$$i(I + K(\lambda - A - K - CDB)^{-1}) = i(I) = 0.$$

Therefore, for any $\|D\| < \varepsilon$

$$I + K(\lambda - A - K - CDB)^{-1} \in \Phi^b(X)$$

and

$$i(I+K(\lambda-A-K-CDB)^{-1})=0.$$

Furthermore, for all $D\in\mathscr{L}(Y,Z)$ satisfying $\|D\|<\varepsilon$, we have

$$\lambda-A-K-CDB\in\Phi(X)$$

and

$$i(\lambda-A-K-CDB)=0.$$

Consequently, for all $\|D\|<\varepsilon$

$$\lambda-A-CDB\in\Phi(X)$$

and

$$i(\lambda-A-CDB)=0.$$

It remains to show fo rall $\|D\|<\varepsilon$,

$$\operatorname{asc}(\lambda-A-CDB)<\infty$$

and

$$\operatorname{desc}(\lambda-A-CDB)<\infty.$$

By using Eq. (9.9), we have

$$(\lambda-A-CDB)^n=\left(I+K(\lambda-A-K-CDB)^{-1}\right)^n(\lambda-A-K-CDB)^n$$

for every $n\in\mathbb{N}$. Since for any $\|D\|<\varepsilon$, we have $(\lambda-A-K-CDB)^n$ is one-to-one, we infer

$$N\left((\lambda-A-CDB)^n\right)\subset N\left((I+K(\lambda-A-K-CDB)^{-1})^n\right).$$

Then, we can write for all $\|D\|<\varepsilon$,

$$\alpha\left((\lambda-A-CDB)^n\right)\leq\alpha\left((I+K(\lambda-A-K-CDB)^{-1})^n\right)$$

and so, by Lemma 2.4.1, we obtain that

$$\alpha\left((\lambda-A-CDB)^n\right)\leq$$

$$\operatorname{asc}\left(I+K(\lambda-A-K-CDB)^{-1}\right)\alpha\left(I+K(\lambda-A-K-CDB)^{-1}\right).$$

In addition, $I \in \mathscr{B}(X)$ and the use both Eq. (9.10) and Theorem 2.5.3 leads to for any $\|D\| < \varepsilon$,

$$I+K(\lambda-A-K-CDB)^{-1} \in \mathscr{B}(X).$$

By Lemma 2.4.2, we can see that for every $\|D\| < \varepsilon$,

$$\operatorname{asc}(\lambda-A-CDB) < \infty.$$

Likewise, Theorem 2.4.1 gives

$$\operatorname{desc}(\lambda-A-CDB) = \operatorname{asc}(\lambda-A-CDB) < \infty,$$

whence

$$\lambda \notin \sigma_{e6}(A,B,C,\varepsilon).$$

Conversely, by Theorem 2.1.8, we may easily observe that

$$\mathscr{K}_\varepsilon(X) \subset \mathscr{G}_\varepsilon(X),$$

where $\mathscr{K}_\varepsilon(X)$ is given in (9.8). Combining this result together with Remark 9.1.2, we obtain

$$\bigcap_{K\in\mathscr{G}_\varepsilon(X)} \sigma(A+K,B,C,\varepsilon) \subset \sigma_{e6}(A,B,C,\varepsilon). \qquad \text{Q.E.D.}$$

Corollary 9.1.5 *Let $A \in \mathscr{C}(X)$ such that $\rho(A) \neq \emptyset$, $B \in \mathscr{L}(X,Y)$, $C \in \mathscr{L}(Z,X)$ and $\varepsilon > 0$. Then,*

$$\sigma_{e6}(A,B,C,\varepsilon) = \bigcap_{K\in\mathscr{G}'_\varepsilon(X)} \sigma(A+K,B,C,\varepsilon),$$

where

$$\mathscr{G}'_\varepsilon(X) = \Big\{K \in \mathscr{L}(X) : \forall D \in \mathscr{L}(Y,Z) \text{ satisfying } \|D\| < \varepsilon, K \text{ commutes with } A+CDB \text{ and } K(\lambda-A)^{-1} \in \mathscr{S}(X),\ \forall\, \lambda \in \rho(A)\Big\}. \quad \diamondsuit$$

Proof. Let $\lambda \notin \bigcap\limits_{K \in \mathscr{G}'_\varepsilon(X)} \sigma(A+K,B,C,\varepsilon)$, then there exists $K \in \mathscr{G}'_\varepsilon(X)$ such that for every $\|D\| < \varepsilon$, we have

$$\lambda \in \rho(A+K+CDB).$$

Let $\mu \in \rho(A)$, we adopt the following identity

$$\mu - A = \left[I + (\mu - \lambda + K + CDB)(\lambda - A - K - CDB)^{-1}\right](\lambda - A - K - CDB).$$

Hence,

$$\begin{aligned} K(\lambda - A - K - CDB)^{-1} &= K(\mu - A)^{-1} \times \\ &\left[I + (\mu - \lambda + K + CDB)(\lambda - A - K - CDB)^{-1}\right]. \quad (9.11) \end{aligned}$$

Thus, we obtain from Theorem 2.1.8 together with Eq. (9.11) that $K \in \mathscr{G}_\varepsilon(X)$, where $\mathscr{G}_\varepsilon(X)$ is given in Theorem 9.1.10. Therefore,

$$\bigcap_{K \in \mathscr{G}_\varepsilon(X)} \sigma(A+K,B,C,\varepsilon) \subset \bigcap_{K \in \mathscr{G}'_\varepsilon(X)} \sigma(A+K,B,C,\varepsilon).$$

On the other hand, we have

$$\mathscr{K}_\varepsilon(X) \subset \mathscr{G}'_\varepsilon(X)$$

and by Remark 9.1.2, we can deduce that

$$\bigcap_{K \in \mathscr{G}'_\varepsilon(X)} \sigma(A+K,B,C,\varepsilon) \subset \sigma_{e6}(A,B,C,\varepsilon). \qquad \text{Q.E.D.}$$

The following theorem gives a relationship between the structured essential pseudospectra of the sum of two bounded operators and the structured essential pseudospectra of one of them.

Theorem 9.1.11 *Let A_1, $A_2 \in \mathscr{L}(X)$, $B \in \mathscr{L}(X,Y)$, $C \in \mathscr{L}(Z,X)$ and $\varepsilon > 0$. Then,*

(i) If for any $D \in \mathcal{L}(Y,Z)$ satisfying $\|D\| < \varepsilon$, we have $A_1(A_2 + CDB) \in \mathcal{F}^b(X)$, then

$$\sigma_{ei}(A_1+A_2,B,C,\varepsilon)\backslash\{0\} \subset \left[\sigma_{ei}(A_1)\bigcup\sigma_{ei}(A_2,B,C,\varepsilon)\right]\backslash\{0\},\ i=4,5.$$

If, further, $(A_2+CDB)A_1 \in \mathcal{F}^b(X)$ whence

$$\sigma_{e4}(A_1+A_2,B,C,\varepsilon)\backslash\{0\} = \left[\sigma_{e4}(A_1)\bigcup\sigma_{e4}(A_2,B,C,\varepsilon)\right]\backslash\{0\}.$$

Furthermore, if $\mathbb{C}\backslash\sigma_{e4}(A_1)$ is connected, then

$$\sigma_{e5}(A_1+A_2,B,C,\varepsilon)\backslash\{0\} = \left[\sigma_{e5}(A_1)\bigcup\sigma_{e5}(A_2,B,C,\varepsilon)\right]\backslash\{0\}.$$

(ii) If the hypothesis of (i) is satisfied and if we have for any $\|D\| < \varepsilon$, $\mathbb{C}\backslash\sigma_{e5}(A_2+CDB)$, $\mathbb{C}\backslash\sigma_{e5}(A_1+A_2+CDB)$ and $\mathbb{C}\backslash\sigma_{e5}(A_1)$ are connected, then

$$\sigma_{e6}(A_1+A_2,B,C,\varepsilon)\backslash\{0\} = \left[\sigma_{e6}(A_1)\cup\sigma_{e6}(A_2,B,C,\varepsilon)\right]\backslash\{0\}. \quad \diamond$$

Proof. By virtue of Theorem 3.1.4, we can assert that for any $\|D\| < \varepsilon$, we have if $A_1(A_2+CDB) \in \mathcal{F}^b(X)$, then

$$\sigma_{ei}(A_1+A_2+CDB)\backslash\{0\} \subset \left[\sigma_{ei}(A_1)\bigcup\sigma_{ei}(A_2+CDB)\right]\backslash\{0\},\ i=4,5.$$

If, further, $(A_2+CDB)A_1 \in \mathcal{F}^b(X)$, then

$$\sigma_{e4}(A_1+A_2+CDB)\backslash\{0\} = \left[\sigma_{e4}(A_1)\bigcup\sigma_{e4}(A_2+CDB)\right]\backslash\{0\}.$$

Furthermore, if $\mathbb{C}\backslash\sigma_{e4}(A_1)$ is connected, then

$$\sigma_{e5}(A_1+A_2+CDB)\backslash\{0\} = \left[\sigma_{e5}(A_1)\bigcup\sigma_{e5}(A_2+CDB)\right]\backslash\{0\}.$$

Then, we can immediately deduce (i).

Using Theorem 3.1.4 again, we obtain for any $\|D\| < \varepsilon$

$$\sigma_{e6}(A_1+A_2+CDB)\backslash\{0\} = \left[\sigma_{e6}(A_1)\bigcup\sigma_{e6}(A_2+CDB)\right]\backslash\{0\}.$$

Therefore,

$$\sigma_{e6}(A_1+A_2,B,C,\varepsilon)\backslash\{0\} = \left[\sigma_{e6}(A_1)\cup\sigma_{e6}(A_2,B,C,\varepsilon)\right]\backslash\{0\}.\ \text{Q.E.D.}$$

9.2 SOME DESCRIPTION OF THE STRUCTURED ESSENTIAL PSEUDOSPECTRA

9.2.1 *Relationship Between Structured Jeribi and Structured Ammar-Jeribi Essential Pseudospectra*

We start our analysis by the following remark which gives some crucial properties of the structured Jeribi essential pseudospectrum.

Remark 9.2.1 *We have the following properties:*

(i) $\sigma_j(A,B,C,\varepsilon) \subset \sigma_{e5}(A,B,C,\varepsilon) \subset \sigma(A,B,C,\varepsilon)$.

(ii) $\bigcap\limits_{\varepsilon>0} \sigma_j(A,B,C,\varepsilon) = \sigma_j(A)$.

(iii) *If* $\varepsilon_1 < \varepsilon_2$, *then*

$$\sigma_j(A) \subset \sigma_j(A,B,C,\varepsilon_1) \subset \sigma_j(A,B,C,\varepsilon_2).$$

(iv) *For all* $K \in \mathscr{W}^*(X)$, *we have*

$$\sigma_j(A+K,B,C,\varepsilon) = \sigma_j(A,B,C,\varepsilon). \qquad \diamondsuit$$

Theorem 9.2.1 *Let* $A \in \mathscr{C}(X)$, $B \in \mathscr{L}(X,Y)$, $C \in \mathscr{L}(Z,X)$ *and* $\varepsilon > 0$. *If* $\mathscr{W}^*(X) = \mathscr{W}(X)$ *and* X *has the Dunford-Pettis property, then*

$$\sigma_{e5}(A,B,C,\varepsilon) = \sigma_j(A,B,C,\varepsilon). \qquad \diamondsuit$$

Proof. By virtue of Theorem 9.1.2, we have

$$\sigma_{e5}(A,B,C,\varepsilon) = \bigcap_{K \in \mathscr{K}(X)} \sigma(A+K,B,C,\varepsilon)$$

and since $\mathscr{K}(X) \subset \mathscr{W}^*(X)$, we obtain

$$\sigma_j(A,B,C,\varepsilon) \subset \sigma_{e5}(A,B,C,\varepsilon).$$

Let $\lambda \notin \sigma_j(A,B,C,\varepsilon)$, then there exists $K \in \mathscr{W}^*(X)$ such that for any $D \in \mathscr{L}(Y,Z)$ satisfying $\|D\| < \varepsilon$, we have

$$\lambda \in \rho(A+K+CDB).$$

Thereby,

$$\lambda - A - K - CDB \in \Phi(X)$$

and

$$i(\lambda - A - K - CDB) = 0.$$

Furthermore, we have for any $\|D\| < \varepsilon$,

$$(\lambda - A - K - CDB)^{-1}K \in \mathcal{W}^*(X).$$

Then, by using Eq. (2.8), we get

$$\left[(\lambda - A - K - CDB)^{-1}K\right]^2 \in \mathcal{K}(X).$$

From Lemma 2.2.5, we can deduce that

$$I + (\lambda - A - K - CDB)^{-1}K$$

is a Fredholm operator and

$$i(I + (\lambda - A - S - CDB)^{-1}K) = 0.$$

Thereby, we can write

$$\lambda - A - CDB = (\lambda - A - K - CDB)\ [I + (\lambda - A - K - CDB)^{-1}K]$$

and, by virtue of Theorem 2.2.19 for all $\|D\| < \varepsilon$,

$$\lambda - A - CDB \in \Phi(X)$$

and

$$i(\lambda - A - CDB) = 0.$$

Consequently,

$$\lambda \notin \sigma_{e5}(A, B, C, \varepsilon). \qquad \text{Q.E.D.}$$

Theorem 9.2.2 *Let $A \in \mathcal{C}(X)$, $B \in \mathcal{L}(X,Y)$, $C \in \mathcal{L}(Z,X)$ and $\varepsilon > 0$. If $\mathcal{W}^*(X) = \mathcal{S}(X)$ and X is isomorphic to one of the spaces $L_p(\Omega)$, $p > 1$, then*

$$\sigma_{e5}(A, B, C, \varepsilon) = \sigma_j(A, B, C, \varepsilon) \qquad \diamondsuit$$

Proof. Since $\mathscr{K}(X) \subset \mathscr{S}(X)$ together with Theorem 9.1.2, we get

$$\sigma_j(A,B,C,\varepsilon) \subset \sigma_{e5}(A,B,C,\varepsilon).$$

It remains to shows that

$$\sigma_{e5}(A,B,C,\varepsilon) \subset \sigma_j(A,B,C,\varepsilon).$$

For this, we reason in the same way as the proof of Theorem 9.2.1. Q.E.D.

Remark 9.2.2 *Under the hypothesis of Theorem 9.2.1 or 9.2.2, we have for all $K \in \mathscr{W}^*(X)$*

$$\sigma_{e5}(A+K,B,C,\varepsilon) = \sigma_{e5}(A,B,C,\varepsilon). \qquad \diamondsuit$$

9.2.2 A Characterization of the Structured Ammar-Jeribi Essential Pseudospectrum

We begin with the following theorem which gives a fine description of the structured Ammar-Jeribi essential pseudospectrum.

Theorem 9.2.3 *Let $A \in \mathscr{C}(X)$, $B \in \mathscr{L}(X,Y)$, $C \in \mathscr{L}(Z,X)$ and $\varepsilon > 0$. Then,*

$$\sigma_{e5}(A,B,C,\varepsilon) = \bigcap_{K \in \mathscr{N}_{n,\varepsilon}(X)} \sigma(A+K,B,C,\varepsilon) = \bigcap_{K \in \mathscr{N}'_{n,\varepsilon}(X)} \sigma(A+K,B,C,\varepsilon),$$

where

$$\mathscr{N}_{n,\varepsilon}(X) = \Big\{K \in \mathscr{L}(X) : \forall D \in \mathscr{L}(Y,Z) \text{ satisfying } \|D\| < \varepsilon, \; \forall \lambda \in \rho(A+K+CDB), \; f\Big([(\lambda - A - K - CDB)^{-1}K]^n\Big) < 1 \text{ for some } n \in \mathbb{N}^*\Big\},$$

$$\mathscr{N}'_{n,\varepsilon}(X) = \Big\{K \in \mathscr{L}(X) : \forall D \in \mathscr{L}(Y,Z) \text{ satisfying } \|D\| < \varepsilon, \; \forall \lambda \in \rho(A+K+CDB), \; f\Big([K(\lambda - A - K - CDB)^{-1}]^n\Big) < 1 \text{ for some } n \in \mathbb{N}^*\Big\},$$

and $f(\cdot)$ is the measure of non-strict-singularity given in (2.22). $\diamondsuit$

Proof. Let $\lambda \notin \bigcap_{K\in\mathcal{N}_{n,\varepsilon}(X)} \sigma(A+K,B,C,\varepsilon)$ (resp. $\lambda \notin \bigcap_{K\in\mathcal{N}'_{n,\varepsilon}(X)} \sigma(A+K,B,C,\varepsilon)$). There exists $K \in \mathcal{N}_{n,\varepsilon}(X)$ (resp. $K \in \mathcal{N}'_{n,\varepsilon}(X)$) such that for every $\|D\| < \varepsilon$, we have

$$\lambda \in \rho(A+K+CDB).$$

Since for some $n \in \mathbb{N}^*$,

$$f\Big([(\lambda - A - K - CDB)^{-1}K]^n\Big) < 1$$

$$\text{(resp. } f\Big([K(\lambda - A - K - CDB)^{-1}]^n\Big) < 1),$$

then the use of Proposition 2.6.4 leads to

$$I + (\lambda - A - K - CDB)^{-1}K \in \Phi^b(X)$$

$$\text{(resp. } I + K(\lambda - A - K - CDB)^{-1} \in \Phi^b(X))$$

and

$$i(I + (\lambda - A - K - CDB)^{-1}K) = 0$$

$$\text{(resp. } i(I + K(\lambda - A - K - CDB)^{-1}) = 0).$$

Using the following identity

$$\lambda - A - CDB = (\lambda - A - K - CDB)(I + (\lambda - A - K - CDB)^{-1}K),$$

$$\text{(resp. } \lambda - A - CDB = (I + K(\lambda - A - K - CDB)^{-1})(\lambda - A - K - CDB)),$$

we get for every $D \in \mathcal{L}(Y,Z)$ satisfying $\|D\| < \varepsilon$

$$\lambda - A - CDB \in \Phi(X)$$

and

$$i(\lambda - A - CDB) = 0.$$

Consequently,

$$\sigma_{e5}(A,B,C,\varepsilon) \subset \bigcap_{K\in\mathcal{N}_{n,\varepsilon}(X)} \sigma(A+K,B,C,\varepsilon)$$

$$\text{(resp. } \sigma_{e5}(A,B,C,\varepsilon) \subset \bigcap_{K \in \mathcal{N}'_{n,\varepsilon}(X)} \sigma(A+K,B,C,\varepsilon)).$$

For the inverse inclusion, we have $\mathcal{K}(X) \subset \mathcal{S}(X)$ and by virtue of Proposition 2.6.3, we get

$$\mathcal{K}(X) \subset \mathcal{N}_{n,\varepsilon}(X)$$

(resp. $\mathcal{K}(X) \subset \mathcal{N}'_{n,\varepsilon}(X)$). Furthermore, by using Theorem 9.1.2, we have

$$\sigma_{e5}(A,B,C,\varepsilon) = \bigcap_{K \in \mathcal{K}(X)} \sigma(A+K,B,C,\varepsilon),$$

then we reach the desired result. Q.E.D.

Corollary 9.2.1 *Let $A \in \mathcal{C}(X)$, $B \in \mathcal{L}(X,Y)$, $C \in \mathcal{L}(Z,X)$, $\varepsilon > 0$ and $\mathcal{T}_\varepsilon(X)$ be any subset of $\mathcal{L}(X)$ satisfying*

$$\mathcal{K}(X) \subset \mathcal{T}_\varepsilon(X) \subset \mathcal{N}_{n,\varepsilon}(X)$$

or

$$\mathcal{K}(X) \subset \mathcal{T}_\varepsilon(X) \subset \mathcal{N}'_{n,\varepsilon}(X).$$

Then,

$$\sigma_{e5}(A,B,C,\varepsilon) = \bigcap_{K \in \mathcal{T}_\varepsilon(X)} \sigma(A+K,B,C,\varepsilon). \qquad \diamondsuit$$

We have not the stability of the structured Ammar-Jeribi essential pseudospectra by perturbations by operators in $\mathcal{N}_{n,\varepsilon}(X)$ or $\mathcal{N}'_{n,\varepsilon}(X)$. For this, we define the following subsets for $n \in \mathbb{N}^*$

$$\mathcal{I}'_n(X) = \{K \in \mathcal{L}(X) \text{ satisfying } f((BK)^n) < 1 \text{ for all } B \in \mathcal{L}(X)\},$$

$$\mathcal{I}_n(X) = \{K \in \mathcal{L}(X) \text{ satisfying } f((KB)^n) < 1 \text{ for all } B \in \mathcal{L}(X)\}$$

and let

$$\mathcal{S}'_n(X) = \{B \in \mathcal{L}(X) \text{ satisfying } B^n \in \mathcal{S}(X)\}.$$

It is proved in [122] that

$$\mathcal{S}(X) \subset \mathcal{I}_n(X) \subset \mathcal{S}'_n(X).$$

The following result is an immediate consequence of Theorem 3.1.1 (ii).

Theorem 9.2.4 *Let $A \in \mathscr{C}(X)$, $B \in \mathscr{L}(X,Y)$, $C \in \mathscr{L}(Z,X)$ and $\varepsilon > 0$. Then, for all $K \in \mathscr{I}_n(X)$, we have*

$$\sigma_{e5}(A+K,B,C,\varepsilon) = \sigma_{e5}(A,B,C,\varepsilon). \qquad \diamondsuit$$

Chapter 10

Structured Essential Pseudospectra and Measure of Noncompactness

This chapter deals with studying the structured Ammar-Jeribi and the structured Browder essential pseudospectra of closed densely defined linear operators via the concept of the measure of noncompactness and the measure of non-strict-singularity.

10.1 NEW DESCRIPTION OF THE STRUCTURED ESSENTIAL PSEUDOSPECTRA

10.1.1 A Characterization of the Structured Ammar-Jeribi Essential Pseudospectrum by Kuratowski Measure of Noncompactness

Theorem 10.1.1 *Let* $n \in \mathbb{N}^*$, $A \in \mathscr{C}(X)$, $B \in \mathscr{L}(X,Y)$, $C \in \mathscr{L}(Z,X)$ *and* $\varepsilon > 0$. *Then,*

$$\sigma_{e5}(A,B,C,\varepsilon) = \bigcap_{K \in \mathscr{M}_{n,\varepsilon}(X)} \sigma(A+K,B,C,\varepsilon),$$

where

$$\mathscr{M}_{n,\varepsilon}(X) = \Big\{K \in \mathscr{L}(X) : \forall D \in \mathscr{L}(Y,Z) \text{ satisfying } \|D\| < \varepsilon,$$
$$\forall \lambda \in \rho(A+K+CDB),\ \gamma\Big([(\lambda - A - K - CDB)^{-1}K]^n\Big) < 1\Big\},$$

and $\gamma(\cdot)$ is the Kuratowski measure of noncompactness given in (2.20). $\diamondsuit$

Proof. Suppose that $\lambda \notin \bigcap_{K \in \mathscr{M}_{n,\varepsilon}(X)} \sigma(A+K,B,C,\varepsilon)$. Then, there exists $K \in \mathscr{M}_{n,\varepsilon}(X)$ such that for every $\|D\| < \varepsilon$, we have

$$\lambda \in \rho(A+K+CDB).$$

Thereby, we can write

$$\lambda - A - CDB = (\lambda - A - K - CDB)\Big[I + (\lambda - A - K - CDB)^{-1}K\Big].$$

Now,

$$\gamma\Big([(\lambda - A - K - CDB)^{-1}K]^n\Big) < 1,$$

and hence

$$\lim_{k \to +\infty} \gamma\Big([(\lambda - A - K - CDB)^{-1}K]^n\Big)^k = 0.$$

Therefore, there exists $k_0 \in \mathbb{N}^*$ such that

$$\gamma\Big([(\lambda - A - K - CDB)^{-1}K]^n\Big)^{k_0} < \frac{1}{2}.$$

It follows from Lemma 2.6.1 (iii) that

$$\gamma\Big([(\lambda - A - K - CDB)^{-1}K]^{nk_0}\Big) < \frac{1}{2}.$$

We apply Theorem 2.6.1 (ii) for $P(z) = z^{nk_0}$ and $Q(z) = 1 - z$, we obtain that

$$I + (\lambda - A - K - CDB)^{-1}K \in \Phi^b(X).$$

Next, for $t \in [0,1]$, we have

$$\gamma\Big([t(\lambda - A - K - CDB)^{-1}K]^{nk_0}\Big) < \frac{1}{2},$$

and so

$$I + t(\lambda - A - K - CDB)^{-1}K \in \Phi^b(X).$$

Furthermore, for $t \in [0,1]$, there is a $\beta > 0$ such that for all $t_1 \in [0,1]$ satisfying $|t_1 - t| < \beta$, we have from Theorem 2.2.16

$$\begin{aligned} & i\Big(I+t(\lambda-A-K-CDB)^{-1}K\Big) \\ = \ & i\Big(I+t_1(\lambda-A-K-CDB)^{-1}K+(t-t_1)(\lambda-A-K-CDB)^{-1}K\Big) \\ = \ & i\Big(I+t_1(\lambda-A-K-CDB)^{-1}K\Big). \end{aligned}$$

Using the Heine-Borel theorem, we get the existence of finite number of sets which cover $[0,1]$. Since the index is constant on each one of these sets which overlaps with at least one another, then we obtain

$$i\Big(I+(\lambda-A-K-CDB)^{-1}K\Big)=i(I)=0.$$

Consequently, for all $D\in\mathscr{L}(Y,Z)$ such that $\|D\|<\varepsilon$, we have

$$\lambda-A-CDB\in\Phi(X)$$

and

$$i(\lambda-A-CDB)=0.$$

Conversely, since $\mathscr{K}(X)\subset\mathscr{M}_{n,\varepsilon}(X)$, then

$$\bigcap_{K\in\mathscr{M}_{n,\varepsilon}(X)}\sigma(A+K,B,C,\varepsilon)\subset\bigcap_{K\in\mathscr{K}(X)}\sigma(A+K,B,C,\varepsilon).$$

From Theorem 9.1.2, we have

$$\sigma_{e5}(A,B,C,\varepsilon)=\bigcap_{K\in\mathscr{K}(X)}\sigma(A+K,B,C,\varepsilon),$$

so we can deduce the result. Q.E.D.

Corollary 10.1.1 *Let $n\in\mathbb{N}^*$, $A\in\mathscr{C}(X)$, $B\in\mathscr{L}(X,Y)$, $C\in\mathscr{L}(Z,X)$, $\mathscr{M}(X)$ be an arbitrary subspace of $\mathscr{L}(X)$ and $\varepsilon>0$. If $\mathscr{K}(X)\subset\mathscr{M}(X)\subset\mathscr{M}_{n,\varepsilon}(X)$, then*

$$\sigma_{e5}(A,B,C,\varepsilon)=\bigcap_{K\in\mathscr{M}(X)}\sigma(A+K,B,C,\varepsilon).$$ ◇

Theorem 10.1.2 *Let $n \in \mathbb{N}^*$, $A \in \mathscr{C}(X)$, $B \in \mathscr{L}(X,Y)$, $C \in \mathscr{L}(Z,X)$ and $\varepsilon > 0$. Then,*

$$\bigcup_{K \in \mathscr{Q}_\varepsilon(X)} \sigma_{e5}(A+K,B,C,\varepsilon) \subset \sigma_{e5}(A,B,C,\varepsilon),$$

where $\mathscr{Q}_\varepsilon(X) = \Big\{ K \in \mathscr{L}(X) : \forall D \in \mathscr{L}(Y,Z)$ *satisfying* $\|D\| < \varepsilon$, *the right quasi-inverse* $N_{\lambda,\varepsilon}$ *of the operator* $\lambda - A - CDB$ *satisfies* $\gamma\Big((N_{\lambda,\varepsilon}K)^n\Big) < 1 \Big\}$. ◇

Proof. We assume that $\lambda \notin \sigma_{e5}(A,B,C,\varepsilon)$, then for every $\|D\| < \varepsilon$, we have

$$\lambda - A - CDB \in \Phi(X) \text{ and } i(\lambda - A - CDB) = 0.$$

Let $K \in \mathscr{Q}_\varepsilon(X)$, we shall prove that $\lambda \notin \sigma_{e5}(A+K,B,C,\varepsilon)$. We have for any $D \in \mathscr{L}(Y,Z)$ satisfying $\|D\| < \varepsilon$ the existence of a right quasi inverse $N_{\lambda,\varepsilon}$ of the operator $\lambda - A - CDB$, this implies that

$$R(N_{\lambda,\varepsilon}) \subset \mathscr{D}(A)$$

and

$$(\lambda - A - CDB)N_{\lambda,\varepsilon} = I - K_1,$$

where $K_1 \in \mathscr{K}(X)$. So, we can write

$$\begin{aligned} \lambda - A - K - CDB &= \lambda - A - CDB - \Big[(\lambda - A - CDB)N_{\lambda,\varepsilon} + K_1\Big]K \\ &= (\lambda - A - CDB)[I - N_{\lambda,\varepsilon}K] - K_1 K. \end{aligned}$$

Since

$$\gamma\Big((N_{\lambda,\varepsilon}K)^n\Big) < 1,$$

then by repeating the same argument used in the proof of Theorem 10.1.1, we obtain

$$I - N_{\lambda,\varepsilon}K \in \Phi^b(X) \text{ and } i(I - N_{\lambda,\varepsilon}K) = 0.$$

Therefore, for every $\|D\| < \varepsilon$, we have

$$\lambda - A - K - CDB \in \Phi(X) \text{ and } i(\lambda - A - K - CDB) = 0. \quad \text{Q.E.D.}$$

We close this subsection by the following result which gives a characterization of the structured Ammar-Jeribi essential pseudospectrum by means of the measure of non-strict-singularity $\Delta_\psi(\cdot)$.

Theorem 10.1.3 *Let $A \in \mathscr{C}(X)$, $B \in \mathscr{L}(X,Y)$, $C \in \mathscr{L}(Z,X)$ and $\varepsilon > 0$. Then,*

$$\sigma_{e5}(A,B,C,\varepsilon) = \bigcap_{K \in \mathscr{H}_\varepsilon(X)} \sigma(A+K,B,C,\varepsilon),$$

where

$$\mathscr{H}_\varepsilon(X) = \Big\{K \in \mathscr{L}(X) : \forall D \in \mathscr{L}(Y,Z) \text{ with } \|D\| < \varepsilon,$$
$$\Delta_\psi\Big((\lambda - A - K - CDB)^{-1}K\Big) < 1 \; \forall\, \lambda \in \rho(A+K+CDB)\Big\},$$

and $\Delta_\psi(\cdot)$ is the measure of non-strict-singularity given in (2.11). $\diamondsuit$

Proof. Let $\lambda \notin \bigcap_{K \in \mathscr{H}_\varepsilon(X)} \sigma(A+K,B,C,\varepsilon)$, then there exists $K \in \mathscr{H}_\varepsilon(X)$ such that for all $\|D\| < \varepsilon$, we have

$$\lambda \in \rho(A+K+CDB).$$

Then, for any $D \in \mathscr{L}(Y,Z)$ satisfying $\|D\| < \varepsilon$, we can see that

$$\lambda - A - K - CDB \in \Phi(X),$$

$$i(\lambda - A - K - CDB) = 0$$

and

$$\Delta_\psi\Big((\lambda - A - K - CDB)^{-1}K\Big) < 1 = \Gamma_\psi(I),$$

where $\Gamma_\psi(\cdot)$ is given in (2.10). Using Lemma 2.2.6, we can conclude that

$$I + (\lambda - A - K - CDB)^{-1}K \in \Phi_+^b(X)$$

and

$$i(I + (\lambda - A - K - CDB)^{-1}K) = i(I) = 0.$$

Therefore, for any $\|D\| < \varepsilon$

$$I + (\lambda - A - K - CDB)^{-1}K \in \Phi^b(X)$$

and

$$i(I+(\lambda-A-K-CDB)^{-1}K)=0.$$

Since we can write

$$\lambda-A-CDB=(\lambda-A-K-CDB)(I+(\lambda-A-K-CDB)^{-1}K),$$

then we obtain that for all $\|D\|<\varepsilon$

$$\lambda-A-CDB\in\Phi(X)$$

and

$$i(\lambda-A-CDB)=0.$$

On the other hand, by Theorem 2.1.8, we can see that

$$\mathscr{K}(X)\subset\mathscr{H}_{\varepsilon}(X).$$

Hence,

$$\bigcap_{K\in\mathscr{H}_{\varepsilon}(X)}\sigma(A+K,B,C,\varepsilon)\subset\sigma_{e5}(A,B,C,\varepsilon). \qquad \text{Q.E.D.}$$

10.1.2 A Characterization of the Structured Browder Essential Pseudospectrum by Means of Measure of Non-Strict-Singularity

Theorem 10.1.4 *Let A, $K\in\mathscr{L}(X)$ such that K commutes with A. Suppose that*

$$f(K^n)<f_M(A^n)$$

for some $n\geq 1$, where $f_M(\cdot)$ (resp. $f(\cdot)$) is given in (2.21) (resp. (2.22)). Then, $A\in\mathscr{B}_+(X)$ if, and only if, $A+K\in\mathscr{B}_+(X)$. ◊

Proof. Let $A \in \mathscr{B}_+(X)$, then $A \in \Phi_+(X)$ and

$$\operatorname{asc}(A) < \infty.$$

Using Lemma 2.6.4, we obtain

$$A + K \in \Phi_+(X).$$

Since $N(A^p)$ is a closed subspace and $\operatorname{asc}(A) = p < \infty$, then

$$\overline{N^\infty(A)} = N^\infty(A).$$

By Proposition 2.5.1, we get

$$N^\infty(A) \bigcap R^\infty(A) = \{0\}.$$

Set $A_\lambda := A + \lambda K$, where $\lambda \in [0,1]$. Since

$$f((\lambda K)^n)) = \lambda^n f(K^n) < f_M(A^n),$$

then

$$A_\lambda \in \Phi_+(X).$$

We can check that

$$\operatorname{asc}(A+K) < \infty$$

in the same way as that of the proof of Theorem 2.5.2. Conversely, let $A + K \in \mathscr{B}_+(X)$. Then, Lemma 2.6.4 proves that

$$A \in \Phi_+(X).$$

It remains to show that

$$\operatorname{asc}(A) < \infty.$$

To do this, we consider

$$(A+K)_\lambda = A + K + \lambda K = A + (\lambda + 1)K.$$

where $\lambda \in [-1,0]$ and we reason in the same way as above. Q.E.D.

Theorem 10.1.5 *Let A, $K \in \mathscr{L}(X)$ such that K commutes with A. Suppose that*

$$f(K^n) < f_M(A^n)$$

for some $n \geq 1$, where $f_M(\cdot)$ (resp. $f(\cdot)$) is given in (2.21) (resp. (2.22)). Then, $A \in \mathscr{B}(X)$ if, and only if, $A+K \in \mathscr{B}(X)$. ◇

Proof. Let $A \in \mathscr{B}(X)$, then $A \in \mathscr{B}_+(X)$ and

$$\operatorname{desc}(A) < \infty.$$

Using Theorem 10.1.4, we infer that $A+K \in \mathscr{B}_+(X)$. So, the use of Theorem 2.4.1 leads to

$$\operatorname{desc}(A+K) = \operatorname{asc}(A+K) < \infty.$$

Conversely, we reason in the same way as above. Q.E.D.

Now, we are ready to state the following result.

Theorem 10.1.6 *Let $A \in \mathscr{C}(X)$, $B \in \mathscr{L}(X,Y)$, $C \in \mathscr{L}(Z,X)$ and $\varepsilon > 0$. Then,*

$$\sigma_{e6}(A,B,C,\varepsilon) = \bigcap_{K \in \mathscr{O}_\varepsilon(X)} \sigma(A+K,B,C,\varepsilon),$$

where

$$\mathscr{O}_\varepsilon(X) = \Big\{K \in \mathscr{L}(X) : \forall D \in \mathscr{L}(Y,Z) \text{ satisfying } \|D\| < \varepsilon, K \text{ commutes with } A+CDB \text{ and } f\Big(\big(K(\lambda - A - K - CDB)^{-1}\big)^n\Big) < 1, \ \forall\, \lambda \in \rho(A+K+CDB) \text{ for some } n \in \mathbb{N}^*\Big\}. \quad ◇$$

Proof. Let $\lambda \notin \bigcap_{K \in \mathscr{O}_\varepsilon(X)} \sigma(A+K,B,C,\varepsilon)$. Then, there exists $K \in \mathscr{O}_\varepsilon(X)$ such that for all $\|D\| < \varepsilon$, we have

$$\lambda \in \rho(A+K+CDB).$$

Thereby, for every $\|D\| < \varepsilon$, we have the following equality

$$\lambda - A - CDB = \big(I + K(\lambda - A - K - CDB)^{-1}\big)(\lambda - A - K - CDB). \tag{10.1}$$

Since $I \in \mathscr{B}(X)$ and

$$f\left(\left(K(\lambda - A - K - CDB)^{-1}\right)^{n}\right) < 1,$$

then by Theorem 10.1.5, we obtain for every $\|D\| < \varepsilon$,

$$I + K(\lambda - A - K - CDB)^{-1} \in \mathscr{B}(X).$$

Thus,

$$I + K(\lambda - A - K - CDB)^{-1} \in \Phi(X)$$

and

$$i(I + K(\lambda - A - K - CDB)^{-1}) = 0.$$

In addition, for every $\|D\| < \varepsilon$, we have

$$\lambda - A - K - CDB \in \Phi(X)$$

and

$$i(\lambda - A - K - CDB) = 0.$$

Therefore, for all $\|D\| < \varepsilon$

$$\lambda - A - CDB \in \Phi(X)$$

and

$$i(\lambda - A - CDB) = 0.$$

By using Eq. (10.1), we have the following identity

$$(\lambda - A - CDB)^{n} = \left(I + K(\lambda - A - K - CDB)^{-1}\right)^{n} (\lambda - A - K - CDB)^{n}$$

for every $n \in \mathbb{N}$. Since we have

$$\alpha\left((\lambda - A - CDB)^{n}\right) \leq \alpha\left((I + K(\lambda - A - K - CDB)^{-1})^{n}\right)$$

for all $\|D\| < \varepsilon$ and, by virtue of Lemma 2.4.1, we can conclude that

$$\alpha\left((\lambda - A - CDB)^{n}\right) \leq \operatorname{asc}\left(I + K(\lambda - A - K - CDB)^{-1}\right) \alpha\left(I + K(\lambda - A - K - CDB)^{-1}\right).$$

Lemma 2.4.2 proves that for every $\|D\| < \varepsilon$,

$$\operatorname{asc}(\lambda - A - CDB) < \infty.$$

This implies by the use of Theorem 2.4.1 that

$$\operatorname{desc}(\lambda - A - CDB) = \operatorname{asc}(\lambda - A - CDB) < \infty.$$

Hence,

$$\lambda \notin \sigma_{e6}(A,B,C,\varepsilon).$$

Conversely, by Remark 9.1.1, we have

$$\sigma_{e6}(A,B,C,\varepsilon) = \bigcap_{K \in \mathscr{K}_\varepsilon(X)} \sigma(A+K,B,C,\varepsilon),$$

where $\mathscr{K}_\varepsilon(X)$ is given in (9.8). Using Proposition 2.6.3 (i), we obtain $\mathscr{K}_\varepsilon(X) \subset \mathscr{O}_\varepsilon(X)$. Consequently,

$$\bigcap_{K \in \mathscr{O}_\varepsilon(X)} \sigma(A+K,B,C,\varepsilon) \subset \sigma_{e6}(A,B,C,\varepsilon). \qquad \text{Q.E.D.}$$

Chapter 11

A Characterization of the Essential Pseudospectra

In this chapter, we investigate some parts of the pseudospectrum of closed densely defined operators on a Banach space from the viewpoint of Fredholms theory by examining the different types of pseudospectra, such as approximation pseudospectrum, defect pseudospectrum, essential approximation pseudospectrum, essential defect pseudospectrum, essential structured approximation pseudospectrum and essential structured defect pseudospectrum.

11.1 APPROXIMATION OF ε-PSEUDOSPECTRUM

In order to obtain the strongest possible results we have to refine the notion of the ε-pseudospectrum of an operator $(A, \mathscr{D}(A))$ defined on the Banach space X. For it turns out that the part $\sigma(A)\backslash\sigma_{ap}(A)$ cannot always be approximated in the general case. For $\varepsilon \geq 0$, we define the ε-approximate spectrum $\sigma_{\varepsilon,ap}(\cdot)$ by

$$\sigma_{\varepsilon,ap}(A) = \{\lambda \in \mathbb{C} \text{ such that } \widetilde{\alpha}(\lambda - A) \leq \varepsilon\},$$

where $\widetilde{\alpha}(\cdot)$ is given in Remark 3.1.1 (iv). In particular $\sigma_{0,ap}(A) = \sigma_{ap}(A)$ as well as

$$\sigma_{\varepsilon,ap}(A) \subset \Sigma_{\varepsilon}(A).$$

The following results, given in this section, come from [161].

Theorem 11.1.1 *Let $A \in \mathscr{C}(X)$. Then, $\sigma_{\varepsilon,ap}(A)$ is always closed.* ◇

Proof. Let $\lambda \notin \sigma_{\varepsilon,ap}(A)$. Then,

$$\widetilde{\alpha}(\lambda - A) > \varepsilon.$$

Now, let $\mu \in \mathbb{C}$ satisfy

$$|\lambda - \mu| < \widetilde{\alpha}(\lambda - A) - \varepsilon.$$

Then,

$$\begin{aligned}
&\widetilde{\alpha}(\mu - A) \\
&\quad = \inf\{\|(\mu - A)x\| \text{ such that } \|x\| = 1 \text{ and } x \in \mathscr{D}(A)\} \\
&\quad \geq \inf\{|\|(\lambda - A)x\| - |\lambda - \mu|\|x\|| \text{ such that } \|x\| = 1 \text{ and } x \in \mathscr{D}(A)\} \\
&\quad > \varepsilon.
\end{aligned}$$

So, the complement of $\sigma_{\varepsilon,ap}(A)$ is open. Q.E.D.

Corollary 11.1.1 *If $A \in \mathscr{C}(X)$ and $\varepsilon > 0$, then $\Sigma_{ap,\varepsilon}(A)$ is closed.* ◇

Now, we present the following simple and useful result:

Proposition 11.1.1 *Let $A \in \mathscr{C}(X)$ and $\varepsilon > 0$. Then,*

(i) $\sigma_{ap,\varepsilon}(A) \subset \sigma_{\varepsilon}(A)$,

(ii) $\sigma_{ap}(A) = \bigcap_{\varepsilon > 0} \sigma_{ap,\varepsilon}(A)$,

(iii) *if* $\varepsilon_1 < \varepsilon_2$, *then* $\sigma_{ap}(A) \subset \sigma_{ap,\varepsilon_1}(A) \subset \sigma_{ap,\varepsilon_2}(A)$,

(iv) *if* $A \in \mathscr{L}(X)$ *and* $\lambda \in \sigma_{ap,\varepsilon}(A)$, *then* $|\lambda| < \varepsilon + \|A\|$,

(v) *if* $\alpha \in \mathbb{C}$ *and* $\varepsilon > 0$, *then* $\sigma_{ap,\varepsilon}(A + \alpha) = \alpha + \sigma_{ap,\varepsilon}(A)$, *and*

(vi) *if* $\alpha \in \mathbb{C}\backslash\{0\}$ *and* $\varepsilon > 0$, *then* $\sigma_{ap,|\alpha|\varepsilon}(\alpha A) = \alpha\, \sigma_{ap,\varepsilon}(A)$. ◇

Proof. (i) If $\lambda \notin \sigma_\varepsilon(A)$, then

$$\|(\lambda - A)^{-1}\| \leq \frac{1}{\varepsilon}.$$

Moreover,

$$\begin{aligned}
\frac{1}{\inf\limits_{x\in\mathscr{D}(A),\ \|x\|=1} \|(\lambda - A)x\|} &= \sup_{x\in\mathscr{D}(A),\ \|x\|=1} \frac{\|x\|}{\|(\lambda - A)x\|} \\
&= \sup_{0\neq x\in\mathscr{D}(A)} \frac{\|x\|}{\|(\lambda - A)x\|} \\
&= \sup_{y\in X\backslash\{0\}} \frac{\|(\lambda - A)^{-1}y\|}{\|y\|} \\
&= \|(\lambda - A)^{-1}\| \\
&\leq \frac{1}{\varepsilon}.
\end{aligned}$$

Hence,

$$\inf_{x\in\mathscr{D}(A),\ \|x\|=1} \|(\lambda - A)x\|) \geq \varepsilon.$$

So,

$$\lambda \notin \sigma_{ap,\varepsilon}(A).$$

(ii) It is clear that $\sigma_{ap}(A) \subset \sigma_{ap,\varepsilon}(A)$, then

$$\sigma_{ap}(A) \subset \bigcap_{\varepsilon>0} \sigma_{ap,\varepsilon}(A).$$

Conversely, if $\lambda \in \bigcap\limits_{\varepsilon>0} \sigma_{ap,\varepsilon}(A)$, then for all $\varepsilon > 0$, we have $\lambda \in \sigma_{ap,\varepsilon}(A)$.
Let

$$\lambda \in \Big\{\lambda \in \mathbb{C} \text{ such that } \inf_{x\in\mathscr{D}(A),\ \|x\|=1} \|(\lambda - A)x\| < \varepsilon\Big\},$$

taking limits as $\varepsilon \to 0^+$, we get for all $x \in \mathscr{D}(A)$ such that $\|x\| = 1$,

$$\inf_{x\in\mathscr{D}(T),\ \|x\|=1} \|(\lambda - A)x\| = 0.$$

We infer that $\lambda \in \sigma_{ap}(A)$.

(iii) Let $\lambda \in \sigma_{ap,\varepsilon_1}(A)$, then

$$\inf_{x\in\mathscr{D}(A),\ \|x\|=1} \|(\lambda - A)x\| < \varepsilon_1 < \varepsilon_2.$$

Hence, $\lambda \in \sigma_{ap,\varepsilon_2}(A)$.

(iv) Let $\lambda \in \sigma_{ap,\varepsilon}(A)$, then

$$\inf_{x\in\mathscr{D}(A),\ \|x\|=1} \|(\lambda - A)x\| < \varepsilon,$$

and also

$$\Big| |\lambda| - \|Ax\| \Big| < \|(\lambda - A)x\|.$$

Hence,

$$|\lambda| < \varepsilon + \|A\|.$$

(v) Let $\lambda \in \sigma_{ap,\varepsilon}(A+\alpha)$, then

$$\inf_{x\in\mathscr{D}(A),\ \|x\|=1} \|((\lambda - \alpha) - A)x\| < \varepsilon.$$

Hence,

$$\lambda - \alpha \in \sigma_{ap,\varepsilon}(A).$$

This yields to

$$\lambda \in \alpha + \sigma_{ap,\varepsilon}(A).$$

The second inclusion follows by using the same reasoning.

(vi) Let $\lambda \in \sigma_{ap,|\alpha|\varepsilon}(\alpha A)$, then

$$\begin{aligned}\inf_{x\in\mathscr{D}(A),\ \|x\|=1} \|(\lambda - \alpha A)x\| &= \inf_{x\in\mathscr{D}(A),\ \|x\|=1} \left\| \alpha\left(\frac{\lambda}{\alpha} - A\right)x \right\| \quad \alpha \neq 0,\\ &= |\alpha| \inf_{x\in\mathscr{D}(A),\ \|x\|=1} \left\| \left(\frac{\lambda}{\alpha} - A\right)x \right\|\\ &< |\alpha|\varepsilon.\end{aligned}$$

Hence,

$$\frac{\lambda}{\alpha} \in \sigma_{ap,\varepsilon}(A).$$

So,

$$\sigma_{ap,|\alpha|\varepsilon}(\alpha A) \subseteq \alpha\ \sigma_{ap,\varepsilon}(A).$$

However, the reverse inclusion is similar. Q.E.D.

Remark 11.1.1 *P. H. Wolff shows that for all* $\varepsilon > 0$ *that* $\Sigma_{ap,\varepsilon}(A) \neq \Sigma_\varepsilon(A)$, *(see [161]).* ◊

Proposition 11.1.2 *Let* $A \in \mathscr{C}(X)$ *and* $\varepsilon > 0$. *Then,*

(i) $\Sigma_{ap,\varepsilon}(A) \subset \Sigma_\varepsilon(A)$,

(ii) $\bigcap_{\varepsilon>0} \Sigma_{ap,\varepsilon}(A) = \sigma_{ap}(A)$, *and*

(iii) *if* $\varepsilon_1 < \varepsilon_2$, *then* $\sigma_{ap}(A) \subset \Sigma_{ap,\varepsilon_1}(A) \subset \Sigma_{ap,\varepsilon_2}(A)$. ◊

Proof. The proof of (i), (ii) and (iii) may be achieved in the same way as the proof of (i), (ii) and (iii) for Proposition 11.1.1. Q.E.D.

11.2 A CHARACTERIZATION OF APPROXIMATION PSEUDOSPECTRUM

In this section, we turn to the problem when the closure of $\sigma_{ap,\varepsilon}(A)$ is equal to $\Sigma_{ap,\varepsilon}(A)$ holds. We consider the following hypothesis for A :

$(H6)$ There is no open set in $\rho_{ap}(A) := \mathbb{C}\backslash\sigma_{ap}(A)$ on which the

$$\lambda \longrightarrow \inf_{x\in\mathscr{D}(A),\ \|x\|=1} \|(\lambda - A)x\|$$

is constant. Our first result is the following.

Theorem 11.2.1 *Let* $A \in \mathscr{C}(X)$ *and* $\varepsilon > 0$. *If* $(H6)$ *holds, then*

$$\overline{\sigma_{ap,\varepsilon}(A)} = \Sigma_{ap,\varepsilon}(A).$$ ◊

Proof. Since $\sigma_{ap,\varepsilon}(A) \subset \Sigma_{ap,\varepsilon}(A)$ and $\Sigma_{ap,\varepsilon}(A)$ is closed, then

$$\overline{\sigma_{ap,\varepsilon}(A)} \subset \Sigma_{ap,\varepsilon}(A).$$

In order to prove the inverse inclusion, we take $\lambda \in \Sigma_{ap,\varepsilon}(A)$. We notice the existence of two cases:

<u>1st *case*</u> : If $\lambda \in \sigma_{ap,\varepsilon}(A)$, then $\lambda \in \overline{\sigma_{ap,\varepsilon}(A)}$.

<u>2nd case</u> : If $\lambda \in \Sigma_{ap,\varepsilon}(A)\backslash\sigma_{ap,\varepsilon}(A)$, then

$$\inf_{x\in\mathscr{D}(A),\ \|x\|=1,} \|(\lambda - A)x\| = \varepsilon.$$

By using Hypothesis $(H6)$, there exists a sequence $\lambda_n \in \rho_{ap}(A)$ such that $\lambda_n \to \lambda$, and

$$\inf_{x\in\mathscr{D}(A),\ \|x\|=1} \|(\lambda_n - A)x\| < \inf_{x\in\mathscr{D}(A),\ \|x\|=1} \|(\lambda - A)x\| = \varepsilon.$$

We deduce that $\lambda_n \in \sigma_{ap,\varepsilon}(A)$ and hence, $\lambda \in \overline{\sigma_{ap,\varepsilon}(A)}$. So,

$$\Sigma_{ap,\varepsilon}(A) \subset \overline{\sigma_{ap,\varepsilon}(A)}.$$ Q.E.D.

Theorem 11.2.2 *Let $A \in \mathscr{C}(X)$ and $\varepsilon > 0$. Then, the following properties are equivalent:*

(i) $\lambda \in \sigma_{ap,\varepsilon}(A)$.

(ii) There exists a bounded operator $D \in \mathscr{L}(X)$ such that $\|D\| < \varepsilon$ and $\lambda \in \sigma_{ap}(A+D)$. ◊

Proof. $(i) \Rightarrow (ii)$ Let $\lambda \in \sigma_{ap,\varepsilon}(A)$. Then, there exists $x_0 \in X$ such that $\|x_0\| = 1$ and

$$\|(\lambda - A)x_0\| < \varepsilon. \tag{11.1}$$

By using the Hahn Banach theorem (see Theorem 2.1.4), there exists $x' \in X^*$ (dual of X) such that $\|x'\| = 1$ and

$$x'(x_0) = \|x_0\|.$$

Consider the operator D defined by the formula

$$D : X \longrightarrow X,\ \ x \longrightarrow Dx := x'(x)(\lambda - A)x_0.$$

Then, D is a linear operator everywhere defined on X. It is bounded, since

$$\begin{aligned} \|Dx\| &= \|x'(x)(\lambda - A)x_0\| \\ &\leq \|x'\|\|x\|\|(\lambda - A)x_0\|, \end{aligned}$$

for $x \neq 0$. Therefore,

$$\frac{\|Dx\|}{\|x\|} \leq \|(\lambda - A)x_0\|.$$

Hence, by using Eq. (11.1), we have

$$\|D\| < \varepsilon.$$

Let $x_0 \in X$, then

$$\begin{aligned} \inf_{x \in \mathscr{D}(A),\ \|x\|=1} \|(\lambda - A - D)x\| &\leq \|(\lambda - A - D)x_0\| \\ &\leq \|(\lambda - A)x_0 - x'(x_0)(\lambda - A)x_0\| = 0. \end{aligned}$$

This prove that

$$\inf_{x \in \mathscr{D}(A),\ \|x\|=1} \|(\lambda - A - D)x\| = 0.$$

$(ii) \Rightarrow (i)$ We assume that there exists a bounded operator $D \in \mathscr{L}(X)$ such that $\|D\| < \varepsilon$ and $\lambda \in \sigma_{ap}(A + D)$, which means that

$$\inf_{x \in \mathscr{D}(A),\ \|x\|=1} \|(\lambda - A - D)x\| = 0.$$

In order to prove that

$$\inf_{x \in \mathscr{D}(A),\ \|x\|=1} \|(\lambda - A)x\| < \varepsilon,$$

we can write

$$\begin{aligned} \|(\lambda - A)x_0\| &= \|(\lambda - A - D + D)x_0\| \\ &\leq \|(\lambda - A - D)x_0\| + \|Dx_0\|. \end{aligned}$$

Then,

$$\inf_{x \in \mathscr{D}(A),\ \|x\|=1} \|(\lambda - A)x\| < \varepsilon.$$

Q.E.D.

We can derive from Theorem 11.2.2 the following result.

Corollary 11.2.1 *Let $A \in \mathscr{C}(X)$ and $\varepsilon > 0$. Then,*

$$\sigma_{ap,\varepsilon}(A) = \bigcup_{\|D\| < \varepsilon} \sigma_{ap}(A + D).$$

$\diamondsuit$

Theorem 11.2.3 *Let $A \in \mathscr{C}(X)$ and $\varepsilon > 0$. Then,*

$$\sigma_{ap,\varepsilon}(A) = \bigcup_{D \in \Theta_\varepsilon(X)} \sigma_{ap}(A+D),$$

where

$$\Theta_\varepsilon(X) := \Big\{ D \in \mathscr{L}(X) \text{ such that } \|D\| < \varepsilon \text{ and } \dim R(D) \leq 1 \Big\}. \quad \diamondsuit$$

Proof. Let $\lambda \in \sigma_{ap,\varepsilon}(A)$, then there exists $x_0 \in X$ such that $\|x_0\| = 1$ and

$$\|(\lambda - A)x_0\| < \varepsilon.$$

Putting $\|x_0\| = \|(\lambda - A)^{-1}(\lambda - A)x_0\|$ implies that

$$\|(\lambda - A)^{-1}\| > \frac{1}{\varepsilon}.$$

Then, we can find $y_0 \in X$ such that $\|y_0\| = 1$ and

$$\|(\lambda - A)^{-1}y_0\| > \frac{1}{\varepsilon}.$$

Hence,

$$\|(\lambda - A)^{-1}y_0\| = \frac{1}{\delta},$$

where $\delta < \varepsilon$. By using the Hahn Banach theorem (see Theorem 2.1.4), there exists $x' \in X^*$ such that $\|x'\| = 1$ and

$$x'\left((\lambda - A)^{-1}y_0\right) = \|(\lambda - A)^{-1}y_0\| = \frac{1}{\delta}.$$

Now, we can define the rank-one operator by,

$$D : X \longrightarrow X, \;\; x \longrightarrow Dx := \delta x'(x) y_0.$$

Clearly, D is a linear operator everywhere defined on X. It is bounded, since

$$\begin{aligned} \|Dx\| &= \|\delta x'(x) y_0\| \\ &\leq \delta \|x'\| \|y_0\| \|x\|. \end{aligned}$$

Then,

$$\|D\| \leq \delta < \varepsilon.$$

Furthermore,

$$\begin{aligned} D\left((\lambda - A)^{-1}y_0\right) &= \delta x'\left((\lambda - A)^{-1}y_0\right)y_0 \\ &= \delta\frac{1}{\delta}y_0 \\ &= y_0. \end{aligned}$$

Putting $x = (\lambda - A)^{-1}y_0$, we will discuss these two cases:

$\underline{1^{st}\ case}$: If $x = x_0$, we obtain

$$\begin{aligned} \inf_{x\in\mathscr{D}(A),\ \|x\|=1} \|(\lambda - A - D)x\| &\leq \|(\lambda - A - D)x_0\| \\ &\leq \|(\lambda - A)x_0 - Dx_0\| \\ &= \|y_0 - y_0\| = 0. \end{aligned}$$

$\underline{2^{nd}\ case}$: If $x \neq x_0$. First, let $x = 0$, then

$$(\lambda - A)^{-1}y_0 = 0,$$

which is a contradiction with

$$\|(\lambda - A)^{-1}y_0\| = \frac{1}{\delta}.$$

Second, let $x \neq 0$, then

$$Dx = y_0 = (\lambda - A)x.$$

Hence,

$$\inf_{x\in\mathscr{D}(A),\ \|x\|=1} \|(\lambda - A - D)x\| = 0.$$

We deduce that $\lambda \in \sigma_{ap}(A+D)$ and $D \in \Theta_\varepsilon(X)$. The second inclusion is clear. Q.E.D.

Theorem 11.2.4 *Let $A \in \mathscr{C}(X)$ and $\varepsilon > 0$. Let $E \in \mathscr{L}(X)$ such that $\|E\| < \varepsilon$. Then,*

$$\sigma_{ap,\varepsilon-\|E\|}(A) \subseteq \sigma_{ap,\varepsilon}(A+E) \subseteq \sigma_{ap,\varepsilon+\|E\|}(A). \qquad \diamondsuit$$

Proof. Let $\lambda \in \sigma_{ap,\varepsilon-\|E\|}(A)$. Then, by using Theorem 11.2.2, there exists a bounded operator $D \in \mathscr{L}(X)$ with $\|D\| < \varepsilon - \|E\|$ such that

$$\lambda \in \sigma_{ap}(A+D) = \sigma_{ap}\Big((A+E)+(D-E)\Big).$$

The fact that

$$\|D-E\| \leq \|D\| + \|E\| < \varepsilon$$

allows us to deduce that $\lambda \in \sigma_{ap,\varepsilon}(A+E)$. Using a similar reasoning to the first inclusion, we deduce that $\lambda \in \sigma_{ap,\varepsilon+\|E\|}(A)$. Q.E.D.

Theorem 11.2.5 *Let $A \in \mathscr{L}(X)$ and $V \in \mathscr{L}(X)$ be invertible. If $B = V^{-1}AV$, then*

$$\sigma_{ap}(B) = \sigma_{ap}(A),$$

and for $\varepsilon > 0$ and $k = \|V^{-1}\|\|V\|$, we have

$$\sigma_{ap,\frac{\varepsilon}{k}}(A) \subseteq \sigma_{ap,\varepsilon}(B) \subseteq \sigma_{ap,k\varepsilon}(A), \text{ and} \tag{11.2}$$

$$\Sigma_{ap,\frac{\varepsilon}{k}}(A) \subseteq \Sigma_{ap,\varepsilon}(B) \subseteq \Sigma_{ap,k\varepsilon}(A). \tag{11.3}$$

◇

Proof. We can write

$$\|\lambda - B\| = \|V^{-1}(\lambda - A)V\| \leq k\|\lambda - A\|, \tag{11.4}$$

$$\|\lambda - A\| = \|V(\lambda - B)V^{-1}\| \leq k\|\lambda - B\|. \tag{11.5}$$

Let $\lambda \in \sigma_{ap}(A)$, which implies that

$$\inf_{\|x\|=1} \|(\lambda - A)x\| = 0.$$

By using relation (11.4), it follows that

$$\inf_{\|x\|=1} \|(\lambda - B)x\| = 0.$$

Hence, $\lambda \in \sigma_{ap}(B)$. The converse is similar, it is sufficient to use relation (11.5). For the second result, if $\lambda \in \sigma_{ap,\frac{\varepsilon}{k}}(A)$, then using (11.4), we obtain

$\lambda \in \sigma_{ap,\varepsilon}(B)$. If $\lambda \in \sigma_{ap,\varepsilon}(B)$, then using relation (11.5), we obtain $\lambda \in \sigma_{ap,k\varepsilon}(B)$. The second formula in (11.2) holds and the proof is similar for the relation (11.3). Q.E.D.

The closure of $\sigma_{ap,\varepsilon}(A)$ is always contained in $\Sigma_{ap,\varepsilon}(A)$, but equality holds if, and only if, A does not have constant infimum norm on any open set. The present part addresses the question on whether or not a similar equality holds in the case of non-strict inequalities:

$$\Sigma_{ap,\varepsilon}(A) \stackrel{?}{=} \bigcup_{\|D\|\leq\varepsilon} \sigma_{ap}(A+D).$$

Theorem 11.2.6 *Let $A \in \mathscr{C}(X)$ and $\varepsilon > 0$. Then,*

$$\bigcup_{\|D\|\leq\varepsilon} \sigma_{ap}(A+D) \subset \Sigma_{ap,\varepsilon}(A). \qquad \diamondsuit$$

Proof. Let $\lambda \notin \Sigma_{ap,\varepsilon}(A)$, then

$$\inf_{x\in\mathscr{D}(A),\ \|x\|=1} \|(\lambda - A)x\| > \varepsilon.$$

It is easy to see

$$\inf_{x\in\mathscr{D}(A),\ \|x\|=1} \|(\lambda - A - D)x\| > 0$$

for all $D \in \mathscr{L}(X)$ such that $\|D\| \leq \varepsilon$. So,

$$\lambda \notin \bigcup_{\|D\|\leq\varepsilon} \sigma_{ap}(A+D),$$

which completes the proof of theorem. Q.E.D.

We first consider the following example: Let $l_1(\mathbb{N})$ the space defined by

$$l_1(\mathbb{N}) = \left\{ (x_j)_{j\geq 1} \text{ such that } x_j \in \mathbb{C} \text{ and } \sum_{j=1}^{+\infty} |x_j| < \infty \right\}$$

which equipped with the following norm

$$\|x\| := \sum_{j=1}^{\infty} |x_j|.$$

Define the operator A by

$$A : l_1(\mathbb{N}) \longrightarrow l_1(\mathbb{N}),$$
$$x \longrightarrow Ax,$$

where

$$Ax = \left((1+2\varepsilon)x_1 - \sum_{j=2}^{\infty} x_j, -\varepsilon_2 x_2, \cdots, -\varepsilon_n x_n, \cdots \right),$$

$x = (x_1, x_2, \cdots, x_n, \cdots) \in l_1(\mathbb{N})$, $\varepsilon > 0$, and ε_n, where $n = 2, 3, \cdots$, is a sequence of positive numbers monotonically decreasing to 0. Then, we have

$$(A - 2\varepsilon)x = \left(x_1 - \sum_{j=2}^{\infty} x_j, -(2\varepsilon + \varepsilon_2)x_2, \cdots, -(2\varepsilon + \varepsilon_n)x_n, \cdots \right),$$

and

$$\|(A - 2\varepsilon)x\| = \left| x_1 - \sum_{j=2}^{\infty} x_j \right| + 2\varepsilon \sum_{j=2}^{\infty} |x_j| + \sum_{j=2}^{\infty} \varepsilon_j |x_j|.$$

Suppose $\|x\| = 1$. Then, we have

$$\|(A - 2\varepsilon)x\| = \left| x_1 - \sum_{j=2}^{\infty} x_j \right| + 2\varepsilon(1 - |x_1|) + \sum_{j=2}^{\infty} \varepsilon_j |x_j|.$$

Let

$$m(x) := \left| x_1 - \sum_{j=2}^{\infty} x_j \right| + 2\varepsilon(1 - |x_1|).$$

If $|x_1| \geq \frac{1}{2}$, then we have

$$\begin{aligned} m(x) &\geq |x_1| - \sum_{j=2}^{\infty} |x_j| + 2\varepsilon(1 - |x_1|) \\ &= 2(1-\varepsilon)|x_1| + 2\varepsilon - 1 \\ &\geq 2(1-\varepsilon)\frac{1}{2} + 2\varepsilon - 1 \\ &= \varepsilon \end{aligned}$$

If $|x_1| \leq \frac{1}{2}$, then we have

$$m(x) \geq 2\varepsilon(1 - |x_1|) \geq \varepsilon.$$

Hence,

$$m(x) \geq \varepsilon, \ \ \|x\| = 1,$$

where the equality is achieved if, and only if, $|x_1| = \frac{1}{2}$ and

$$\sum_{j=2}^{\infty} x_j = x_1.$$

Consequently,

$$\|(A-2\varepsilon)x\| = m(x) + \sum_{j=2}^{\infty} \varepsilon_j |x_j| > \varepsilon, \ \ \|x\| = 1, \tag{11.6}$$

and

$$\|(A-2\varepsilon)x^{(k)}\| \to \varepsilon \text{ as } k \to \infty,$$

where

$$x^{(k)} = \Big(\frac{1}{2}, \underbrace{0, \cdots, 0}_{k \text{ zeros}}, \frac{1}{2}, 0, \cdots\Big), \ \ k \in \mathbb{N}.$$

Thus,

$$\inf_{\|x\|=1} \|(A-2\varepsilon)x\| = \varepsilon. \tag{11.7}$$

It is easy to see that $A-2\varepsilon$ is invertible. It follows from (11.7) that

$$\|(A-2\varepsilon)^{-1}\| = \frac{1}{\varepsilon}.$$

Hence, $2\varepsilon \in \Sigma^{\varepsilon}(A)$. Let $D \in \mathcal{L}(X)$ such that $\|D\| \leq \varepsilon$. Then, $D-2\varepsilon$ is invertible and $A+D-2\varepsilon$ is a Fredholm operator of index 0. Suppose that its kernel is not trivial. Then, there exists $x \in l_1(\mathbb{N})$ such that $\|x\| = 1$ and

$$(A-2\varepsilon)x = -Dx.$$

It follows from (11.6) that the left-hand side of the last equality has a norm strictly larger than ε. The norm of the right-hand side, on the other hand, is less than or equal to ε. This contradiction implies that the kernel of $A+D-$

2ε is trivial. $A+D-2\varepsilon$ is therefore invertible and we have $2\varepsilon \in \rho(A+D)$. It follows that $2\varepsilon \in \Sigma_{ap,\varepsilon}(A)$ and $2\varepsilon \notin \sigma_{ap}(A+D)$ for all $\|D\| \leq \varepsilon$. Then,

$$2\varepsilon \notin \bigcup_{\|D\|\leq\varepsilon} \sigma_{ap}(A+D).$$

Hence,

$$\bigcup_{\|D\|\leq\varepsilon} \sigma_{ap}(T+D) \subsetneq \Sigma_{ap,\varepsilon}(A).$$

$\diamondsuit$

Theorem 11.2.7 *Let $A \in \mathscr{C}(X)$ and $\varepsilon > 0$. If $(H6)$ holds, then*

$$\Sigma_{ap,\varepsilon}(A) = \bigcup_{\|D\|\leq\varepsilon} \sigma_{ap}(A+D). \tag{11.8}$$

$\diamondsuit$

Proof. It follows from both Theorems 11.2.6 and 11.2.2 that

$$\overline{\sigma_{ap,\varepsilon}(A)} = \overline{\bigcup_{\|D\|\leq\varepsilon} \sigma_{ap}(A+D)} \subseteq \Sigma_{ap,\varepsilon}(A).$$

So, by using the hypothesis $(H6)$, we have

$$\overline{\sigma_{ap,\varepsilon}(A)} = \Sigma_{ap,\varepsilon}(A).$$

Hence,

$$\overline{\bigcup_{\|D\|\leq\varepsilon} \sigma_{ap}(A+D)} = \Sigma_{ap,\varepsilon}(A).$$

It follows from both Theorems 11.2.6 and 11.2.2 that Eq. (11.8) is an equality if, and only if, the level set

$$\left\{\lambda \in \mathbb{C} \text{ such that } \inf_{x\in\mathscr{D}(A),\ \|x\|=1} \|(\lambda - A)x\| = \varepsilon\right\}$$

is a subset of

$$\bigcup_{\|D\|=\varepsilon} \sigma_{ap}(A+D).$$

Q.E.D.

11.3 ESSENTIAL APPROXIMATION PSEUDOSPECTRUM

In this section, we have the following useful stability result for the essential approximation pseudospectrum.

Proposition 11.3.1 *Let* $A \in \mathscr{C}(X)$ *and* $\varepsilon > 0$. *Then,*

(i) $\bigcap_{\varepsilon>0} \sigma_{eap,\varepsilon}(A) = \sigma_{e7}(A)$,

(ii) *if* $\varepsilon_1 < \varepsilon_2$, *then* $\sigma_{e7}(A) \subset \sigma_{eap,\varepsilon_1}(A) \subset \sigma_{eap,\varepsilon_2}(A)$, *and*

(iii) $\sigma_{eap,\varepsilon}(A+F) = \sigma_{eap,\varepsilon}(A)$ *for all* $F \in \mathscr{K}(X)$. ◊

Proof. (i) $\sigma_{e7}(A) \subset \sigma_{eap,\varepsilon}(A)$. Indeed, Let $\lambda \notin \sigma_{eap,\varepsilon}(A)$. Then, there exists $K \in \mathscr{K}(X)$ such that

$$\inf_{x\in\mathscr{D}(X),\ \|x\|=1} \|(\lambda - A - K)x\| > \varepsilon > 0.$$

Hence, $\lambda \notin \sigma_{e7}(A)$. So,

$$\sigma_{e7}(A) \subset \bigcap_{\varepsilon>0} \sigma_{eap,\varepsilon}(A).$$

Conversely, let

$$\lambda \in \bigcap_{\varepsilon>0} \sigma_{eap,\varepsilon}(A).$$

Then, for all $\varepsilon > 0$, we have $\lambda \in \sigma_{eap,\varepsilon}(A)$. Hence, for every $K \in \mathscr{K}(X)$, we obtain $\lambda \in \sigma_{ap,\varepsilon}(A+K)$. This implies that

$$\inf_{x\in\mathscr{D}(X),\ \|x\|=1} \|(\lambda - A - K)x\| < \varepsilon.$$

Taking limits as $\varepsilon \to 0^+$, we infer that $\lambda \in \sigma_{e7}(A)$.

(ii) Let $\lambda \in \sigma_{eap,\varepsilon_1}(A)$, then there exists $K \in \mathscr{K}(X)$, such that

$$\inf_{x\in\mathscr{D}(X),\ \|x\|=1} \|(\lambda - A - K)x\| < \varepsilon_1 < \varepsilon_2.$$

So, $\lambda \in \sigma_{eap,\varepsilon_2}(A)$.

(iii) It follows immediately from the definition of $\sigma_{eap,\varepsilon}(A)$ that

$$\sigma_{eap,\varepsilon}(A+F) = \sigma_{eap,\varepsilon}(A)$$

for all $F \in \mathscr{K}(X)$. Q.E.D.

Theorem 11.3.1 *Let $A \in \mathscr{C}(X)$ and $\varepsilon > 0$. Then, the following properties are equivalent:*

(i) $\lambda \notin \sigma_{eap,\varepsilon}(A)$.

(ii) For all $D \in \mathscr{L}(X)$ such that $\|D\| < \varepsilon$, we have $\lambda - A - D \in \Phi_+(X)$ and

$$i(\lambda - A - D) \leq 0. \qquad \diamondsuit$$

Proof. $(i) \Rightarrow (ii)$ Let $\lambda \notin \sigma_{eap,\varepsilon}(A)$. Then, there exists a compact operator K on X such that $\lambda \notin \sigma_{ap,\varepsilon}(A+K)$. By using Theorem 11.2.2, we notice that $\lambda \notin \sigma_{ap}(A+D+K)$, for all $D \in \mathscr{L}(X)$ such that $\|D\| < \varepsilon$. So,

$$\lambda - A - D - K \in \Phi_+(X) \text{ and } i(\lambda - A - D - K) \leq 0,$$

for all $D \in \mathscr{L}(X)$ such that $\|D\| < \varepsilon$. Using Lemma 2.3.1, we get for all $D \in \mathscr{L}(X)$ such that $\|D\| < \varepsilon$,

$$\lambda - A - D \in \Phi_+(X) \text{ and } i(\lambda - A - D) \leq 0.$$

$(ii) \Rightarrow (i)$ We assume that for all $D \in \mathscr{L}(X)$ such that $\|D\| < \varepsilon$, we have

$$\lambda - A - D \in \Phi_+(X) \text{ and } i(\lambda - A - D) \leq 0.$$

Based on Lemma 2.2.7, $\lambda - A - D$ can be expressed in the form

$$\lambda - A - D = S + K,$$

where $K \in \mathscr{K}(X)$ and $S \in \mathscr{C}(X)$ is an operator with closed range and $\alpha(S) = 0$. So,

$$\lambda - T - D - K = S \text{ and } \alpha(\lambda - A - D - K) = 0.$$

By using Theorem 2.1.3, there exists a constant $c > 0$ such that

$$\|(\lambda - A - D - K)x\| \geq c\|x\|, \text{ for all } x \in \mathscr{D}(A).$$

This proves that

$$\inf_{x \in \mathscr{D}(T),\ \|x\|=1} \|(\lambda - A - D - K)x\| \geq c > 0.$$

Thus, $\lambda \notin \sigma_{ap}(A + D + K)$, and therefore $\lambda \notin \sigma_{eap,\varepsilon}(A)$. Q.E.D.

Remark 11.3.1 *It follows immediately from Theorem 11.3.1 that* $\lambda \notin \sigma_{eap,\varepsilon}(A)$ *if, and only if, for all* $D \in \mathscr{L}(X)$ *such that* $\|D\| < \varepsilon$, *we obtain*

$$\lambda - A - D \in \Phi_+(X)$$

and

$$i(\lambda - A - D) \leq 0.$$

This is equivalent to

$$\sigma_{eap,\varepsilon}(A) = \bigcup_{\|D\|<\varepsilon} \sigma_{e7}(A + D). \qquad \diamondsuit$$

From the definition of $\sigma_{eap,\varepsilon}(A)$ and Proposition 11.3.1, we get the following corollary:

Corollary 11.3.1 *Let* $A \in \mathscr{C}(X)$ *and* $\varepsilon > 0$. *Then,*

$$\sigma_{e7}(T) = \lim_{\varepsilon \to 0} \overline{\bigcap_{K \in \mathscr{K}(X)} \sigma_{ap,\varepsilon}(A + K)} = \bigcap_{\varepsilon > 0} \left(\bigcup_{\|D\|<\varepsilon} \sigma_{e7}(A + D) \right). \qquad \diamondsuit$$

Theorem 11.3.2 *Let* $A \in \mathscr{C}(X)$ *and* $\varepsilon > 0$. *Then,*

$$\sigma_{eap,\varepsilon}(A) = \bigcap_{F \in \mathscr{F}_+(X)} \sigma_{ap,\varepsilon}(A + F). \qquad \diamondsuit$$

Proof. Let $\lambda \notin \bigcap\limits_{F \in \mathscr{F}^{+}(X)} \sigma_{ap,\varepsilon}(A+F)$, then there exists $F \in \mathscr{F}_{+}(X)$ such that

$$\lambda \notin \sigma_{ap,\varepsilon}(A+F).$$

In view of Theorem 11.2.2, we have

$$\lambda \notin \sigma_{ap}(A+F+D),$$

for all $D \in \mathscr{L}(X)$ such that $\|D\| < \varepsilon$. Therefore,

$$\lambda - A - F - D \in \Phi_{+}(X) \text{ and } i(\lambda - A - F - D) \leq 0.$$

Using Lemma 2.3.1, we conclude that for all $D \in \mathscr{L}(X)$ such that $\|D\| < \varepsilon$,

$$\lambda - A - D \in \Phi_{+}(X) \text{ and } i(\lambda - A - D) \leq 0.$$

Finally, Theorem 11.3.1 shows that $\lambda \notin \sigma_{eap,\varepsilon}(A)$. For the second inclusion, it is clear that

$$\bigcap_{F \in \mathscr{F}_{+}(X)} \sigma_{ap,\varepsilon}(A+F) \subset \bigcap_{F \in \mathscr{K}(X)} \sigma_{ap,\varepsilon}(A+F) =: \sigma_{eap,\varepsilon}(A),$$

because $\mathscr{K}(X) \subset \mathscr{F}_{+}(X)$, which completes the proof. Q.E.D.

Remark 11.3.2 *Let $A \in \mathscr{C}(X)$ and $\varepsilon > 0$.*

(i) Using Theorem 11.3.2, we infer that $\sigma_{eap,\varepsilon}(A+F) = \sigma_{eap,\varepsilon}(A)$ for all $F \in \mathscr{F}_{+}(X)$.

(ii) Let $\mathfrak{I}(X)$ be any subset of $\mathscr{L}(X)$. If $\mathscr{K}(X) \subset \mathfrak{I}(X) \subset \mathscr{F}_{+}(X)$, then

$$\sigma_{eap,\varepsilon}(A) = \bigcap_{M \in \mathfrak{I}(X)} \sigma_{ap,\varepsilon}(A+M)$$

and

$$\sigma_{eap,\varepsilon}(A+J) = \sigma_{eap,\varepsilon}(A)$$

for all $J \in \mathfrak{I}(X)$.

Lemma 11.3.1 *Let $\varepsilon > 0$, A and B be two elements of $\mathscr{C}(X)$. Assume that for a bounded operator D such that $\|D\| < \varepsilon$, the operator B is $(A+D)$-compact, then*

$$\sigma_{eap,\varepsilon}(A) = \sigma_{eap,\varepsilon}(A+B). \qquad \diamondsuit$$

Proof. Let $\lambda \notin \sigma_{eap,\varepsilon}(A)$, then for all $D \in \mathscr{L}(X)$ such that $\|D\| < \varepsilon$, we have $\lambda - A - D \in \Phi_+(X)$ and

$$i(\lambda - A - D) \leq 0.$$

Since B is $(A+D)$-compact and applying Theorem 2.3.5, we get

$$\lambda - A - B - D \in \Phi_+(X) \text{ and } i(\lambda - A - B - D) \leq 0.$$

Therefore,

$$\lambda \notin \sigma_{eap,\varepsilon}(T+D).$$

We conclude that

$$\sigma_{eap,\varepsilon}(A+B) \subset \sigma_{eap,\varepsilon}(A).$$

Conversely, let $\lambda \notin \sigma_{eap,\varepsilon}(A+B)$. Then, for all $D \in \mathscr{L}(X)$ such that $\|D\| < \varepsilon$, we have

$$\lambda - A - B - D \in \Phi_+(X)$$

and

$$i(\lambda - A - B - D) \leq 0.$$

On the other hand, B is $(A+D)$-compact. Using Theorem 2.3.7, we deduce that B is $(A+B+D)$-compact. Hence,

$$\lambda - A - D \in \Phi_+(X)$$

and

$$i(\lambda - A - D) \leq 0.$$

Therefore, $\lambda \notin \sigma_{eap,\varepsilon}(A)$. This proves that

$$\sigma_{eap,\varepsilon}(A) \subset \sigma_{eap,\varepsilon}(A+B). \qquad \text{Q.E.D.}$$

11.4 PROPERTIES OF ESSENTIAL PSEUDOSPECTRA

In this section, we give the results of stability and sum for the essential pseudospectra.

Theorem 11.4.1 *Let X be a Banach space, $\varepsilon > 0$ and consider $A \in \mathscr{C}(X)$. Then,*

(i) If $J \in \mathscr{F}(X)$, then

$$\sigma_{ei,\varepsilon}(A) = \sigma_{ei,\varepsilon}(A+J),\ i = 4,5.$$

(ii) If $J \in \mathscr{F}_+(X)$, then

$$\sigma_{ei,\varepsilon}(A) = \sigma_{ei,\varepsilon}(A+J),\ i = 1, ap.$$

(iii) If $J \in \mathscr{F}_-(X)$, then

$$\sigma_{ei,\varepsilon}(A) = \sigma_{ei,\varepsilon}(A+J),\ i = 2, \delta.$$

(iv) If $J \in \mathscr{F}_+(X) \bigcap \mathscr{F}_-(X)$, then

$$\sigma_{e3,\varepsilon}(A) = \sigma_{e3,\varepsilon}(A+J).$$ ◇

Proof. (i) Let $\lambda \notin \sigma_{e5,\varepsilon}(A)$, then $\lambda - A - D \in \Phi(X)$ and

$$i(\lambda - A - D) = 0$$

for all $||D|| < \varepsilon$. Hence, using Lemma 2.3.1, we have $\lambda - A - D - J \in \Phi(X)$ and

$$i(\lambda - A - D - J) = 0$$

for $||D|| < \varepsilon$. Therefore, $\lambda \notin \sigma_{e5,\varepsilon}(A+J)$, i.e.,

$$\sigma_{e5,\varepsilon}(A+J) \subset \sigma_{e5,\varepsilon}(A).$$

The opposite inclusion is obtained by symmetry:

$$\sigma_{e5,\varepsilon}(A) = \sigma_{5,\varepsilon}(A+J-J) \subset \sigma_{e5,\varepsilon}(A+J).$$

(ii) Let $\lambda \notin \sigma_{eap,\varepsilon}(A)$, then $\lambda - A - D \in \Phi_+(X)$ and $i(\lambda - A - D) \leq 0$ for all $||D|| < \varepsilon$. Hence, using Lemma 2.3.1, we have

$$\lambda - A - D - J \in \Phi_+(X)$$

and

$$i(\lambda - A - D - J) \leq 0$$

for all $||D|| < \varepsilon$. Therefore,

$$\lambda \notin \sigma_{eap,\varepsilon}(A+J),$$

i.e.,

$$\sigma_{eap,\varepsilon}(A+J) \subset \sigma_{eap,\varepsilon}(A).$$

Analogously, by using Lemma 2.3.1 and arguing as above, we can deriving easily the opposite inclusion

$$\sigma_{eap,\varepsilon}(A) \subset \sigma_{eap,\varepsilon}(A+J).$$

So,

$$\sigma_{ei,\varepsilon}(A+J) = \sigma_{ei,\varepsilon}(A),\ i = 1, ap.$$

Statements (iii), (iv) can be checked in the same way (ii). Q.E.D.

Theorem 11.4.2 *Let X be a Banach space, $\varepsilon > 0$ and consider A, $B \in \mathscr{C}(X)$. Assume that there are A_0, $B_0 \in \mathscr{L}(X)$ and F_1, $F_2 \in \mathscr{F}^b(X)$ such that*

$$AA_0 = I - F_1, \tag{11.9}$$

$$BB_0 = I - F_2. \tag{11.10}$$

(i) If $0 \in \Phi_A \bigcap \Phi_B$, $A_0 - B_0 \in \mathscr{F}_+^b(X)$ and $i(A) = i(B)$, then

$$\sigma_{eap,\varepsilon}(A) = \sigma_{eap,\varepsilon}(B). \tag{11.11}$$

(ii) If $0 \in \Phi_A \bigcap \Phi_B$, $A_0 - B_0 \in \mathscr{F}_-(X)$ and $i(A) = i(B)$, then

$$\sigma_{e\delta,\varepsilon}(A) = \sigma_{e\delta,\varepsilon}(B).$$

$\diamondsuit$

Proof. Let λ be a complex number, Eqs. (11.9) and (11.10) imply that

$$(\lambda-A-D)A_0-(\lambda-B-D)B_0=F_1-F_2+(\lambda-D)(A_0-B_0). \quad (11.12)$$

(i) Let $\lambda \notin \sigma_{eap,\varepsilon}(B)$, then

$$\lambda-B-D\in\Phi_+(X)$$

and

$$i(\lambda-B-D)\leq 0$$

for all $D\in\mathscr{L}(X)$ such that $||D||<\varepsilon$. Since $B+D$ is closed and $\mathscr{D}(B+D)=\mathscr{D}(B)$ endowed with the graph norm is a Banach space denoted by X_{B+D}. We can regard $B+D$ as operator from X_{B+D} into X. This will be denoted by $\widehat{B+D}$. Using Eq. (2.6), we can show that

$$\lambda-(\widehat{B+D})\in\Phi_+^b(X,X_{B+D})$$

and

$$i(\lambda-(\widehat{B+D}))\leq 0.$$

Moreover, since $F_2\in\mathscr{F}^b(X)$, Eq. (11.10), Lemma 2.3.1, and Theorem 2.2.22, we find that $B_0\in\Phi^b(X,X_{B+D})$ and consequently,

$$(\lambda-(\widehat{B+D}))B_0\in\Phi_+^b(X).$$

Since $A_0-B_0\in\mathscr{F}_+^b(X)$, and by using both Eq. (11.12) and Lemma 2.3.1, we can prove that

$$(\lambda-A-D)A_0\in\Phi_+^b(X)$$

and

$$i[(\lambda-A-D)A_0]=i[(\lambda-B-D)B_0]. \quad (11.13)$$

A similar reasoning show that

$$A_0\in\Phi^b(X,X_{A+D}),$$

where $X_{A+D}=(\mathscr{D}(A),||.||_{A+D})$. By Theorem 2.2.3, we can write

$$A_0S=I-F \text{ on } X_{A+D}, \quad (11.14)$$

where $S \in \mathscr{L}(X_{A+D}, X)$ and $F \in \mathscr{K}(X_A)$. Hence, by Eq. (11.14), we have

$$(\lambda - (\widehat{A+D}))A_0 S = (\lambda - (\widehat{A+D})) - (\lambda - (\widehat{A+D}))F. \qquad (11.15)$$

Since $S \in \Phi^b(X_{A+D}, X)$, using Theorem 2.2.5, we show that

$$(\lambda - (\widehat{A+D}))A_0 S \in \Phi_+^b(X, X_{A+D}).$$

By using Eq. (11.15) and, by applying Lemma 2.3.1, we prove that

$$\lambda - (\widehat{A+D}) \in \Phi_+^b(X, X_{A+D}),$$

and in view of Eq. (2.6), we have

$$\lambda - A - D \in \Phi_+(X).$$

It follows that

$$\lambda - A \in \Phi_+^{\varepsilon}(X). \qquad (11.16)$$

Since, $F_1, F_2 \in \mathscr{F}^b(X)$, Eqs. (11.9), (11.10) and Lemma 2.3.1 give

$$i(A) + i(A_0) = i(I - F_1) = 0$$

and

$$i(B) + i(B_0) = i(I - F_2) = 0.$$

Since $i(A) = i(B)$, then

$$i(A_0) = i(B_0).$$

Using both Eq. (11.13) and Theorem 2.2.5, we can write

$$i(\lambda - A - D) + i(A_0) = i(\lambda - B - D) + i(B_0).$$

Therefore,

$$i(\lambda - A - D) \leq 0. \qquad (11.17)$$

Using both Eqs. (11.16) and (11.17), we conclude that

$$\lambda \notin \sigma_{eap,\varepsilon}(A).$$

Therefore,

$$\sigma_{eap,\varepsilon}(A) \subset \sigma_{eap,\varepsilon}(B).$$

The opposite inclusion follows by symmetry and we obtain (11.11).

(ii) Similarly, we can prove the statement

$$\sigma_{e\delta,\varepsilon}(A) = \sigma_{e\delta,\varepsilon}(B). \qquad \text{Q.E.D.}$$

Theorem 11.4.3 *Let X be a Banach space, $\varepsilon > 0$ and consider $A, B \in \mathcal{L}(X)$. Then,*

(i) If for all bounded operator D such that $||D|| < \varepsilon$ and $A(B+D) \in \mathcal{F}^b(X)$, then

$$\sigma_{e4,\varepsilon}(A+B)\backslash\{0\} \subset [\sigma_{e4}(A) \bigcup \sigma_{e4,\varepsilon}(B)]\backslash\{0\}.$$

If, further, $(B+D)A \in \mathcal{F}^b(X)$, then

$$\sigma_{e4,\varepsilon}(A+B)\backslash\{0\} = [\sigma_{e4}(A) \bigcup \sigma_{e4,\varepsilon}(B)]\backslash\{0\}.$$

(ii) If for all bounded operator D such that $||D|| < \varepsilon$ and $A(B+D) \in \mathcal{F}_+^b(X)$, then

$$\sigma_{e1,\varepsilon}(A+B)\backslash\{0\} \subset [\sigma_{e1}(A) \bigcup \sigma_{e1,\varepsilon}(B)]\backslash\{0\}.$$

If, further, $(B+D)A \in \mathcal{F}_+^b(X)$, then

$$\sigma_{e1,\varepsilon}(A+B)\backslash\{0\} = [\sigma_{e1}(A) \bigcup \sigma_{e1,\varepsilon}(B)]\backslash\{0\}. \tag{11.18}$$

(iii) If for all bounded operator D such that $||D|| < \varepsilon$ and $A(B+D) \in \mathcal{F}_-^b(X)$, then

$$\sigma_{e2,\varepsilon}(A+B)\backslash\{0\} \subset [\sigma_{ei}(A) \bigcup \sigma_{e2,\varepsilon}(B)]\backslash\{0\}.$$

If, further, $(B+D)A \in \mathcal{F}_-^b(X)$, then

$$\sigma_{e2,\varepsilon}(A+B)\backslash\{0\} = [\sigma_{e2}(A) \bigcup \sigma_{e2,\varepsilon}(B)]\backslash\{0\}. \tag{11.19}$$

(iv) If for all bounded operator D such that $||D|| < \varepsilon$ and $A(B+D) \in \mathcal{F}_+^b(X) \bigcap \mathcal{F}_-^b(X)$, then

$$\sigma_{e3,\varepsilon}(A+B)\backslash\{0\} \subset$$
$$[(\sigma_{e3}(A) \bigcup \sigma_{e3,\varepsilon}(B)) \bigcup (\sigma_{e3}(A) \bigcap \sigma_{e2,\varepsilon}(B)) \bigcup (\sigma_{e2}(A) \bigcap \sigma_{e1,\varepsilon}(B))]\backslash\{0\}.$$

Moreover, if $(B+D)A \in \mathcal{F}_+^b(X) \bigcap \mathcal{F}_-^b(X)$, then

$$\sigma_{e3,\varepsilon}(A+B)\backslash\{0\} =$$
$$[(\sigma_{e3}(A) \bigcup \sigma_{e3,\varepsilon}(B)) \bigcup (\sigma_{e3}(A) \bigcap \sigma_{e2,\varepsilon}(B)) \bigcup (\sigma_{e2}(A) \bigcap \sigma_{e1,\varepsilon}(B))]\backslash\{0\}. \ \diamond$$

Proof. (*i*) For $\lambda \in \mathbb{C}$, we can write

$$(\lambda - A)(\lambda - B - D) = A(B+D) + \lambda(\lambda - A - B - D) \tag{11.20}$$

and

$$(\lambda - B - D)(\lambda - A) = (B+D)A + \lambda(\lambda - A - B - D). \tag{11.21}$$

Let $\lambda \notin \left[\sigma_{e4}(A) \bigcup \sigma_{e4,\varepsilon}(B)\right] \bigcup \{0\}$. Then, $\lambda - A \in \Phi^b(X)$ and for all $\|D\| < \varepsilon$, $\lambda - B - D \in \Phi^b(X)$. It follows from Theorem 2.2.5 that

$$(\lambda - A)(\lambda - B - D) \in \Phi^b(X).$$

Since $A(B+D) \in \mathscr{F}^b(X)$, applying Eq. (11.20), we have $\lambda - A - B - D \in \Phi^b(X)$, then $\lambda \notin \sigma_{e4,\varepsilon}(A+B)$. Therefore,

$$\sigma_{e4,\varepsilon}(A+B)\backslash\{0\} \subseteq \left[\sigma_{e4}(A) \bigcup \sigma_{e4,\varepsilon}(B)\right]\backslash\{0\}. \tag{11.22}$$

Now, we prove the inverse inclusion of Eq. (11.22). Suppose $\lambda \notin \sigma_{e4,\varepsilon}(A+B)\backslash\{0\}$, then for all $D \in \mathscr{L}(X)$ such that $\|D\| < \varepsilon$, we have $\lambda - A - B - D \in \Phi^b(X)$. In view of $A(B+D) \in \mathscr{F}^b(X)$ and $(B+D)A \in \mathscr{F}^b(X)$, and by using both Eqs. (11.20) and (11.21), we have

$$(\lambda - A)(\lambda - B - D) \in \Phi^b(X)$$

and

$$(\lambda - B - D)(\lambda - A) \in \Phi^b(X).$$

Applying Lemma 2.2.3, it is clear that $\lambda - A \in \Phi^b(X)$ and for all $\|D\| < \varepsilon$, we have $\lambda - B - D \in \Phi^b(X)$. Therefore, $\lambda \notin \sigma_{e4}(A) \bigcup \sigma_{e4,\varepsilon}(B)$. This proves that

$$\sigma_{e4,\varepsilon}(A+B)\backslash\{0\} = \left[\sigma_{e4}(A) \bigcup \sigma_{e4,\varepsilon}(B)\right]\backslash\{0\}.$$

(*ii*) Suppose that $\lambda \notin \sigma_{e1}(A) \bigcup \sigma_{e1,\varepsilon}(B) \bigcup \{0\}$, then

$$\lambda - A \in \Phi_+^b(X)$$

and

$$\lambda - B \in \Phi_+^\varepsilon(X).$$

Hence, $\lambda - A \in \Phi_+^b(X)$ and $\lambda - B - D \in \Phi_+^b(X)$ for all $||D|| < \varepsilon$. Using Theorem 2.2.5, we have

$$(\lambda - A)(\lambda - B - D) \in \Phi_+^b(X).$$

Since $A(B+D) \in \mathscr{F}_+^b(X)$, and by applying Eq. (11.20), we infer that $\lambda - A - B - D \in \Phi_+^b(X)$. Hence, $\lambda - A - B \in \Phi_+^\varepsilon(X)$. So, $\lambda \notin \sigma_{e1,\varepsilon}(A+B)$. Therefore,

$$\sigma_{e1,\varepsilon}(A+B)\backslash\{0\} \subset [\sigma_{e1}(A)\bigcup\sigma_{e1,\varepsilon}(B)]\backslash\{0\}.$$

Suppose $\lambda \notin \sigma_{e1,\varepsilon}(A+B)]\bigcup\{0\}$, then $\lambda - A - B \in \Phi_+^\varepsilon(X)$, so $\lambda - A - B - D \in \Phi_+^b(X)$ for all bounded operator D such that $||D|| < \varepsilon$. Since $A(B+D) \in \mathscr{F}_+^b(X)$ and $(B+D)A \in \mathscr{F}_+^b(X)$, then by Eqs. (11.20), (11.21) and Lemma 2.3.1, we have

$$(\lambda - A)(\lambda - B - D) \in \Phi_+^b(X) \text{ and } (\lambda - B - D)(\lambda - A) \in \Phi_+^b(X). \quad (11.23)$$

By using (11.23) and Theorem 2.2.10, it is clear that $\lambda - A \in \Phi_+^b(X)$ and $\lambda - B \in \Phi_+^\varepsilon(X)$. Hence, $\lambda \notin \sigma_{e1}(A)\bigcup\sigma_{e1,\varepsilon}(B)$. Therefore,

$$\sigma_{e1}(A)\bigcup\sigma_{e1,\varepsilon}(B) \subset \sigma_{e1,\varepsilon}(A+B).$$

This prove that

$$\sigma_{e1}(A)\bigcup\sigma_{e1,\varepsilon}(B) = \sigma_{e1,\varepsilon}(A+B).$$

(iii) can be checked in the same way as (ii).

(iv) Since the equalities

$$\sigma_{e3}(A) = \sigma_{e1}(A)\bigcap\sigma_{e2}(A),$$
$$\sigma_{e3,\varepsilon}(B) = \sigma_{e1,\varepsilon}(B)\bigcap\sigma_{e2,\varepsilon}(B),$$

and

$$\sigma_{e3,\varepsilon}(A+B) = \sigma_{e1,\varepsilon}(A+B)\bigcap\sigma_{e2,\varepsilon}(A+B)$$

and, in view of both $A(B+D) \in \mathscr{F}_+^b\bigcap\mathscr{F}_-^b(X)$ and $(B+D)A \in \mathscr{F}_+^b(X)\bigcap\mathscr{F}_-^b(X)$, one can infer from both Eqs. (11.18) and (11.19) that

$$\sigma_{e3,\varepsilon}(A+B)\backslash\{0\} =$$
$$[(\sigma_{e3}(A)\bigcup\sigma_{e3,\varepsilon}(B))\bigcup(\sigma_{e3}(A)\bigcap\sigma_{e2,\varepsilon}(B))\bigcup(\sigma_{e2}(A)\bigcap\sigma_{e1,\varepsilon}(B))]\backslash\{0\}.$$

Q.E.D.

11.5 PSEUDOSPECTRUM OF BLOCK OPERATOR MATRICES

Let X_1 and X_2 be two Banach spaces and consider the 2×2 block operator matrices defined on $X_1 \times X_2$ by

$$\mathcal{L} := \begin{pmatrix} A & B \\ C & D \end{pmatrix},$$

where $A \in \mathcal{L}(X_1)$, $B \in \mathcal{L}(X_2 \times X_1)$, $C \in \mathcal{L}(X_1 \times X_2)$, and $D \in \mathcal{L}(X_2)$. Let

$$T = \begin{pmatrix} T_1 & 0 \\ 0 & T_2 \end{pmatrix},$$

where $T_i \in \mathcal{L}(X_i)$, $i = 1, 2$. Defining the norm of operator matrix T as

$$||T|| = \max\left\{\|T_1\|, \|T_2\|\right\}.$$

Lemma 11.5.1 *Let $A \in \mathcal{L}(X_1)$, $B \in \mathcal{L}(X_2)$ and consider the 2×2 block operator matrices*

$$M_C := \begin{pmatrix} A & C \\ 0 & B \end{pmatrix},$$

where $C \in \mathcal{L}(X_2, X_1)$. Then,

(i) if $A \in \Phi^b(X_1)$ and $B \in \Phi^b(X_2)$, then $M_C \in \Phi^b(X_1 \times X_2)$ for every $C \in \mathcal{L}(X_2, X_1)$,

(ii) if $A \in \Phi^b_+(X_1)$ and $B \in \Phi^b_+(X_2)$, then $M_C \in \Phi^b_+(X_1 \times X_2)$ for every $C \in \mathcal{L}(X_2, X_1)$, and

(iii) if $A \in \Phi^b_-(X_1)$ and $B \in \Phi^b_-(X_2)$, then $M_C \in \Phi^b_-(X_1 \times X_2)$ for every $C \in \mathcal{L}(X_2, X_1)$. ◇

Proof. (i) Let us write M_C in the form

$$M_C = \begin{pmatrix} I & 0 \\ 0 & B \end{pmatrix}\begin{pmatrix} I & C \\ 0 & I \end{pmatrix}\begin{pmatrix} A & 0 \\ 0 & I \end{pmatrix}. \tag{11.24}$$

Since $A \in \Phi^b(X_1)$ and $B \in \Phi^b(X_2)$, then

$$\begin{pmatrix} A & 0 \\ 0 & I \end{pmatrix}$$

and

$$\begin{pmatrix} I & 0 \\ 0 & B \end{pmatrix}$$

are both Fredholm operators. So, M_C is a Fredholm operator since

$$\begin{pmatrix} I & C \\ 0 & I \end{pmatrix}$$

is invertible for every $C \in \mathscr{L}(X_2, X_1)$.

(ii) and (iii) can be checked in the same way as (i). Q.E.D.

Remark 11.5.1 *Using the same reasoning as in the proof of Lemma 11.5.1, we can show that:*

(i) if $A \in \Phi^b(X_1)$ and $B \in \Phi^b(X_2)$, then

$$M_D := \begin{pmatrix} A & 0 \\ D & B \end{pmatrix}$$

is a Fredholm operator on $X_1 \times X_2$ for every $D \in \mathscr{L}(X_1, X_2)$,

(ii) if $A \in \Phi_+^b(X_1)$ and $B \in \Phi_+^b(X_2)$, then $M_D \in \Phi_+^b(X_1 \times X_2)$ for every $D \in \mathscr{L}(X_1, X_2)$, and

(iii) if $A \in \Phi_-^b(X_1)$ and $B \in \Phi_-^b(X_2)$, then $M_D \in \Phi_-^b(X_1 \times X_2)$ for every $D \in \mathscr{L}(X_1, X_2)$. ◇

Lemma 11.5.2 *Let $A \in \mathscr{L}(X_1)$, $B \in \mathscr{L}(X_2)$ and consider the 2×2 block operator matrices*

$$M_C := \begin{pmatrix} A & C \\ 0 & B \end{pmatrix},$$

where $C \in \mathscr{L}(X_2, X_1)$. Then,

(i) if $M_C \in \Phi_+^b(X_1 \times X_2)$, then $A \in \Phi_+^b(X_1)$, and

(ii) if $M_C \in \Phi_-^b(X_1 \times X_2)$, then $B \in \Phi_-^b(X_2)$. ◇

Proof. The result follows immediately from Eq. (11.24). Q.E.D.

Remark 11.5.2 *(i) It follows from Lemma 11.5.2 that, if $M_C \in \Phi^b(X_1 \times X_2)$, then $A \in \Phi_+^b(X_1)$ and $B \in \Phi_-^b(X_2)$.*

(ii) Using the same reasoning as in the proof of Lemma 11.5.1, we can show that, if the operator

$$\begin{pmatrix} A & 0 \\ D & B \end{pmatrix} \text{ is in } \Phi^b(X_1 \times X_2)$$

for some $D \in \mathscr{L}(X_1, X_2)$, then $A \in \Phi_-^b(X_1)$ and $B \in \Phi_+^b(X_2)$. ◇

Theorem 11.5.1 *(A. Jeribi, N. Moalla, and S. Yengui [102]) Let*

$$F := \begin{pmatrix} F_{11} & F_{12} \\ F_{21} & F_{22} \end{pmatrix},$$

where $F_{ij} \in \mathscr{L}(X_j, X_i)$, with $i, j = 1, 2$. Then, $F \in \mathscr{F}^b(X_1 \times X_2)$ if, and only if, $F_{ij} \in \mathscr{F}^b(X_j, X_i)$, with $i, j = 1, 2$. ◇

Proof. In order to prove the second implication, let us consider the following decomposition:

$$F = \begin{pmatrix} F_{11} & 0 \\ 0 & 0 \end{pmatrix} + \begin{pmatrix} 0 & F_{12} \\ 0 & 0 \end{pmatrix} + \begin{pmatrix} 0 & 0 \\ F_{21} & 0 \end{pmatrix} + \begin{pmatrix} 0 & 0 \\ 0 & F_{22} \end{pmatrix}.$$

It is sufficient to prove that if $F_{ij} \in \mathscr{F}^b(X_j, X_i)$, with $i, j = 1, 2$, then each operator in the right side of the previous equality is a Fredholm perturbation on $X_1 \times X_2$. For example, we will prove the result for the first operator. The proofs for the other operators will be similarly achieved. Consider

$$L = \begin{pmatrix} A & B \\ C & D \end{pmatrix} \in \Phi^b(X_1 \times X_2)$$

and let us denote

$$\widetilde{F} := \begin{pmatrix} F_{11} & 0 \\ 0 & 0 \end{pmatrix}.$$

From Theorem 2.2.3, it follows that there exist

$$L_0 = \begin{pmatrix} A_0 & B_0 \\ C_0 & D_0 \end{pmatrix} \in \mathscr{L}(X_1 \times X_2)$$

and

$$K = \begin{pmatrix} K_{11} & K_{12} \\ K_{21} & K_{22} \end{pmatrix} \in \mathscr{K}(X_1 \times X_2),$$

such that

$$LL_0 = I - K \text{ on } X_1 \times X_2.$$

Then,

$$(L+\widetilde{F})L_0 = I - K + \widetilde{F}L_0 = \begin{pmatrix} I - K_{11} + F_{11}A_0 & -K_{12} + F_{11}B_0 \\ -K_{21} & I - K_{22} \end{pmatrix}.$$

Since $F_{11} \in \mathscr{F}^b(X_1)$, and using Theorem 2.3.1, we will have

$$I - K_{11} + F_{11}A_0 \in \Phi^b(X_1).$$

This, together with the fact that $I - K_{22} \in \Phi^b(X_2)$, allows us to deduce, from Lemma 11.5.1 (i), that

$$(L+\widetilde{F})L_0 - \begin{pmatrix} 0 & 0 \\ -K_{21} & 0 \end{pmatrix}$$

is a Fredholm operator on $X_1 \times X_2$. The fact that K_{21} is a compact operator and $L_0 \in \Phi^b(X_1 \times X_2)$ leads, by Theorem 2.2.6, to $L + \widetilde{F} \in \Phi^b(X_1 \times X_2)$. Conversely, assume that $F \in \mathscr{F}^b(X_1 \times X_2)$. We will prove that $F_{11} \in \mathscr{F}^b(X_1)$. Let $A \in \Phi^b(X_1)$ and let us define the operator

$$L_1 := \begin{pmatrix} A & -F_{12} \\ 0 & I \end{pmatrix}.$$

From Lemma 11.5.1 (i), it follows that

$$L_1 \in \Phi^b(X_1 \times X_2).$$

Hence,

$$F+L_1=\begin{pmatrix} A+F_{11} & 0 \\ F_{21} & I+F_{22} \end{pmatrix}\in\Phi^b(X_1\times X_2).$$

The use of Remark 11.5.2 (ii) leads to

$$A+F_{11}\in\Phi_-^b(X_1). \tag{11.25}$$

In the same way, we may consider the Fredholm operator

$$\begin{pmatrix} A & 0 \\ -F_{21} & I \end{pmatrix}.$$

Using Remarks 11.5.1 (i) and 11.5.2 (i), it is easy to deduce that

$$A+F_{11}\in\Phi_+^b(X_1). \tag{11.26}$$

From Eqs. (11.25) and (11.26), it follows that $F_{11}\in\mathscr{F}^b(X_1)$. In the same way, we can prove that

$$F_{22}\in\mathscr{F}^b(X_2).$$

Now, we have to prove that $F_{12}\in\mathscr{F}^b(X_2,X_1)$ and $F_{21}\in\mathscr{F}^b(X_1,X_2)$. For this, let us consider $A\in\Phi^b(X_2,X_1)$ and $B\in\Phi^b(X_1,X_2)$. Then,

$$\begin{pmatrix} 0 & A \\ B & 0 \end{pmatrix}\in\Phi^b(X_1\times X_2).$$

Using the facts that $F_{11}\in\mathscr{F}^b(X_1)$, that $F_{22}\in\mathscr{F}^b(X_2)$, as well as the result of the second implication, we can deduce that

$$F+\begin{pmatrix} -F_{11} & 0 \\ 0 & -F_{22} \end{pmatrix}\in\mathscr{F}^b(X_1\times X_2).$$

Hence,

$$\begin{pmatrix} 0 & A+F_{12} \\ B+F_{21} & 0 \end{pmatrix}\in\Phi^b(X_1\times X_2).$$

So,

$$A+F_{12}\in\Phi^b(X_2,X_1)$$

and

$$B+F_{21}\in\Phi^b(X_1,X_2).$$

Q.E.D.

Lemma 11.5.3 *Let $A \in \mathcal{L}(X_1)$, $B \in \mathcal{L}(X_2 \times X_1)$, $\varepsilon > 0$ and consider the block operator matrices*

$$M_C = \begin{pmatrix} A & C \\ 0 & B \end{pmatrix},$$

where $C \in \mathcal{L}(X_2 \times X_1)$. Then,

(*i*) $\sigma_{ei,\varepsilon}(M_C) \subset \sigma_{ei,\varepsilon}(A) \bigcup \sigma_{ei,\varepsilon}(B)$, $i = 4, 5$,

(*ii*) $\sigma_{ei,\varepsilon}(M_C) \subset \sigma_{ei,\varepsilon}(A) \bigcup \sigma_{e,\varepsilon}(B)$ $i = 1, ap$, *and*

(*iii*) $\sigma_{ei,\varepsilon}(M_C) \subset \sigma_{ei,\varepsilon}(A) \bigcup \sigma_{ei,\varepsilon}(B)$ $i = 2, \delta$. ◇

Proof. Let $T_1 \in \mathcal{L}(X_1)$, $T_2 \in \mathcal{L}(X_2)$ and consider

$$T = \begin{pmatrix} T_1 & 0 \\ 0 & T_2 \end{pmatrix}$$

such that $||T|| < \varepsilon$. The operator $\lambda - M_C - T$ can be written as the following form

$$\lambda - M_C - T = \begin{pmatrix} I & 0 \\ 0 & \lambda - B - T_2 \end{pmatrix} \begin{pmatrix} I & -C \\ 0 & I \end{pmatrix} \begin{pmatrix} \lambda - A - T_1 & 0 \\ 0 & I \end{pmatrix}. \quad (11.27)$$

If $\lambda \notin \sigma_{e4,\varepsilon}(A) \bigcup \sigma_{e4,\varepsilon}(B)$, then

$$\lambda - A \in \Phi^{\varepsilon}(X_1)$$

and

$$\lambda - B \in \Phi^{\varepsilon}(X_2).$$

Since $||T|| < \varepsilon$, then

$$||T_1|| < \varepsilon$$

and

$$||T_2|| < \varepsilon.$$

So, $\lambda - A - T_1$ and $\lambda - B - T_2$ are Fredholm operators. Therefore,

$$\begin{pmatrix} \lambda - A - T_1 & 0 \\ 0 & I \end{pmatrix} \in \Phi^b(X_1 \times X_2)$$

and
$$\begin{pmatrix} I & 0 \\ 0 & \lambda - B - T_2 \end{pmatrix} \in \Phi^b(X_1 \times X_2).$$

Since
$$\begin{pmatrix} I & -C \\ 0 & I \end{pmatrix}$$

is invertible, then
$$\lambda - M_C - T \in \Phi^b(X_1 \times X_2)$$

for all $||T|| < \varepsilon$. Therefore,
$$\lambda - M_C \in \Phi^\varepsilon(X_1 \times X_2).$$

Hence,
$$\sigma_{e4,\varepsilon}(M_C) \subset \sigma_{e4,\varepsilon}(A) \bigcup \sigma_{e4,\varepsilon}(B).$$

On the other hand, Eq. (11.27), the invertibility of
$$\begin{pmatrix} I & -C \\ 0 & I \end{pmatrix}$$

and Theorem 2.2.19 leads to
$$i(\lambda - M_C - T) = i\left[\begin{pmatrix} I & 0 \\ 0 & \lambda - B - T_2 \end{pmatrix}\right] + i\left[\begin{pmatrix} \lambda - A - T_1 & 0 \\ 0 & I \end{pmatrix}\right].$$

So,
$$i(\lambda - M_C - T) = i(\lambda - A - T_1) + i(\lambda - B - T_2).$$

Hence,
$$\sigma_{e5,\varepsilon}(M_C) \subset \sigma_{e5,\varepsilon}(A) \bigcup \sigma_{e5,\varepsilon}(B).$$

(ii) and (iii) can be checked in the same way as (i). Q.E.D.

In all that follows we will make the following assumptions:

$(H7)$ $CB \in \mathscr{F}^b(X_2)$ and $C(A + T_1) \in \mathscr{F}^b(X_1 \times X_2)$,

$(H8)$ $CB \in \mathscr{K}(X_2)$ and $C(A + T_1) \in \mathscr{K}(X_1 \times X_2)$,

and

$(H9)$ $BC \in \mathscr{K}(X_2)$ and $(D + T_2)C \in \mathscr{K}(X_1 \times X_2)$.

Theorem 11.5.2 *(i) If (H7) holds, then*

$$\sigma_{ei,\varepsilon}(\mathscr{L})\backslash\{0\} \subseteq [\sigma_{ei,\varepsilon}(A)\bigcup\sigma_{ei,\varepsilon}(D)]\backslash\{0\},\ 4,5.$$

(ii) If (H8) holds, then

$$\sigma_{e1,\varepsilon}(\mathscr{L})\backslash\{0\} \subseteq [\sigma_{e1,\varepsilon}(A)\bigcup\sigma_{e1,\varepsilon}(D)]\backslash\{0\}.$$

(iii) If (H9) holds, then

$$\sigma_{e2,\varepsilon}(\mathscr{L})\backslash\{0\} \subseteq [\sigma_{e2,\varepsilon}(A)\cup\sigma_{e2,\varepsilon}(D)]\backslash\{0\}. \qquad \diamondsuit$$

Proof. Let $T_1 \in \mathscr{L}(X_1)$, $T_2 \in \mathscr{L}(X_2)$ and consider

$$T = \begin{pmatrix} T_1 & 0 \\ 0 & T_2 \end{pmatrix}$$

such that $||T|| < \varepsilon$. Then, for all $\lambda \in \mathbb{C}\backslash\{0\}$, we have

$$\lambda - \mathscr{L} - T = \\ \frac{1}{\lambda}\begin{pmatrix} 0 & 0 \\ -C(A+T_1) & -CB \end{pmatrix} + \begin{pmatrix} I & 0 \\ -\frac{C}{\lambda} & I \end{pmatrix}\begin{pmatrix} \lambda - A - T_1 & -B \\ 0 & \lambda - D - T_2 \end{pmatrix}. \tag{11.28}$$

(i) Suppose that $\lambda \notin \sigma_{e5,\varepsilon}(A)\cup\sigma_{e5,\varepsilon}(D)\bigcup\{0\}$, then by Lemma 11.5.3,

$$\begin{pmatrix} \lambda - A - T_1 & -B \\ 0 & \lambda - D - T_2 \end{pmatrix} \in \Phi^b(X_1 \times X_2)$$

and

$$i\left[\begin{pmatrix} \lambda - A - T_1 & -B \\ 0 & \lambda - D - T_2 \end{pmatrix}\right] = 0.$$

Since

$$\begin{pmatrix} I & 0 \\ -\frac{C}{\lambda} & I \end{pmatrix}$$

is invertible, using Theorem 2.2.19

$$\begin{pmatrix} I & 0 \\ -\frac{C}{\lambda} & I \end{pmatrix}\begin{pmatrix} \lambda - A - T_1 & -B \\ 0 & \lambda - D - T_2 \end{pmatrix}$$

is a Fredholm operator and

$$i\left[\begin{pmatrix} I & 0 \\ -\frac{C}{\lambda} & I \end{pmatrix}\begin{pmatrix} \lambda - A - T_1 & -B \\ 0 & \lambda - D - T_2 \end{pmatrix}\right] =$$
$$i\left[\begin{pmatrix} I & 0 \\ -\frac{C}{\lambda} & I \end{pmatrix}\right] + i\left[\begin{pmatrix} \lambda - A - T_1 & -B \\ 0 & \lambda - D - T_2 \end{pmatrix}\right] = 0.$$

On the other hand, it follows from the hypothesis $(H7)$ and Theorem 11.5.1 that

$$\begin{pmatrix} 0 & 0 \\ -C(A+T_1) & -CB \end{pmatrix} \in \mathscr{F}^b(X_1 \times X_2).$$

So, using Eq. (11.28), $\lambda - \mathscr{L} \in \Phi^{\varepsilon}(X_1 \times X_2)$ and

$$i(\lambda - \mathscr{L} - T) = 0$$

for all $||T|| < \varepsilon$. Thus,

$$\lambda \notin \sigma_{e5,\varepsilon}(\mathscr{L}).$$

Hence,

$$\sigma_{ei,\varepsilon}(\mathscr{L})\backslash\{0\} \subseteq [\sigma_{ei,\varepsilon}(A) \bigcup \sigma_{ei,\varepsilon}(D)]\backslash\{0\},\ i = 4,5.$$

(ii) If $\lambda \notin \sigma_{e1,\varepsilon}(A) \bigcup \sigma_{e1,\varepsilon}(D) \bigcup \{0\}$, then

$$\lambda - A \in \Phi_{+}^{\varepsilon}(X_1)$$

and

$$\lambda - D \in \Phi_{+}^{\varepsilon}(X_2).$$

Using Lemma 11.5.3, we have

$$\begin{pmatrix} \lambda - A - T_1 & -B \\ 0 & \lambda - D - T_2 \end{pmatrix} \in \Phi_{+}^{b}(X_1 \times X_2)$$

for all $||T|| < \varepsilon$. Since

$$\begin{pmatrix} I & 0 \\ -\frac{C}{\lambda} & I \end{pmatrix}$$

is invertible, then

$$\begin{pmatrix} I & 0 \\ -\frac{C}{\lambda} & I \end{pmatrix} \begin{pmatrix} \lambda - A - T_1 & -B \\ 0 & \lambda - D - T_2 \end{pmatrix} \in \Phi_+^b(X_1 \times X_2).$$

Hence, it follows from both hypothesis $(H8)$ and Eq. (11.28), that

$$\lambda - \mathscr{L} - T \in \Phi_+^b(X_1 \times X_2)$$

for all $||T|| < \varepsilon$. Thus,

$$\lambda \notin \sigma_{e1,\varepsilon}(\mathscr{L}).$$

Hence,

$$\sigma_{e1,\varepsilon}(\mathscr{L})\backslash\{0\} \subseteq [\sigma_{e1,\varepsilon}(A)\bigcup\sigma_{e1,\varepsilon}(D)]\backslash\{0\}.$$

(iii) The proof of (iii) may be checked in the same way as the proof of the item (ii). Q.E.D.

Chapter 12

Conditional Pseudospectra

Our aim in this chapter is to show some properties of condition pseudospectra of a linear operator A in Banach spaces and reveal the relation between their condition pseudospectrum.

12.1 SOME PROPERTIES OF $\Sigma^{\varepsilon}(A)$

In the next lemma, we give some properties of the condition pseudospectral radius $r_{\varepsilon}(\cdot)$.

Lemma 12.1.1 *Let $A \in \mathscr{L}(X)$ and $0 < \varepsilon < 1$. Then,*

(i) $r_{\sigma}(A) \leq r_{\varepsilon}(A) \leq \left(\frac{1+\varepsilon}{1-\varepsilon}\right) \|A\|$.

(ii) If $\|A^k\| \leq \delta < \frac{1}{\varepsilon}$ for all $k \geq 0$, then

$$r_{\varepsilon}(A) \leq \frac{1+\delta^2\varepsilon}{1-\delta\varepsilon}.$$ ◇

Proof. (i) Since

$$\sigma(A) \subseteq \Sigma^{\varepsilon}(A)$$

for $0 < \varepsilon < 1$, we obtain that

$$r_{\sigma}(A) \leq r_{\varepsilon}(A).$$

Now, we prove that

$$r_\varepsilon(A) \leq \Big(\frac{1+\varepsilon}{1-\varepsilon}\Big)\|A\|.$$

Let $\lambda \in \Sigma^\varepsilon(A)$. We will discuss these two cases:

<u>1st *case*</u> : If $|\lambda| \leq \|A\|$, then it is clear that

$$r_\varepsilon(A) \leq \Big(\frac{1+\varepsilon}{1-\varepsilon}\Big)\|A\|.$$

<u>2nd *case*</u> : If $|\lambda| > \|A\|$, then $\lambda - A$ is an invertible operator and

$$\|(\lambda - A)^{-1}\| \leq \frac{1}{|\lambda| - \|A\|}.$$

So,

$$1 < \varepsilon\|(\lambda - A)^{-1}\|\|\lambda - A\| \leq \Big(\frac{|\lambda| + \|A\|}{|\lambda| - \|A\|}\Big)\varepsilon.$$

After a simple computation, we have

$$|\lambda| \leq \Big(\frac{1+\varepsilon}{1-\varepsilon}\Big)\|A\|.$$

Hence,

$$r_\varepsilon(A) \leq \Big(\frac{1+\varepsilon}{1-\varepsilon}\Big)\|A\|.$$

(ii) Since $\lambda \in \Sigma^\varepsilon(A)$ and

$$\begin{aligned} r_\sigma(A) &= \lim_{k\to\infty} \|A^k\|^{\frac{1}{k}} \\ &= \lim_{k\to\infty} \delta^{\frac{1}{k}} \\ &= 1, \end{aligned}$$

then there are two possible cases:

1st case : If $|\lambda| \leq 1$, then it is clear that

$$|\lambda| \leq \frac{1+\delta^2\varepsilon}{1-\delta\varepsilon}.$$

2nd case : If $|\lambda| > 1$, then

$$\lambda \notin \sigma(A).$$

We get $\lambda - A$ is an invertible operator. Therefore,

$$\begin{aligned}\|(\lambda - A)^{-1}\| &\leq \frac{1}{|\lambda|}\sum_{k=0}^{\infty}\frac{\|A^k\|}{|\lambda|^k}\\ &\leq \frac{\delta}{|\lambda|}\sum_{k=0}^{\infty}\frac{1}{|\lambda|^k}\\ &= \frac{\delta}{|\lambda|-1}.\end{aligned}$$

Hence,

$$1 < \varepsilon\|(\lambda - A)^{-1}\|\|\lambda - A\| \leq \varepsilon(|\lambda| + \delta)\frac{\delta}{|\lambda|-1},$$

which implies that

$$|\lambda| \leq \frac{1+\delta^2\varepsilon}{1-\delta\varepsilon}.$$

So,

$$r_\varepsilon(A) \leq \frac{1+\delta^2\varepsilon}{1-\delta\varepsilon}. \qquad \text{Q.E.D.}$$

The following theorem establish the relationship between condition pseudospectrum and pseudospectrum of a bounded linear operator $A \in \mathscr{L}(X)$. We set

$$\delta_A := \inf\Big\{\|\lambda - A\| : \lambda \in \mathbb{C}\Big\}.$$

Theorem 12.1.1 *Let $A \in \mathscr{L}(X)$ such that $A \neq I$ and $0 < \varepsilon < 1$. Then,*

$$\Sigma^{\varepsilon}(A) \subseteq \sigma_{\gamma_\varepsilon}(A) \subseteq \Sigma^{\upsilon_\varepsilon}(A),$$

where

$$\gamma_\varepsilon = \frac{2\varepsilon\|A\|}{1-\varepsilon} \in]0,1[,$$

$$\upsilon_\varepsilon = \frac{2\varepsilon\|A\|}{(1-\varepsilon)\delta_A} \in]0,1[$$

and $\sigma_{\gamma_\varepsilon}(A)$ is the pseudospectrum of A. ◊

Proof. Let $\lambda \in \Sigma^{\varepsilon}(A)$, then

$$|\lambda| \leq \left(\frac{1+\varepsilon}{1-\varepsilon}\right)\|A\|$$

and

$$\|\lambda - A\|\|(\lambda - A)^{-1}\| > \frac{1}{\varepsilon}.$$

Thus,

$$\begin{aligned}\|(\lambda - A)^{-1}\| &> \frac{1}{\varepsilon\|\lambda - A\|} \\ &> \frac{1}{\varepsilon(|\lambda| + \|A\|)} \\ &\geq \frac{1-\varepsilon}{2\varepsilon\|A\|}.\end{aligned}$$

Hence,

$$\lambda \in \sigma_{\gamma_\varepsilon}(A).$$

For the second inclusion, let $\lambda \in \sigma_{\gamma_\varepsilon}(A)$. Then,

$$\|(\lambda - A)^{-1}\| > \frac{1-\varepsilon}{2\varepsilon\|A\|}.$$

Also, we have

$$\|\lambda - A\| \geq \inf\left\{\|\lambda - A\| : \lambda \in \mathbb{C}\right\} := \delta_A.$$

Hence,

$$\|\lambda - A\|\|(\lambda - A)^{-1}\| > \delta_A \frac{1-\varepsilon}{2\varepsilon\|A\|}.$$

Therefore,

$$\lambda \in \Sigma^{\upsilon_\varepsilon}(A).$$ Q.E.D.

In the following we gives a precise information about the condition pseudospectrum of bounded linear operator under linear transformation.

Proposition 12.1.1 *Let* $A \in \mathcal{L}(X)$ *and* $0 < \varepsilon < 1$. *Then,*

(i) $\sigma(A) = \bigcap_{0<\varepsilon<1} \Sigma^{\varepsilon}(A)$.

(ii) If $0 < \varepsilon_1 < \varepsilon_2 < 1$*, then*

$$\sigma(A) \subset \Sigma^{\varepsilon_1}(A) \subset \Sigma^{\varepsilon_2}(A).$$

(iii) If $\alpha \in \mathbb{C}$*, then*

$$\Sigma^{\varepsilon}(A+\alpha I) = \alpha + \Sigma^{\varepsilon}(A).$$

(iv) If $\alpha \in \mathbb{C}\backslash\{0\}$*, then*

$$\Sigma^{\varepsilon}(\alpha A) = \alpha \Sigma^{\varepsilon}(A). \qquad \diamondsuit$$

Proof. (i) It is clear that

$$\sigma(A) \subset \Sigma^{\varepsilon}(A)$$

for all $0 < \varepsilon < 1$, then

$$\sigma(A) \subset \bigcap_{0<\varepsilon<1} \Sigma^{\varepsilon}(A).$$

Conversely, if

$$\lambda \in \bigcap_{0<\varepsilon<1} \Sigma^{\varepsilon}(A),$$

then for all $0 < \varepsilon < 1$, we have

$$\lambda \in \Sigma^{\varepsilon}(A).$$

We will discuss these two cases:

$\underline{1^{st}\ case}$: If $\lambda \in \sigma(A)$, then we get the desired result.

$\underline{2^{nd}\ case}$: If $\lambda \in \left\{\lambda \in \mathbb{C} : \|\lambda - A\| \|(\lambda - A)^{-1}\| > \frac{1}{\varepsilon}\right\}$, then taking limits as $\varepsilon \to 0^+$, we get

$$\|\lambda - A\| \|(\lambda - A)^{-1}\| = \infty.$$

Thus,

$$\lambda \in \sigma(A).$$

(ii) Let $\lambda \in \Sigma^{\varepsilon_1}(A)$, then

$$\|\lambda - A\| \|(\lambda - A)^{-1}\| > \frac{1}{\varepsilon_1} > \frac{1}{\varepsilon_2}.$$

Hence,

$$\lambda \in \Sigma^{\varepsilon_2}(A).$$

(iii) Let $\lambda \in \Sigma^{\varepsilon}(A+\alpha I)$, then

$$\|(\lambda-\alpha)I-A\|\|((\lambda-\alpha)I-A)^{-1}\| > \frac{1}{\varepsilon}.$$

Hence,

$$\lambda-\alpha \in \Sigma^{\varepsilon}(A).$$

This yields to

$$\lambda \in \alpha+\Sigma^{\varepsilon}(A).$$

The opposite inclusion follows by symmetry.

(iv) Let $\lambda \in \Sigma^{\varepsilon}(\alpha A)$ such that $\alpha \neq 0$, then

$$\begin{aligned} \frac{1}{\varepsilon} < \|\lambda-\alpha A\| \; \|(\lambda-\alpha A)^{-1}\| &= \left\|\alpha\left(\frac{\lambda}{\alpha}-A\right)\right\|\left\|\alpha^{-1}\left(\frac{\lambda}{\alpha}-A\right)^{-1}\right\|, \\ &= \left\|\left(\frac{\lambda}{\alpha}-A\right)\right\|\left\|\left(\frac{\lambda}{\alpha}-A\right)^{-1}\right\|. \end{aligned}$$

Hence,

$$\frac{\lambda}{\alpha} \in \Sigma^{\varepsilon}(A),$$

so

$$\Sigma^{\varepsilon}(\alpha A) \subseteq \alpha\Sigma^{\varepsilon}(A).$$

However, the opposite inclusion follows by symmetry. Q.E.D.

Theorem 12.1.2 *Let $A \in \mathscr{L}(X)$, $k=\|A\|\|A^{-1}\|$ and $0<\varepsilon<1$. Then,*

(i) $\lambda \in \Sigma^{\varepsilon}(A)$ if, and only if, $\overline{\lambda} \in \Sigma^{\varepsilon}(A^*)$.

(ii) If $\lambda \in \Sigma^{\varepsilon}(A^{-1})\backslash\{0\}$, then

$$\frac{1}{\lambda} \in \Sigma^{\varepsilon k}(A)\backslash\{0\}.$$

◊

Proof. (i) Using the identity

$$\|\lambda-A\|\|(\lambda-A)^{-1}\| = \|\overline{\lambda}-A^*\|\|(\overline{\lambda}-A^*)^{-1}\|,$$

it is easy to see that the condition pseudospectrum of A^* is given by the mirror image of $\Sigma^{\varepsilon}(A)$ with respect to the real axis.

(ii) Let $\lambda \in \Sigma^{\varepsilon}(A^{-1})\backslash\{0\}$, then

$$\frac{1}{\varepsilon} < \|\lambda - A^{-1}\|\|(\lambda - A^{-1})^{-1}\|$$

and

$$\begin{aligned}
&\|\lambda - A^{-1}\|\|(\lambda - A^{-1})^{-1}\| \\
&\quad = \left\|-\lambda\left(\frac{1}{\lambda} - A\right)A^{-1}\right\|\left\|-\lambda^{-1}A\left(\frac{1}{\lambda} - A\right)^{-1}\right\|, \\
&\quad \leq \|A\|\|A^{-1}\|\left\|\frac{1}{\lambda} - A\right\|\left\|\left(\frac{1}{\lambda} - A\right)^{-1}\right\|.
\end{aligned}$$

Hence,

$$\frac{1}{\lambda} \in \Sigma^{\varepsilon k}(A)\backslash\{0\}. \qquad \text{Q.E.D.}$$

Theorem 12.1.3 *Let* $A \in \mathcal{L}(X)$ *and* $V \in \mathcal{L}(X)$ *be invertible. Let* $B = V^{-1}AV$. *Then, for all* $0 < \varepsilon < 1$ *and* $k = \|V^{-1}\|\|V\|$ *such that* $0 < k^2\varepsilon < 1$, *we have*

$$\Sigma^{\frac{\varepsilon}{k^2}}(A) \subseteq \Sigma^{\varepsilon}(B) \subseteq \Sigma^{k^2\varepsilon}(A). \qquad \diamondsuit$$

Proof. We can write

$$\begin{aligned}
\|\lambda - B\|\|(\lambda - B)^{-1}\| &= \|V^{-1}(\lambda - A)V\|\|V^{-1}(\lambda - A)^{-1}V\| \\
&\leq \left(\|V\|\|V^{-1}\|\right)^2\|\lambda - A\|\|(\lambda - A)^{-1}\| \\
&\leq k^2\|\lambda - A\|\|(\lambda - A)^{-1}\|, \qquad (12.1)
\end{aligned}$$

and

$$\begin{aligned}
\|\lambda - A\|\|(\lambda - A)^{-1}\| &= \|V(\lambda - B)V^{-1}\|\|V(\lambda - B)^{-1}V^{-1}\| \\
&\leq \left(\|V\|\|V^{-1}\|\right)^2\|\lambda - B\|\|(\lambda - B)^{-1}\| \\
&\leq k^2\|\lambda - B\|\|(\lambda - B)^{-1}\|. \qquad (12.2)
\end{aligned}$$

Let $\lambda \in \Sigma^{\frac{\varepsilon}{k^2}}(A)$, then using relation (12.2), we obtain

$$\lambda \in \Sigma^{\varepsilon}(B).$$

If $\lambda \in \Sigma^{\varepsilon}(B)$, then using relation (12.1), we obtain

$$\lambda \in \Sigma^{k^2\varepsilon}(A). \qquad \text{Q.E.D.}$$

12.2 CHARACTERIZATION OF CONDITION PSEUDOSPECTRUM

In this section, we give a characterization of the condition pseudospectrum of linear operators on a Banach space. Our first result is the following.

Lemma 12.2.1 *Let $A \in \mathscr{L}(X)$ and $0 < \varepsilon < 1$. Then, $\lambda \in \Sigma^{\varepsilon}(A)\backslash\sigma(A)$ if, and only if, there exists $x \in X$, such that*

$$\|(\lambda - A)x\| < \varepsilon\|\lambda - A\|\|x\|. \qquad \diamondsuit$$

Proof. Let $\lambda \in \Sigma^{\varepsilon}(A)\backslash\sigma(A)$, then

$$\|\lambda - A\|\|(\lambda - A)^{-1}\| > \frac{1}{\varepsilon}.$$

Thus,

$$\|(\lambda - A)^{-1}\| > \frac{1}{\varepsilon\|\lambda - A\|}.$$

Moreover,

$$\sup_{y \in X\backslash\{0\}} \frac{\|(\lambda - A)^{-1}y\|}{\|y\|} > \frac{1}{\varepsilon\|\lambda - A\|}.$$

Then, there exists a nonzero $y \in X$, such that

$$\|(\lambda - A)^{-1}y\| > \frac{\|y\|}{\varepsilon\|\lambda - A\|}.$$

Putting $x = (\lambda - A)^{-1}y$. We have the result.

Conversely, we assume that there exists $x \in X$ such that

$$\|(\lambda - A)x\| < \varepsilon\|\lambda - A\|\|x\|.$$

Let $\lambda \notin \sigma(A)$ and $x = (\lambda - A)^{-1}y$, then

$$\|x\| \leq \|(\lambda - A)^{-1}\|\|y\|.$$

Moreover,

$$1 < \varepsilon\|\lambda - A\|\|(\lambda - A)^{-1}\|.$$

So,

$$\lambda \in \Sigma^{\varepsilon}(A)\backslash\sigma(A). \qquad \text{Q.E.D.}$$

In the following theorem, we investigate the relation between the condition pseudospectrum and the usual spectrum in a complex Banach space.

Theorem 12.2.1 *Let $A \in \mathscr{L}(X)$, $\lambda \in \mathbb{C}$, and $0 < \varepsilon < 1$. If there is $D \in \mathscr{L}(X)$ such that $\|D\| \leq \varepsilon\|\lambda - A\|$ and $\lambda \in \sigma(A+D)$, then $\lambda \in \Sigma^{\varepsilon}(A)$.* $\diamondsuit$

Proof. We assume that there exists D such that $\|D\| < \varepsilon\|\lambda - A\|$ and $\lambda \in \sigma(A+D)$. If $\lambda \notin \Sigma^{\varepsilon}(A)$, then

$$\|\lambda - A\|\|(\lambda - A)^{-1}\| \leq \frac{1}{\varepsilon}.$$

Now, we define the operator $S : X \longrightarrow X$ by

$$S := \sum_{n=0}^{\infty}(\lambda - A)^{-1}\left(D(\lambda - A)^{-1}\right)^{n}.$$

Since

$$\|D(\lambda - A)^{-1}\| < 1,$$

we can write

$$S = (\lambda - A)^{-1}\left(I - D(\lambda - A)^{-1}\right)^{-1}.$$

Then, there exists $y \in X$ such that

$$S\left(I - D(\lambda - A)^{-1}\right)^{-1} y = (\lambda - A)^{-1}y.$$

Let $x = (\lambda - A)^{-1}y$. Then,

$$S(\lambda - A - D)x = x$$

for every $x \in X$. Similarly, we can prove that

$$(\lambda - A - D)Sy = y$$

for all $y \in X$. Hence, $\lambda - A - D$ is invertible, so

$$\lambda \in \Sigma^{\varepsilon}(A). \qquad \text{Q.E.D.}$$

Theorem 12.2.2 *Let X be a complex Banach space satisfying the following property:*

for all bounded operator A with $0 \in \rho(A)$, there exists a non invertible bounded operator D such that

$$\|A-D\| = \frac{1}{\|A^{-1}\|}.$$

If $\lambda \in \Sigma^{\varepsilon}(A)$, then there exists $D \in \mathscr{L}(X)$ such that $\|D\| \leq \varepsilon\|\lambda - A\|$ and $\lambda \in \sigma(A+D)$. ♢

Proof. Suppose $\lambda \in \Sigma^{\varepsilon}(A)$. We will discuss these two cases:

1st case : If $\lambda \in \sigma(A)$, then it is sufficient to take $D = 0$.

2nd case : If $\lambda \in \Sigma^{\varepsilon}(A)\backslash\sigma(A)$, then by assumption, there exists an element $B \in \mathscr{L}(X)$ such that

$$\|\lambda - A - B\| = \frac{1}{\|(\lambda - A)^{-1}\|}.$$

Let $D = \lambda - A - B$. Then

$$\|D\| = \frac{1}{\|(\lambda - A)^{-1}\|} \leq \varepsilon\|\lambda - A\|.$$

Also $B = \lambda - (A+D)$, is not invertible. So, $\lambda \in \sigma(A+D)$. Q.E.D.

The next corollary is an immediate consequence of Theorems 12.2.1 and 12.2.2.

Corollary 12.2.1 *Let X be a complex Banach space satisfying the hypothesis of Theorem 12.2.2. Then, $\lambda \in \Sigma^{\varepsilon}(A)$ if, and only if, there exists $D \in \mathscr{L}(X)$ such that $\|D\| \leq \varepsilon\|\lambda - A\|$ and $\lambda \in \sigma(A+D)$.* ♢

In the sequel of this section, we consider the Hilbert space X, we use the notation $conv(\Sigma^{\varepsilon}(A))$, the convex hull in $\mathbb{C}$ of a set $\Sigma^{\varepsilon}(A)$, $\mathbb{B}(a,r)$ the open ball with center at a and radius r and we define the distance between two nonempty subsets U and V by the formula

$$\inf\{\|u-v\| : u \in U,\ v \in V\}.$$

The next theorem gives a relation between the condition pseudospectrum and the numerical range of A, given by

$$W(A) := \{\langle Ax, x\rangle \text{ such that } x \in \mathbb{S}_X\},$$

where

$$\mathbb{S}_X := \{x \in X \text{ such that } \|x\| = 1\}$$

is the unit sphere in X. It is well known that $W(A)$ is a convex set whose closure contains the spectrum $\sigma(A)$ of A.

Theorem 12.2.3 *Let $A \in \mathscr{L}(X)$ and $0 < \varepsilon < 1$. Then,*

$$conv(\Sigma^{\varepsilon}(A)) \subseteq W(A) + \mathbb{B}\left(0, \frac{2\varepsilon}{1-\varepsilon}\|A\|\right). \qquad \diamondsuit$$

Proof. Let $\lambda \in \mathbb{C}$ such that

$$\text{dist}(\lambda, W(A)) > 0,$$

then $\lambda - A$ is invertible and, we have

$$\|(\lambda - A)^{-1}\| \leq \frac{1}{\text{dist}(\lambda, W(A))}.$$

Assume that $\lambda \in \Sigma^{\varepsilon}(A)$. There are two possible cases:

$\underline{1^{st}\ case}$: If $\lambda \in \overline{W(A)}$, then the result is trivial.

$\underline{2^{nd}\ case}$: If $\lambda \in \Sigma^{\varepsilon}(A)\backslash\overline{W(A)}$. Then,

$$\frac{1}{\varepsilon} < \|\lambda - A\|\|(\lambda - A)^{-1}\| \leq \frac{\|\lambda - A\|}{\text{dist}(\lambda, W(A))}.$$

It follows that

$$\begin{aligned}\operatorname{dist}(\lambda, W(A)) &< \varepsilon\|\lambda - A\| \\ &\leq \varepsilon\left(|\lambda| + \|A\|\right).\end{aligned}$$

Using Lemma 12.1.1, we obtain that

$$\begin{aligned}\operatorname{dist}(\lambda, W(A)) &\leq \varepsilon\left(\frac{1+\varepsilon}{1-\varepsilon}\|A\| + \|A\|\right) \\ &= \left(\frac{2\varepsilon}{1-\varepsilon}\right)\|A\|.\end{aligned}$$

Hence,

$$\Sigma^{\varepsilon}(A) \subseteq W(A) + \mathbb{B}\left(0, \frac{2\varepsilon}{1-\varepsilon}\|A\|\right).$$

The result follows from the fact that $W(A)$ is convex set. Q.E.D.

Corollary 12.2.2 *Let $A \in \mathscr{L}(X)$ and $0 < \varepsilon < 1$. Then,*

$$\left\{\lambda \in \mathbb{C} \text{ such that } dist(\lambda, W(A)) < \varepsilon\|\lambda - A\|\right\} \subseteq \Sigma^{\varepsilon}(A). \qquad \diamondsuit$$

Proof. Let $\lambda \notin \Sigma^{\varepsilon}(A)$. By using Theorem 12.2.3, we have

$$\begin{aligned}\frac{1}{\operatorname{dist}(\lambda, W(A))} &\leq \|(\lambda - A)^{-1}\| \\ &\leq \frac{1}{\varepsilon\|\lambda - A\|}.\end{aligned}$$

Hence,

$$\operatorname{dist}(\lambda, W(A)) \geq \varepsilon\|\lambda - A\|. \qquad \text{Q.E.D.}$$

Bibliography

1. Jeribi, A., (2015). *Spectral Theory and Applications of Linear Operators and Block Operator Matrices*, Springer-Verlag, New York.

2. Jeribi, A., (2018). *Spectral Theory of Linear Non-Self-Adjoint Operators in Hilbert and Banach Spaces: Denseness and Bases with Applications*, preprint.

3. Jeribi, A., (2018). *Denseness, Bases and Frames in Banach Spaces and Applications*, de Gruyter, Berlin.

4. Jeribi, A., & Krichen, B., (2016). Nonlinear Functional Analysis in Banach Spaces and Banach Algebras. Fixed point theory under weak topology for nonlinear operators and block operator matrices with applications. (English) *Monographs and Research Notes in Mathematics.* Boca Raton, FL: CRC Press (ISBN 978-1-4987-3388-5/hbk; 978-1-4987-3389-2/ebook). xvi, 355 p.

5. Jeribi, A., Hammami, M. A., & Masmoudi, A., (eds.), (2015). *Applied Mathematics in Tunisia.* International Conference on Advances in Applied Mathematics (ICAAM), Hammamet, Tunisia, December 1619, 2013. Springer proceedings in mathematics and statistics 131. Cham: Springer (ISBN 978-3-319-18040-3/hbk; 978-3-319-18041-0/ebook). xix, 397 p.

6. Abdelmoumen, B., Dehici, A., Jeribi, A., & Mnif, M., (2008). Some new properties of fredholm theory, Schechter essential spectrum, and application to transport theory, *J. Inequal. Appl.,* Art. ID 852676, 1–14.

7. Abdelmoumen, B., Jeribi, A., & Mnif, M., (2011). Invariance of the Schechter essential spectrum under polynomially compact operators perturabation, *Extr. Math., 26*(1), 61–73.

8. Abdelmoumen, B., Jeribi, A., & Mnif, M., (2012). On graph measures in banch spaces and description of essential spectra of multidimensional transport equation, *Acta Math. Sci. Ser. B Engl. Ed., 32*(5), 2050–2064.

9. Abdelmoumen, B., Jeribi, A., & Mnif, M., (2012). Measure of weak noncompactness, some new properties in fredholm theory, characterization of the Schechter essential spectrum and application to transport operators, *Ricerche Mat., 61*, 321–340.

10. Abdmouleh, F., Ammar, A., & Jeribi, A., (2013). *Stability of the S-Essential Spectra on a Banach Space*, Math. Slovaca, *63*(2), 299–320.

11. Abdmouleh, F., Ammar, A., & Jeribi, A., A (2015). Characterization of the pseudo-browder essential spectra of linear operators and application to a transport equation, *J. Comput. Theor. Transp., 44*(3), 141–153.

12. Abdmouleh, F., Charfi, S., & Jeribi, A., (2012). On a characterization of the essential spectra of the sum and the product of two operators, *J. Math. Anal. Appl., 386*(1), 83–90.

13. Abdmouleh, F., & Jeribi, A., (2010). Symmetric family of Fredholm operators of indices zero, stability of essential spectra and application to transport operators, *J. Math. Anal. Appl., 364*, 414–223.

14. Abdmouleh, F., & Jeribi, A., (2011). Gustafson, Weidmann, Kato, Wolf, Schechter, Browder, Rakočević and Schmoger essential spectra of the sum of two bounded operators and application to a transport operator, *Math. Nachr., 284*(2–3), 166–176.

15. Aiena, P., (1989). *Riesz Operators and Perturbation Ideals*, Note Mat., *9*(1), 1–27.

16. Aiena, P., (2007). *Semi-Fredholm Operators*, Perturbation Theory and Localized SVEP, Caracas, Venezuela.

17. Akhmerov, R. R., Kamenskii, M., I., Potapov, A. S., Rodkina, A., E., & Sadovskii, B. N., (1992). *Measures of Noncompactness and Condensing Operators*, Birkhäuser Verlag, Basel Boston Berlin.

18. Alvarez, T., (1998). Perturbation and coperturbation functions characterizing semi-Fredholm type operators, *Math. Proc. R. Ir. Acad.*, *98 A*(1), 41–26.

19. Ammar, A., Boukattaya, B., & Jeribi, A., (2014). Stability of the S-left and S-right essential spectra of a linear operator, *Acta Math. Scie., Ser. B Engl.* Ed., *34*(5), 1–13.

20. Ammar, A., Boukettaya, B., & Jeribi, A., (2014). Stability of the S-left and S-right essential spectra of a linear operator, *Acta Math. Scie. Ser. B Engl.* Ed., *34*(6), 1922–1934.

21. Ammar, A., Boukettaya, B., & Jeribi, A., (2016). A note on the essential pseudospectra and application, *Linear and Multilinear Algebra, 64*(8), 1474–1483.

22. Ammar, A., Bouzayeni, F., & Jeribi, A., (2017). Perturbation of unbounded linear operators by γ-relative boundedness, Ricerche Mat., DOI 10.1007/s11587-017-0341-0, 1–13.

23. Ammar, A., Dhahri, M. Z., & Jeribi, A., (2015). Stability of essential approximate point spectrum and essential defect spectrum of linear operator, *Filomat, 29*(9), 1983–1994.

24. Ammar, A., Dhahri, M. Z., & Jeribi, A., (2015). Some properties of the M-essential spectra of closed linear operator on a Banach space, *Functional Analysis, Approximation and Computation, 7*(1), 15–28.

25. Ammar, A., Dhahri, M. Z., & Jeribi, A., (2015). A Characterization of S-essential spectrum by mean of measure of non-strict-singularity and application, *Azerbaijan J. Math., V. 5*(1), 2218–6816.

26. Ammar, A., Dhahri, M. Z., & Jeribi, A., (2015). Some properties of upper triangular 3×3-block matrices of linear relations, *Boll. Unione Mat. Ital., 8*(3), 189–204.

27. Ammar, A., Djaidja, N., & Jeribi, A., (2017). The essential spectrum of a sequence of linear operators in Banach spaces, *International J. Anal. Appl.,* 1–7.

28. Ammar, A., Heraiz, T., & Jeribi, A., (2017). Essential approximate point spectrum and essential defect spectrum of a sequence of linear operators in Banach spaces, preprint.

29. Ammar, A., & Jeribi, A., (2013). A characterization of the essential pseudospectra on a Banach space, *Arab. J. Math., 2*(2), 139–145.

30. Ammar, A., & Jeribi, A., (2013). A characterization of the essential pseudospectra and application to a transport equation, *Extr. Math., 28,* 95–112.

31. Ammar, A., & Jeribi, A., (2014). Measures of noncompactness and essential pseudo-spectra on Banach Space, *Math. Methods Appli. Sci., 37*(3), 447–452.

32. Ammar, A., & Jeribi, A., (2016). The Weyl essential spectrum of a sequence of linear operators in Banach spaces, *Indag. Math. (N.S.), 27*(1), 282–295.

33. Ammar, A., Jeribi, A., & Mahfoudhi, K., (2017). A characterization of the essential approximation pseudospectrum on a Banach space, *Filomat 31*(11), 3599–3610.

34. Ammar, A., Jeribi, A., & Mahfoudhi, K., (2017). Global bifurcation from the real leading eigenvalue of the transport operator, *J. Comput. Theor. Transp., 46*(4), 229–241.

35. Ammar, A., Jeribi, A., & Mahfoudhi, K., (2017). *A Characterization of the Condition Pseudospectrum,* preprint.

36. Ammar, A., Jeribi, A., & Moalla, N., (2013). On a characterization of the essential spectra of a 3×3 operator matrix and application to three-group transport operators, *Ann. Funct. Anal. AFA 4*(2), 153–170, electronic only.

37. Ammar, A., Jeribi, A., & Saadaoui, B., (2017). Frobenius-Schur factorization for multivalued 2×2 matrices linear operator, *Mediterr. J. Math., 14*(1), Paper No. 29, 29 p.

38. Astala, K., (1980). On measure of noncompactness and ideal variations in Banach spaces, *Ann. Acad. Sci. Fenn. Ser. A.I. Math. Diss., 29.*

39. Atkinson, F. V., Langer, H., Mennicken, R., & Shkalikov, A. A., (1994). The essential spectrum of some matrix operators, *Math. Nachr., 167*, 5–20.

40. Baloudi, H., & Jeribi, A., (2014). Left-Right Fredholm and Weyl spectra of the sum of two bounded operators and application, *Mediterr. J. Math., 11*, 939–953.

41. Baloudi, H., & Jeribi, A., (2016). Holomorphically Weyl-decomposably regular. (English) *Funct. Anal. Approx. Comput., 8*(2), 13–22.

42. Banaś, J., & Geobel, K., (1980). Measures of noncompactness in Banach spaces, *Lecture Notes in Pure and Applied Mathematics,* Volume 60, Marcel Dekker, New York, 259–262.

43. Ben Ali, N., Jeribi, A., & Moalla, N., (2010). Essential spectra of some matrix operators, *Math. Nachr., 283*(9), 1245–1256.

44. Ben Amar, A., Jeribi, A., & Mnif, M., (2010). Some applications of the regularity and irreducibility on transport theory, *Acta Appl. Math., 110*, 431–448.

45. Ben Amar, A., Jeribi, A., & Krichen, B., (2010). Essential spectra of a 3×3 operator matrix and application to three-group transport equation, *Inte. Equa. Oper. Theo., 68*, 1–21.

46. Ben Amar, A., Jeribi, A., & Mnif, M., (2014). Some results on Fredholm and semi-Fredholm operators, *Arab. J. Math. (Springer) 3*(3), 313–323.

47. Benharrat, M., Ammar, A., Jeribi, A., & Messirdi, B., (2014). On the Kato, semi-regular and essentially semi-regular spectra, *Functional Analysis, Approximation and Computation* 6(2), 9–22.

48. Browder, F. E., (1961). On the spectral theory of elliptic differential operators, I, *Math. Ann., 142*, 22–130.

49. Caradus, S. R., (1966). Operators of Riesz type, *Pacific J. Math., 18*, 61–71.

50. Caradus, S. R., (1966). Operators with finite ascent and descent, *Pacific J. Math., 18*, 437–449.

51. Chaker, W., Jeribi, A., & Krichen, B., (2015). Demicompact linear operators, essential spectrum and some perturbation results, *Math. Nachr., 288*(13), 1476–1486.

52. Charfi, S., & Jeribi, A., (2009). On a characterization of the essential spectra of some matrix operators and applications to two-group transport operators, *Math. Z., 262*(4), 775–794.

53. Charfi, S., Jeribi, A., & Moalla, N., (2013). Time asymptotic behavior of the solution of an abstract Cauchy problem given by a one-velocity transport operator with Maxwell boundary condition, *Collectanea Mathematica, 64,* 97–109.

54. Charfi, S., Jeribi, A., & Moalla, R., (2014). Essential spectra of operator matrices and applications, *Methods Appli. Sci. 37*(4), 597–608.

55. Charfi, S., Jeribi, A., & Walha, I., (2010). Essential Spectra, Matrix Operator and Applications, Acta Appl. Math., 111, (3), 319–337.

56. Damak, M., & Jeribi, A., (2007). On the essential spectra of some matrix operators and applications, *Electron. J. Differential Equations 11*, 1–16.

57. Danes, J., (1972). On the istratescu measure of noncompactness, *Bull. Math. Soc. R.S. Roumanie, 16*(64), 403–406.

58. Davies, E. B., (2007). *Linear Operators and Iheir Spectra,* Cambridge University Press, New York.

59. Dehici, A., Latrach, K., & Jeribi, A., (2002). On a transport operator arising in growing cell populations I. *Spectral analysis, Advances in Mathematics Research*, pp. 149–175.

60. Dehici, A., Jeribi, A., & Latrach, K., Spectral analysis of a transport operator arising in growing cell populations, Acta Appl. Math., 92(1), 37–62 (2006).

61. Diagana, T., (2015). Perturbations of unbounded Fredholm linear operators in Banach spaces, in: *Handbook on Operator Theory,* 875–880, Springer Basel.

62. Diestel, J., (1980). A survey of results related of the Dunford-Pettis property, Amer. Math. Soc. of Conf. on integration, Topology and Geometric in Linear Spaces, 15–60.

63. Dominguez Benavides, T., (1986). Some properties of the set and ball measures of noncompactness and applications, *J. London Math. Soc., 34*(2), 120–128.

64. Dunford, N., & Pettis, B. J., (1940). Linear operators on summable functions, *Trans. Amer. Math. Soc., 47*, 323–392.

65. Dunford, N., & Schwartz, J. T., (1971). *Linear Operators*, Part III, Wiley-Interscience, New York.

66. Edmum, D. E., & Evans, W. D., (1987). *Spectral Theory and Differential Operators,* Oxford Science Publications.

67. Elleuch, A., & Jeribi, A., (2016). On a characterization of the structured Wolf, Schechter and Browder essential pseudospectra, *Indag. Math., 27,* 212–224.

68. Elleuch, A., & Jeribi, A., (2016). New description of the structured essential pseudospectra, *Indag. Math., 27,* 368–382.

69. Faierman, M., Mennicken, R., & Muller, M., (1995). A boundary eigenvalue problem for a system of partial differential operators occurring in magnetohydrodynamics, *Math. Nachr., 173*, 141–167.

70. Fakhfakh, F., & Mnif, M., (2009). Browder and semi-Browder operators and perturbation function, *Extr. Math., 24*(3), 219–241.

71. Gethner, R. M., & Shapiro, J. H., (1987). Universal vectors for operators on space of holomorphic functions, *Proc. Am. Math. Soc., 100*, 281–288.

72. Gohberg, I. C., & Krein, G., (1960). Fundamental theorems on deficiency numbers, root numbers and indices of linear operators, *Amer. Math. Soc. Transl., Ser. 2*, 13, 185–264.

73. Gohberg, I. C., & Krein, M. G., (1969). Introduction to the theory of linear non-self-adjoint operators in Hilbert space, *Amer. Math. Soc., Providence.*

74. Gohberg, I., Markus, A., ad Feldman, I. A., (1967). Normally solvable operators and ideals associated with them, *Proc. Amer. Math. Soc. Tran. Ser.* 2(61), 63–84.

75. Goldberg, S., (1966). *Unbounded Linear Operators*, New York: McGraw-Hill.

76. Goldberg, S., (1974). Perturbations of semi-Fredholm operators by operators converging to zero compactly, *Proc. Amer. Math. Soc., 45*, 93–98.

77. Gonzalez, M., & Onieva, M. O., (1985).On Atkinson operators in locally convex spaces, *Math. Z., 190*, 505–517.

78. Grothendieck, B., (1953). Sur les applications linéaires faiblement compactes d'espaces du type $C(K)$, *Canad. J. Math., 5*, 129–173.

79. Gustafson, K., & Weidmann, J., (1969). On the essential spectrum, *J. Math. Anal. Appl., 25*, 121–127.

80. Harrabi, A., (1998). Pseudospectre d'une suite d'opérateurs bornés, RAIRO Modél. Math. Anal. Numér., 32, 671–680.

81. Hinrichsen, D., & Kelb, B., (1993). Spectral value sets: a graphical tool for robustness analysis, *Systems Control Lett., 21*(2), 127–136.

82. Jeribi, A., (1997). Quelques remarques sur les oprateurs de Fredholm et application l'quation de transport, *C. R. Acad. Sci., Paris, Srie, I 325*, 43–48.

83. Jeribi, A., (1998). Quelques remarques sur le spectre de Weyl et applications, *C. R. Acad. Sci., Paris Sr. I Math., 327*(5), 485–490.

84. Jeribi, A., (1998). Développement de certaines propriétés fines de la théorie spectrale et applications à des modèles monocinétiques et à des modèles de Reggeons, Thesis of Mathematics. University of Corsica, Frensh (16 Janvier 1998).

85. Jeribi, A., (2000). Une nouvelle caractrisation du spectre essentiel et application, *C. R. Acad. Sci. Paris, t. 331*, Srie I, pp. 525–530.

86. Jeribi, A., (2002). A characterization of the Schechter essential spectrum on Banach spaces and applications, *J. Math. Anal. Appl., 271*(2), 343–358.

87. Jeribi, A., (2002). Some remarks on the Schechter essential spectrum and applications to transport equations, *J. Math. Anal. Appl., 275*(1), 222–237.

88. Jeribi, A., (2002). A characterization of the essential spectrum and applications, *Boll. dell. Unio. Mate. Ital., 8 B-5,* 805–825.

89. Jeribi, A., (2002). On the Schechter essential spectrum on Banach spaces and applications, *Facta Univ. Ser. Math. Inform., 17*, 35–55.

90. Jeribi, A., (2003). Time asymptotic behavior for unbounded linear operator arising in growing cell populations, *Nonl. Anal. Real Worl. Appli., 4*, 667–688.

91. Jeribi, A., (2004). Fredholm operators and essential spectra, *Arch. Inequal. Appl., 2*(2–3), 123–140.

92. Jeribi, A., Krichen, B., & Zarai Dhahri, M. (2016). Essential spectra of some of matrix operators involving γ-relatively bounded entries and an applications, *Linear and Multilinear Algebra*, 1654–1668.

93. Jeribi, A., & Latrach, K., (1996). Quelques remarques sur le spectre essentiel et application l'quation de transport, *C. R. Acad. Sci. Paris, t. 323*, Srie I, 469–474.

94. Jeribi, A., Latrach, K., & Megdiche, H., (2005). Time asymptotic behavior of the solution to a Cauchy problem governed by a transport operator, *J. Inte. Equa. Appl., Volume 17, Number 2,* 121–139.

95. Jeribi, A., Megdiche, H., & Moalla, N., (2005). On a transport operator arising in growing cell populations II. Cauchy problem, *Math. Meth. Appl. Scie., 28*, 127–145.

96. Jeribi, A., & Mnif, M., (2005). Fredholm operators, essential spectra and application to transport equations, *Acta Appl. Math., 89*(1–3), 155–176.

97. Jeribi, A., & Moalla, N., (2006). Fredholm operators and Riesz theory for polynomially compact operators, *Acta Appl. Math., 90*(3), 227–245.

98. Jeribi, A., & Moalla, N., (2009). A characterization of some subsets of Schechter's essential spectrum and application to singular transport equation, *J. Math. Anal. Appl., 358*, 434–444.

99. Jeribi, A., Ould Ahmed Mahmoud, S. A., & Sfaxi, R., (2007). Time asymptotic behavior for a one-velocity transport operator with Maxwell boundary condition, *Acta Appl. Math., 3*, 163–179.

100. Jeribi, A., Moalla, N., & Walha, I., (2009). Spectra of some block operator matrices and application to transport operators, *J. Math. Anal. Appl., 351*(1), 315–325.

101. Jeribi, A., Moalla, N., & Yengui, S., (2014). S-essential spectra and application to an example of transport operators, *Math. Methods Appl. Sci., 37*(16), 2341–2353.

102. Jeribi, A., Moalla, N., & Yengui, S., (2017). Some results on perturbation theory of matrix operators, M-essential spectra of matrix operators and application to an example of transport operators, preprint.

103. Jeribi, A., & Walha, I., (2011). Gustafson, Weidmann, Kato, Wolf, Schechter and Browder Essential Spectra of Some Matrix Operator and Application to Two-Group Transport Equation, *Math. Nachr., 284*(1), 67–86.

104. Kaashoek, M. A., (1967). Ascent, descent, nullity and defect, a note on a paper by A.E. Taylor, *Math. Ann., 172*, 105–115.

105. Kaashoek, M. A., & Lay, D. C., (1972). Ascent, descent and commuting perturbations, *Trans. Amer. Math. Soc., 169,* 35–47.

106. Karow, M., (2006). Eigenvalue condition numbers and a formula of Burke, Lewis and Overton, *Electron. J. Linear Algebra 15,* 143–153.

107. Kato, T., (1966). *Perturbation Theory for Linear Operators,* Die Grundlehren der mathematischen Wissenschaften, Band 132, Springer-Verlag New York, Inc., New York.

108. Kato, T., (1980). *Perturbation Theory for Linear Operators*, Second edition. Grundlehren der Mathematischen Wissenschaften, Band 132, Springer-Verlag, Berlin Heidelberg, New York.

109. Kulkarni, S. H., & Sukumar, D., (2008). The condition spectrum, *Acta Sci. Math. (Szeged), 74*, 625–641.

110. Kumar, G. K., & Lui, S. H., (2015). Pseudospectrum and condition spectrum, *Operators and Matrices, 1*, 121–145.

111. Langer, H., Markus, A., Matsaev, V., & Tretter, C., (2003). Self-adjoint block operator matrices with non-separated diagonal entries and their Schur complements, *J. Funct. Anal., 199*(2), 427–451.

112. Latrach, K., & Dehici, A., (2001). Fredholm, semi-Fredholm perturbations, and essential spectra, *J. Math. Anal. Appl., 259*(1), 277–301.

113. Latrach, K., & Jeribi, A., (1996). On the essential spectrum of transport operators on L_1-spaces, *J. Math. Phys., 37*(12), 6486–6494.

114. Latrach, K., & Jeribi, A., (1997). Sur une quation de transport intervenant en dynamique des populations, *C. R. Acad. Sci. Paris, t. 325,* Série I, pp. 1087–1090.

115. Latrach, K., & Jeribi, A., (1998). Some results on Fredholm operators, essential spectra and application, *J. Math. Anal. Appl., 225*(2), 461–485.

116. Latrach, K., Megdiche, H., & Jeribi, A., (2005). Time asymptotic behavior of the solution to a Cauchy problem governed by a transport operator, *J. Inte. Equa. Appl. 17*(2), 121–140.

117. Lebow, A., & Schechter, M., (1971). Semigroups of operators and measures of noncompactness, *J. Func. Anal., 7,* 1–26.

118. Leon-Saavedra, F., & Müller, V., (2004). Rotations of hypercyclic operators, *Inte. Equa. Oper. Theo., 50*(3), 385–391.

119. Markus, A. S., (1988). Introduction to the spectral theory of polynomial operator pencils, *Amer. Math. Soc., Providence*, RI . iv+250 pp. ISBN: 0-8218-4523-3.

120. Mennicken, R., & Motovilov, A. K., (1999). Operator interpretation of resonances arising in spectral problems for 2×2 operator matrices, *Math. Nachr., 201*, 117–181.

121. M. M. Milovanović-Arandjelović, (2001). Measures of noncompactness on uniform spaces—the axiomatic approach, IMC *Filomat 2001*, Niš, 221–225.

122. Moalla, N., (2012). A characterization of Schechter's essential spectrum by mean of measure of non-strict-singularity and application to matrix operator, *Acta Math. Sci. Ser. B* Engl. Ed., *32*(6), 2329–2340.

123. Moalla, N., Damak, M., & Jeribi, A., (2006). Essential spectra of some matrix operators and application to two-group transport operators with general boundary conditions, *J. Math. Anal. Appl., 323*(2), 1071–1090.

124. Müller, V., (2003). *Spectral Theory of Linear Operator and Spectral Systems in Banach Algebras*, Birkhäuser Verlag.

125. Müller, V., (2007). Spectral theory of linear operators and spectral system in Banach algebras, Second edition, *Operator Theory Advance and Application, vol. 139*, Birkhäuser Verlag, Basel.

126. Murray, F. J., (1937). On complementary manifolds and projections in spaces L_p and l_p. *Trans. Amer. Math. Soc., 41,* 138–152.

127. Pelczynski, A., (1965). On strictly singular and strictly cosingular operators. II. Strictly singular and strictly cosingular operators in $L(\mu)-$spaces, *Bull. Acad. Polon. Sci. Sr. Sci. Math. Astronom. Phys., 13*, 37–41.

128. Pelczynski, A., (1965). On strictly singular and strictly cosingular operators. I. Strictly singular and strictly cosingular operators in $\mathscr{C}(\Omega)$-spaces, *Bull. Acad. Polon. Sci. Sr. Sci. Math. Astronom. Phys., 13,* 31–36.

129. Rakočević, V., (1981). On one subset of M. Schester's essential spectrum, *Math. Vesnik., 5*(18), *33*(4), 389–391.

130. Rakočević, V. R., (1983). Measures of non-strict-singularity of operators, *Math. Vesnik., 35*, 79–82.

131. Rakočević, V., (1984). On the essential approximate point spectrum, II, *Math. Vesnik., 36*, 89–97.

132. Rakočević, V., (1995). Semi-Fredholm operators with finite ascent or descent and perturbations, *Amer. Math. Soci., 123*(12), December.

133. Rakočević, V., (1998). Measures of noncompactness and some applications, *Filomat, 12*(2), 87–120.

134. Ransford, T. J., (1984). Generalized spectra and analytic multivalued functions, *J. London Math. Soc., 29,* 306–322.

135. Reddy, S. C., & Trefethen, L. N., (1990). Lax-stability of fully discrete spectral methods via stability regions and pseudo-eigenvalues, *Comput. Meth. Appl. Mech. and Engrg., 80*(1), 147–164.

136. Riesz, F., & Nagy, B. S., (1952). Leçons d'analyse fonctionnelle, Acadmie des Sciences de Hongrie, Akadmiai Kiadó, Budapest.

137. Schechter, M., (1966). On the essential spectrum of an arbitrary operator, I. *J. Math. Anal. Appl., 13,* 205–215.

138. Schechter, M., (1967). Basic theory of Fredholm operators, *Anna. Scuola Norm. Sup., Pisa 21*(3), 261–280.

139. Schechter, M., (1968). Riesz operators and Fredholm perturbations, *Bull. Amer. Math. Soc., 74,* 1139–1144.

140. Schechter, M., (1971). *Spectra of Partial Differential Operators,* North-Holland, Amsterdam.

141. Schechter, M., (1971). *Principles of Functional Analysis,* New Work: Academic Press.

142. Schechter, M., (1972). Quantities related to strictly singular operators, *Ind. Univ. Math. J., 21*(11), 1061–1071.

143. Schmoeger, C., (1997). The spectral mapping theorem for the essential approximate point spectrum, *Colloq. Math. 74*(2) 167–176.

144. Shapiro, J., & Snow, M., (1974). The Fredholm spectrum of the sum and product of two operators, *Trans. Amer. Math. Soc., 191,* 387–393.

145. Shargorodsky, E., (2009). On the definition of pseudospectra, *Bull. London Math. Soc., 41*, 524–534.

146. Shkalikov, A. A., (1995). On the essential spectrum of some matrix operators, *Math. Notes, 58*(5–6), 1359–1362.

147. Shkalikov, A. A., & Tretter, C., (1996). Spectral analysis for linear pencils $N - \lambda P$ of ordinary differential operators, *Math. Nachr. 179,* 275–305.

148. Taylor, A. E., (1951). Spectral theory of closed distributive operators, *Acta Math., 84*, MR 12, 717, 189–224.

149. Taylor, A. E., (1966). Theorems on ascent, descent, nullity, and defect of linear operators, *Math. Annalen, 163*, 18–49.

150. Trefethen, L. N., (1990). Approximation theory and numerical linear algebra, *Algorithms for Approximation II*, Mason, J. C., & Cox, M. G., eds., Chapman and Hall, London, 336–360.

151. Trefethen, L. N., (1992). Pseudospectra of matrices, numerical analysis. In: *Pitman Research notes in Mathematics Series,* Longman Science and Technology, Harlow, *260*, 234–266.

152. Trefethen, L. N., (1997). Pseudospectra of linear operators, *SAIM rev., 39*(3), 383–406.

153. Trefethen, L. N., & Embree, M., (2005). *Spectra and Pseudospectra* (The behavior of non normal matrices and operators), Princeton University Press.

154. Tretter, C., (2008). *Spectral Theory of Block Operator Matrices and Applications*, Imperial College Press, London.

155. Varah, J. M., (1967). The computation of bounds for the invariant subspaces of a general matrix operator, PhD thesis, Stanford University Pro. Quest LLC, Ann. Arbor.

156. Vladimirskii, Ju. I., (1967). Strictly cosingular operators, *Soviet. Math. Dokl. 8*, 739–740.

157. West, T. T., (1987). A Riesz-Schauder theorem for semi-Fredholm operators, *Proc. Roy. Irish Acad. Sect. A, 87(*2), 137–146.

158. Weyl, H., (1909). Uber beschrankte quadratische Formen, deren differenz vollstelig ist., *Rend. Circ. Mat. Palermo 27,* 373–392.

159. Wilkinson, J. H., (1986). Sensitivity of eigenvalues II, Util. Math., 30, 243–286.

160. Wolf, F., (1959). On the invariance of the essential spectrum under a change of boundary conditions of partial differential boundary operators, *Indag. Math., 21,* 142–147.

161. Wolff, M. P. H., (2001). Discrete approximation of unbounded operators and approximation of their spectra, *J. Approx. Theo., 113*, 229–244.

162. Yood, B., (1951). Properties of linear transformations preserved under addition of a completely continuous transformation, *Duke Math. J., 18*, 599–612.

163. Živković-Zlatanović, S. Č., Djordjević, D. S., & Harte, R. E., (2011). On left and right Browder operators, *J. Korean Math. Soc., 48*, 1053–1063.

164. Živković-Zlatanović, S. Č., Djordjević, D. S., & Harte, R. E., (2011). Left-right Browder and left-right Fredholm operators, *Inte. Equa. Oper. Theo., 69*, 347–363.

165. Živković-Zlatanović, S. Č., Djordjević, D. S., Harte, R. E., & Duggal, B. P., (2014). On polynomially Riesz operators, *Filomat, 28*, 197–205.

Index

A

M

N

O

P

Q

R

T

U

W

X

Y